Springer Series in **Materials Science** 4

Springer Series in **Materials Science**

Editors: Aaram Mooradian   Morton B. Panish

Microclusters

Proceedings of the First NEC Symposium,
Hakone and Kawasaki, Japan, October 20–23, 1986

Editors: S. Sugano,
Y. Nishina, and S. Ohnishi

With 187 Figures

Springer-Verlag Berlin Heidelberg New York
London Paris Tokyo

Professor Dr. Satoru Sugano

Institute for Solid State Physics, The University of Tokyo,
7-22-1 Roppongi, Minato-ku, Tokyo 106, Japan

Professor Dr. Yuichiro Nishina

The Research Institute of Iron, Steel and Other Metals,
Tohoku University, Katahira, Sendai 980, Japan

Dr. Shuhei Ohnishi

Fundamental Research Labs., NEC Corp., 1-1 Miyazaki 4,
Miyamae-ku, Kawasaki, Kanagawa 213, Japan

Series Editors:
Dr. *Aaram Mooradian*

Leader of the Quantum Electronics Group, MIT,
Lincoln Laboratory, P.O. Box 73, Lexington, MA 02173, USA

Dr. *Morton B. Panish*

AT&T Bell Laboratories, 600 Mountain Avenue, Murray Hill, NJ 07974, USA

ISBN-13: 978-3-642-83066-2   e - ISBN-13:978-3-642-83064-8
DOI: 10.1007/ 978-3-642-83064-8

© Springer-Verlag Berlin Heidelberg 1987
Softcover reprint of the hardcover 1st edition 1987

2153/3150-543210

Preface

This volume contains the proceedings of the first in a series of biennial NEC Symposia on Fundamental Approaches to New Material Phases sponsored by the NEC Corporation, Tokyo, Japan. The symposium was held October 20–22, 1986, at Hakone Kanko Hotel in Hakone near Mt. Fuji, and on October 23 at NEC Laboratories in Kawasaki, Japan. About 40 participants stayed together at the symposium sites during this period. They enjoyed intense and wide-ranging discussions in a conference room facing Mt. Fuji and the beautiful lake Ashinoko extending from the foot of the slope in the old crater.

The title of the volume, *Microclusters*, means microscopic aggregates consisting of a few tens through a few hundreds of atoms. Microclusters, which are too big to be described as inorganic molecules but too small to have translational symmetry, are expected to show exotic properties which can be found in neither molecules nor solids. In the past few years the research field of microclusters has shown rapid and epoch-making development. This is partly due to rapid development of the experimental techniques which have enabled the production of relatively dense, noninteracting microclusters of various sizes in the form of cluster beams, thus allowing measurement of the properties of a free microcluster of a given size. The progress in the field is also due to rapid development of computers and computational techniques which have made it possible to perform ab initio calculations of the atomic and electronic structure of smaller microclusters, as well as to carry out computer simulations of their dynamics. This volume deals with these experimental and theoretical developments. The microclusters discussed here are those consisting of atoms of metallic, semiconducting, and insulating materials. Their electronic and structural properties are also compared with nuclear properties, since both microclusters and nuclei are small many-body systems. A similarity to nuclei is clearly seen in the application of the shell model to metal clusters. Some of the papers here suggest the need for new concepts of melting and phase transitions for small many-body systems represented by microclusters.

The editors are indebted to the following members of the Steering Committee of the NEC Symposium for establishing the symposium and choosing microclusters as the key subject of the first symposium: Dr. M. Uenohara, Executive Vice President of NEC Corporation; Dr. D. Shinoda, Vice President of NEC Corporation; Dr. H. Kamimura, Professor of the University of Tokyo; Dr. T. Shinjo, Professor of Kyoto University; and Dr. A. Yoshimori, Professor of Osaka University. The editors are also indebted to the following advisory members of the Organizing Committee of the first NEC Symposium for their valuable advice: Dr. R. Kubo, Professor of Keio University, and Dr. R. Uyeda, Professor Emeritus of Nagoya University. The editors thank the following members of the Organizing Committee for organizing the first NEC Symposium and for contributing to this volume: Professor K. Terakura, Professor M. Tsukada, Dr. S. Iijima, Dr. C. Satoko, Dr. N. Hamada, and Dr. A. Oshiyama. Finally, sincere thanks should be expressed to all the participants and those who helped us through their kind cooperation.

Tokyo, Japan   *S. Sugano*
January 1, 1987   *Y. Nishina*
                                                                 S. Ohnishi

Contents

Part I

Shell Model

Shell Structure in Clusters

M.L. Cohen

Department of Physics, University of California, and
Materials and Molecular Research Division, Lawrence Berkeley Laboratory,
Berkeley, CA 94720, USA

Using jellium and pseudopotential models, it is possible to make realistic calculations of the properties of small metal clusters. For simple metals, it is found that the electronic energy is a large fraction of the total energy of a cluster and can be dominant in determining the cluster size. Experimental mass abundance spectra show discontinuities which can be attributed to shell structure in the electronic energy levels.

The jellium calculations for the alkali metal clusters are done self-consistently within a local density approximation (LDA). Since the ground state properties are determined accurately in LDA, the limiting approximation for the total energies is the jellium potential. To test this approximation, pseudopotential calculations have been done. It is found that the shell structure remains and should influence static properties.

For dynamic properties of the electrons, the usual limitations of the LDA found for calculations on bulk solids remain. Excitation energies are not given accurately, and theoretical estimates of the polarizabilities and ionization potentials are limited by the same uncertainties as for bulk properties. Despite these limitations, the jellium approach has been extremely useful in determining physical properties.

The discussion presented here will include a description of the theoretical bases for the calculations and results of applications.

1. Introduction

Significant motivation for studying clusters comes from both basic and applied science. The transitions involved in grouping atoms to form extended solids are not understood in detail; hence the study of the relationships between the properties of atoms, molecules, clusters, solid surfaces, and bulk solids has attracted considerable attention in materials science. On the applied side, the field of microelectronics has continually sought smaller structures for device applications. It is likely that a detailed understanding of future devices will require a knowledge of the role of restricted geometries with sizes comparable to clusters.

Physical scientists have responded to the increased need for research in this area. Refined experimental techniques now yield reproducible results on cluster properties such as the abundance spectra--the number of clusters as a function of their size. Other reproducibly measured properties include the polarizabilities of clusters and the first ionization potentials for electron removal. Theoretical work has come from solid state physics and quantum chemistry. The former have used techniques previously employed for bulk solids, while the latter have extended schemes originally devised for molecules. Close collaborations have developed between experimental and theoretical researchers in both physics and chemistry, resulting in some new views of the origins of the observed properties of clusters. Several reviews and conference proceedings are now available [1-5] which document recent activity.

The discussion here will focus on a solid-state physics approach, but many of the ideas have been borrowed from nuclear physics models of the shell structure of nuclei. Specific applications will be made to clusters of alkali metals. These metals are known to be the materials which are closest in nature to a free-electron gas, and the models described here for clusters of metal atoms will build on that property. Although some experimental results will be discussed, the emphasis here will be on the theoretical models and calculational results.

2. Electronic Models and Shell Structure

Current *ab initio* techniques for solids are capable of yielding structural, vibrational, and electronic properties. In some cases, such as the pseudo-potential approach, the only required inputs are the atomic numbers and atomic masses of the constituent atoms. The pseudopotential total energy approach [6] uses a local density approximation (LDA) [7] to model electron-electron interactions and pseudopotentials to describe the electron-core potentials. Although structural properties are reproduced to within a few percent, excited state properties require modifications [8] beyond the LDA. Hence the pseudopotential approach can be used in general, but corrections to LDA are required except for properties of the ground state.

Ab initio calculations of the type described above have been done for alkali metal clusters [9,10,11], but because of their complexity, these calculations are limited to small numbers of atoms. If the core potentials are "smoothed out" into a structureless positive "jelly," then calculations for large N are possible. This jellium model is still limited by the restrictions of the LDA, but it can describe the behavior of the electrons in a free-electron-like approximation. Jellium approximations to the bulk and surface properties of metals have yielded considerable information [12] about the properties of these systems. Applications to metal clusters have also been fruitful [13-17].

Most of the jellium calculations assume spherical clusters, but deformations can be examined [18] using a one-electron spheroidal potential which is similar to the approach for nuclei introduced by NILSSON [19]. The spherical symmetry of the cluster model introduces shell structure. Electronic shells occur in the following series, 1s, 1p, 1d, 2s, 1f, 2p, 1g, 2d, 3s, 1h..., but the order can be altered somewhat by the potential, and spheroidal corrections can introduce subshells. Shell closings occur when there are a sufficient number of electrons to fill a shell. For monovalent atoms, like alkali metals, the number of electrons equals the number of atoms in a cluster, N, and shell closings occur at N = 2,8,18,20,34,40,58,68,70,92.... It is straightforward to calculate the total energy per atom for a cluster as a function of N within the jellium approximation. This is shown in Fig. 1. The curve is generally smooth with small discontinuities at the shell closings. A bulk limit $N \rightarrow \infty$ is expected to be around -2.2 eV for Na (Fig. 1).

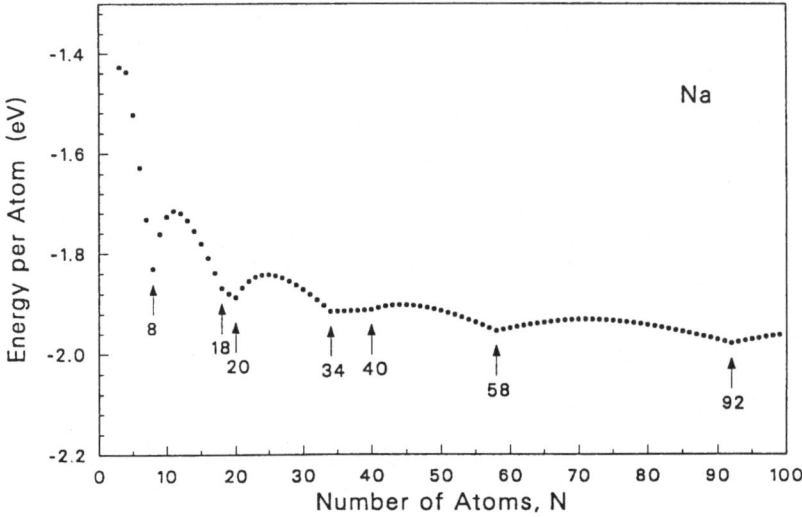

Fig. 1 Total energy per atom for Na clusters versus the number of atoms in the cluster

3. Abundance Spectra

When N corresponds to a closed shell, this cluster is expected to be particularly stable. The lowered energy at shell closings and the reduced probability of evaporation or accretion are expected to cause a larger abundance of clusters of this type. A function $\Delta_2(N)$ was introduced [15] to illustrate the special stability of closed shell clusters

$$\Delta_2(N) = E(N+1) + E(N-1) - 2E(N) \qquad (1)$$

which is related to the second derivative with respect to N of the total energy of a cluster E(N). The function $\Delta_2(N)$ is independent of the reference energy of the free atoms, and it is a measure of the relative stability of clusters. This function is expected to have a peak for stable clusters whose highest electron energy is separated by an energy gap from the energy of the next cluster with one more electron.

A plot of $\Delta_2(N)$ for Na appears in Fig. 2, and a comparison between the $\Delta_2(N)$ curves for Li, Na, and K is given in Fig. 3. Figure 2 also contains a plot of the experimental abundance spectrum for Na [15] taken with argon-carried seeded beams. The correspondences between the discontinuities in the experimental and theoretical curves are attributed to cluster stability when N corresponds to a closed electronic shell configuration.

The experimental data is sensitive to the conditions associated with the production and propagation of the clusters in the beam. During the supersonic expansion of the Na, a quasi-equilibrium size distribution is established and maintained. If the nozzle temperature or carrier gas cooling are not optimum, predictable distortions of the abundance spectra result [20].

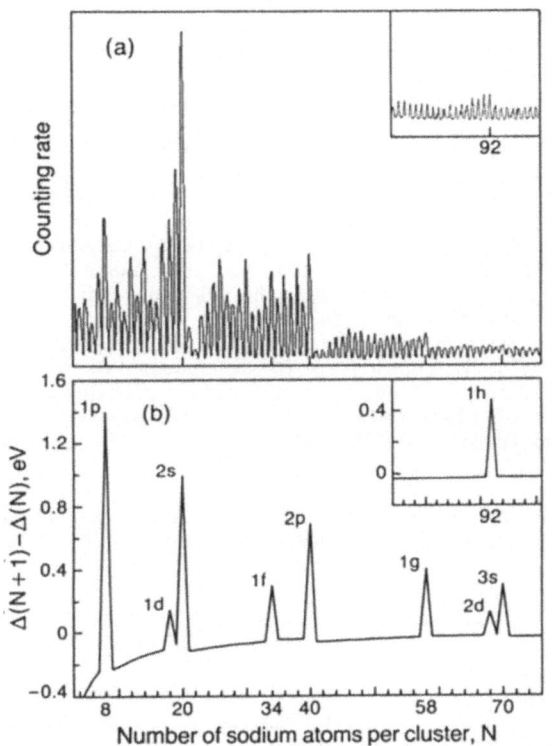

Fig. 2 (a) Experimentally measured abundance spectrum of Na. (b) Theoretical calculation of the function $\Delta_2(N)$ (see text). The $\Delta_2(N)$ function is labeled $\Delta(N+1)-\Delta(N)$ here

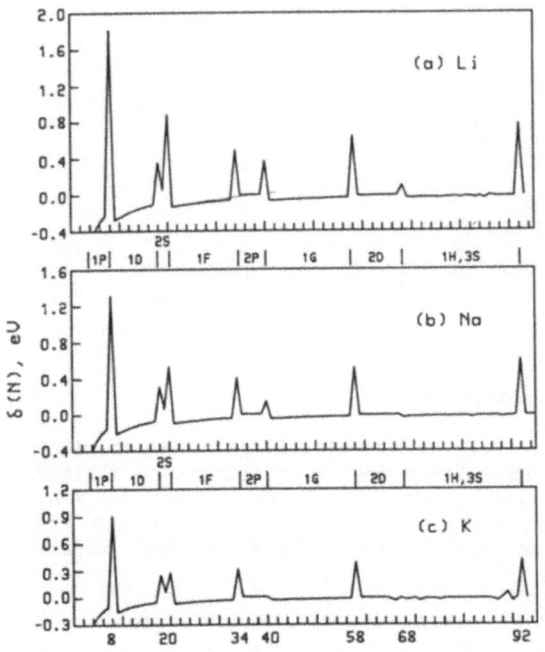

Fig. 3 Calculated $\Delta_2(N)$ spectra (labeled $\delta(N)$ here) for (a) Li, (b) Na, and (c) K

For Li and K, the theoretical results give peaks in $\Delta_2(N)$ at the same N as for Na but with differences in magnitude. For example, the small N = 40 peak in K results from the features of the K jellium potential. This predicted [17] reduction relative to Na has been verified [21].

The jellium model can also be applied to divalent and trivalent atoms. The $\Delta_2(N)$ spectra are given for Mg and Al and compared to Na in Fig. 4. Again, the peaks can be interpreted in terms of shell closings. The numbers in parentheses correspond to the number of electrons in a stable structure. Because of the higher valencies of Mg and Al, relatively small structures can have a sufficient number of electrons for a shell closing. For example, only six Al atoms contain 18 electrons which is near the 20 required to close a spherical shell. Because the jellium model is less appropriate for small clusters where structure must play a larger role, the results here for the smaller clusters should be considered to be suggestive of possible stabilities. Experimentally in addition to alkali metal clusters, shell structure is seen in clusters of other metals: noble [22], divalent [23], trivalent [24], lead [25], and transition [26] metals.

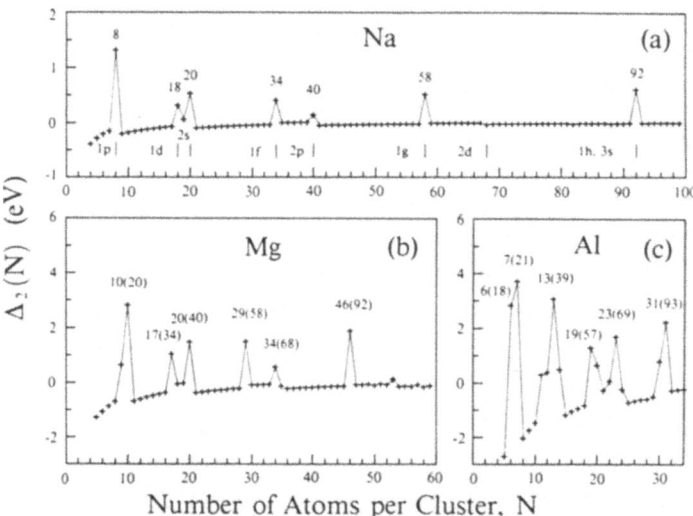

Fig. 4 Calculated $\Delta_2(N)$ spectra for (a) Na, (b) Mg, and (c) Al. The number of electrons in the stable clusters is given in parenthesis for Mg and Al

4. Ionization Energies and Polarizabilities

Studies of the ionization potentials (IP) of metal clusters are important for understanding cluster properties. Since neutral clusters are ionized to make detection possible, the ionization process is central to many experimental measurements. In addition, a knowledge of the IP and its dependence on cluster size can be used to test the theoretical models for electronic properties.

Theoretically the IP of a cluster can be viewed as similar to the IP for a single atom at one extreme or the work function for an extended solid at the other. In the latter case [12], it is known that the important energy contributions are the electrostatic, exchange-correlation, and kinetic or Fermi energy. LDA jellium calculations give work functions which are higher by about 10% than the experimental values. When structure is included using pseudopotentials, the calculated values decrease by ~10%.

In the case of atoms, the LDA results do not agree with the measured IP. Since structure is not relevant here, attention has been given to corrections of the LDA. One useful approach is to remove self-interaction effects. When this self-interaction correction (SIC) is included [27] to calculate the IP of atoms, there is an improvement in the agreement with experiment. Recently ISHII *et al.* [28] examined the effect of the SIC on the ionization energy of Na clusters and found better overall agreement with measured values than for the LDA calculations. These calculations can be classified as spin-polarized LDA with SIC, and they estimate the ionization energy from the energy of the highest occupied orbital. Another approach is to calculate the difference in energy in the usual LDA between the Nth and the ionized (N-1)th cluster. This should give results consistent with those using SIC corrections.

Although the theoretical focus has been on the magnitude, overall shape, and N → 1 or ∞ limits of the IP, there is general agreement that shell structure is seen in the calculated IP versus N curve. The IP generally drops after a closed shell where it is easier to remove a single added electron. For alkalai metal clusters, differences in magnitude and shape are evident when the element is changed. This can be simulated in the jellium model by using different electron densities to represent Li, Na, and K [16]. However, shell structure is evident in all cases.

Experimentally shell structure is also observed in the IP spectra [1,21]. Some measurements [29] are featureless near prominent shell closings, but this may be the result of the method used to ionize the clusters. Data of this kind on homogeneous and mixed clusters have motivated criticisms [30] of the jellium model and shell structure, however the definitive observations of shell structure [1,21] in the IP spectra appear conclusive.

Despite the inclusion of the SIC, the theoretical discontinuities are larger than those observed. Further modifications of the electron-electron energies can be done, as in the case of excited states for solids [8] or the inclusion of pseudopotential effects [31] may give changes in the IP which are similar to those discussed above for the work function. However, it is highly unlikely that effects of this kind will mask the discontinuities at the shell closings. These effects can change the overall magnitude, as in the case of SIC correction where a shift of ~1 eV was found, and they may decrease the structure now seen in the LDA calculations.

Another measured property of metal clusters showing shell structure [7,32] is the polarizabilities. In the measurements, beams are deflected in an inhomogeneous electric field, and the polarizabilities (normalized to the atomic polarizability) can be examined [1,32] as a function of N. There is an overall decrease with N from the atomic limit of 1 to the bulk value of ~1/2. Local minima appear near shell closings at N = 2,8, and 18-20, but not much structure is seen above N = 20.

Theoretically, the classical limit for the polarizability is R^3 for a metal sphere of radius R. It is known that the electron density protrudes beyond the positive jellium edge at surfaces. This effect has been analyzed both for bulk solid surfaces [12] and for closed shell metal clusters. The resulting increase in the effective radius of a cluster increases its polarizability. Other effects which have been examined are the elastic deformation of the jellium background [33] and nonspherical contributions [18]. Although these corrections improve the agreement between theory and experiment, the theoretical values are still lower by ≲20%. Again, structural effects and corrections to the electron-electron potentials can modify the energy gaps and hence the polarizabilities. Although further refinements of both the experimental and theoretical results are needed, there is reasonable agreement and strong support for the existence of shell structure.

5. Conclusions

Evidence for electronic shell structure in the measured abundance spectra, ionization potentials, and polarizabilities are well established. The most compelling data comes from measurements of the abundance spectra. In developing a theory of the abundance spectra and other properties, a central feature is the stability of clusters of certain sizes. The observation [15] that the total electronic energy of a jellium-like system will change with cluster size has resulted in a "giant atom" model [1,15] of a cluster.

In the jellium approximation, the giant atom model has no core structure and the abrupt changes in measured quantities with cluster size must originate from discontinuous changes in the electron energy. These occur at shell closings. To examine the influence of structure on this approach, the energy levels have been calculated [11] for fixed structures. For example, calculations for a body-centered cubic cluster of 15 atoms and a face-centered cubic cluster of 13 atoms reveal charge densities similar to those found in jellium. In addition, the energy levels do not differ significantly from the jellium results; hence the shell closing model is appropriate. A perturbation on jellium with pseudopotentials gives similar conclusions [31]. Although the structural energy can be comparable with the electronic energy, features arising from the shell closings still dominate.

Pseudopotential corrections and treatments of the electron-electron interactions beyond the LDA will also affect the results for the ionization potentials and polarizabilities. However in these cases, as for abundance spectra, the shell closing features should remain.

Based on these arguments, the "giant atom" model used to explain the data on alkali metal clusters [1,15] appears to be generally appropriate for free electron-like systems. It is not likely to work well for very small clusters or for clusters formed by covalent and Van der Waals interactions. In addition, transition metals with strong directional bonding may be out of the domain of this model.

It is likely that more detailed calculations will be done with *ab initio* methods. Unless dramatic progress is made in the theoretical techniques, these will be limited to modest cluster sizes. At this point, there are still interesting modifications to make on the jellium approach and new applications. There may be more ways to use the conceptually and computationally simple jellium model to test additional features of the giant atom model of a cluster. Studies of this kind may lead to new models which like the giant atom model and shell structure will provide new insights for understanding the physical properties of clusters.

6. Acknowledgements

Most of the theoretical work was done in collaboration with M. Y. Chou. Close collaboration with the experimental group of W. D. Knight, W. A. de Heer, and W. A. Saunders is also acknowledged. This work was supported by National Science Foundation Grant No. 8319024 and by the Director, Office of Energy Reserach, Office of Basic Energy Sciences, Materials Sciences Division of the U.S. Department of Energy under Contract No. DE-AC03-76SF00098.

7. References

1. W. de Heer, W. D. Knight, M. Y. Chou, and M. L. Cohen: Solid State Physics (in press)
2. J. C. Phillips: Chem. Rev. 86, 619 (1986)
3. M. D. Morse: Chem. Rev. (in press)
4. J. Physique Colloque 38, C2 (1977)
5. Surf. Sci. 156 (1985)
6. M. L. Cohen: Phys. Scr. T1, 5 (1982)
7. W. Kohn, L. J. Sham: Phys. Rev. 140, A1333 (1965)
8. M. S. Hybertsen, S. G. Louie: Phys. Rev. Lett. 55, 1418 (1985)
9. J. L. Martins, J. Buttet, R. Car: Phys. Rev. B31, 1804 (1985)
10. B. K. Rao, S. N. Karma, P. Jena: Solid State Comm. 56, 731 (1985)
11. A. N. Cleland, M. L. Cohen: Solid State Comm. 55, 35 (1985)
12. N. D. Lang: Solid State Physics 28, 225 (1973)
13. D. E. Beck: Solid State Comm. 49, 381 (1984)
14. W. Ekardt: Phys. Rev. B29, 1558 (1984)
15. W. D. Knight, K. Clemenger, W. A. de Heer, W. A. Saunders, M. Y. Chou, M. L. Cohen: Phys. Rev. Lett. 52, 2141 (1984)
16. M. Y. Chou, A. N. Cleland, M. L. Cohen: Solid State Comm. 52, 645 (1984)
17. M. Y. Chou, M. L. Cohen: Phys. Lett. 113A, 420 (1986)
18. K. Clemenger: Phys. Rev. B32, 1359 (1985)
19. S. G. Nilsson: K. Dan. Vidensk. Selsk. Mat. Fys. Medd. 29, No. 16 (1955)
20. W. D. Knight, W. A. de Heer, W. A. Saunders: private communication
21. W. A. Saunders, K. Clemenger, W. A. de Heer, W. D. Knight: Phys. Rev. B32, 1366 (1986)
22. I. Katakuse, I. Ichihara, Y. Fujita, T. Matsuo, T. Sakurai, H. Matsuda: Int. J. Mass Spect. Ion Proc. 67, 229 (1985)
23. I. Katakuse, I. Ichihara, Y. Fujita, T. Matsuo, H. Matsuda: Int. J. Mass Spect. Ion Proc. 69, 153 (1986)
24. T. P. Martin: J. Chem. Phys. 83, 78 (1985)
25. J. Muhlbach, P. Pfau, E. Recknagel, K. Sattler: Z. Phys. B47, 233 (1982)
26. R. L. Whetten, D. M. Cox, A. Kaldor: Surf. Sci. 156, 8 (1985)
27. J. P. Perdew, A. Zunger: Phys. Rev. B23, 5048 (1981)
28. Y. Ishii, S. Ohnishi, S. Sugano: Phys. Rev. B33, 5271 (1986)
29. A. Herrmann, E. Schumacher, L. Woste: J. Chem. Phys. 68, 2327 (1978)
30. M. Kappes, M. Schar, P. Radi, E. Schumacher: J. Chem. Phys. 84, 1863 (1985)
31. M. Manninen: Solid State Comm. (in press)
32. W. D. Knight, K. Clemenger, W. A. De Heer, W. A. Saunders: Phys. Rev. B31, 2539 (1985)
33. P. Sheng, M. Y. Chou, M. L. Cohen: Phys. Rev. B34, 732 (1986)

SIMS Experiments on Metal Cluster Ions

I. Katakuse

Department of Physics, Faculty of Science, Osaka University,
1-1 Machikaneyama-cho, Toyonaka-shi, Osaka 560, Japan

Mass distributions of positive and negative cluster ions of copper, silver, gold, zinc and cadmium have been investigated. The clusters were produced by bombardment with Xe ions with 10keV energy and mass-analyzed with a single focusing mass spectrometer.

The mass spectra of the noble metals are quite similar. Two types of anomalies are observed. One is the odd-even alternation of the ion intensity, in which the intensity of the odd-n clusters is generally greater than that of even-n. This can be understood by the pairing energy of s-valence electrons in the clusters. The other anomaly is discontinuous variation of cluster ion intensity at certain values of n (magic number). This behaviour can be explained by a one-electron shell model in which the s-valence electrons are bound in a spherically symmetric potential well.[1]

The mass spectra of zinc and cadmium are also quite similar to each other. Almost all intense peaks occur at cluster sizes equal to half of the magic numbers of the noble metals. This behaviour is presumably due to the shell structure, because the number of the free electrons of the m-mer of zinc or cadmium is 2m, while that of a noble metal is m.

The metastable decay of the boble metal cluster ions in flight in the mass spectrometer was investigated. The excited clusters decay by emitting some constituent atoms; however, they do not decay beyond magic numbers.

1. Experimental

Mass spectra were obtained using a single focusing mass spectrometer (magnet radius r_m=260cm). Secondary ions were produced by bombardment with Xe ions with 10keV energy from a discharge-type primary gun[2]. The discharge volage and current were 13kV and 0.8mA respectively. The primary ion current was about 7μA and the spot size at the target was about 1mm in diameter. The accelerating potential for the secondary ions was 6kV. The mass-analyzed ions were amplified by a conventional sixteen-stage Cu/Be secondary electron multiplier and a d.c. amplifier. In order to increase the conversion efficiency from ions to electrons at the conversion dynode of the electron multiplier, the whole detection system was kept at +10kV and -10kV potential in the case of the detection of negative ions and positive ions respectively. The mass spectra were recorded with a conventional pen-recorder or UV-recorder by scanning the magnet current in a linear mode. Mass calibration was performed using $[Cs(CsI)_n]^+$ (for positive mass spectra) and $[I(CsI)_n]^-$ (for negative mass spectra).

In order to investigate the metastable decay of the cluster ions in the flight time of the mass spectrometer, the 7-, 8-, and 9-mers were studied in detail using a Matsuda-type double focusing mass spectrometer[3] with normal geometry. In a double focusing mass spectrometer of normal geometry, those ions which decompose in flight before entering the electric sector cannot pass because of their energy losses. Those ions can, however, be detected, if the acceleration voltage is raised by an amount corresponding to the energy loss due to the emission of the atoms. The experimental procedure was as follows; first, the acceleration voltage (V_a) was adjusted so as to maximize the ion current of clusters for a fixed potential (V_d) of the electric sector. Then the accelerating potential was changed in 10V step, and for each acceleration voltage, a mass spectrum was obtained by scanning the magnet current. By this procedure of raising the acceleration potential, the peaks with full energy gradually disappear, then the peaks of ions which have emitted one atom appear and next, peaks of clusters which have lost two atoms appear. By plotting the ion intensity of the n-mer with respect to the different acceleration voltages V_a, its behaviour with respect to unimolecular decomposition during the flight between the ion source and the electric sector was characterized.

2. Results and discussion

Mass spectra of silver clusters were obtained up to n=250. The peaks corresponding to successive clusters were resolved up to n=100 for the positive clusters and n=80 for the negative clusters. Figure 1 shows the size distributions of positive and negative cluster ions of silver up to n=65 plotted on a logarithmic scale. The abscissa of the mass spectrum of $(Ag)_n^-$ is shifted by two cluster sizes towards the right-hand side in order to show the similarity between the two traces. In each trace, two types of anomalies are observed. One is the odd-even alternation observed up to about n=40 in the mass spectrum of positive clusters and to about n=30 in the negative clusters. The other anomaly is the discontinuous variation of ion intensity at certain cluster sizes, so-called magic numbers. They are observed at n=3, 9, 19, 21, 35, 41, and 59 in the mass spectrum of positive clusters and at n=1, 7, 17, 19, 33, 39, and 57 in the mass spectrum of nega-

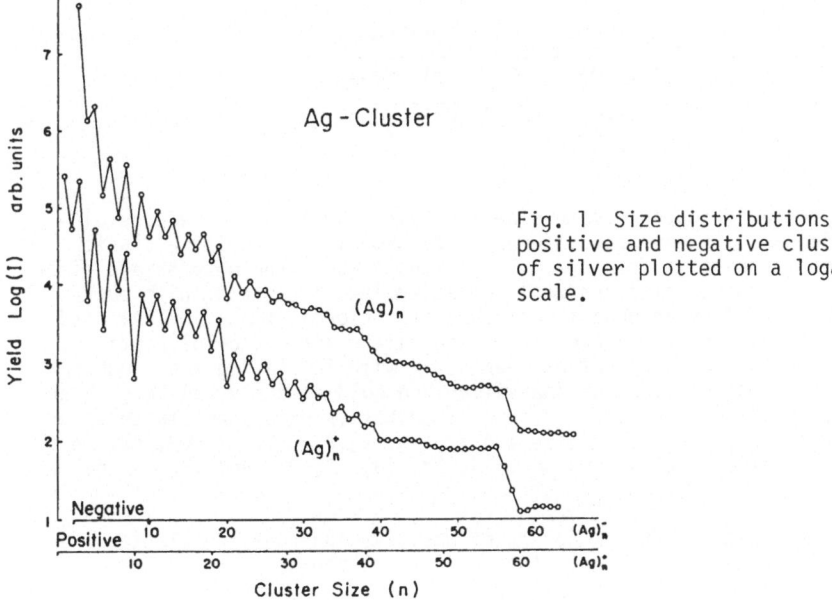

Fig. 1 Size distributions of positive and negative cluster ions of silver plotted on a logarithmic scale.

tive clusters. In the range of larger cluster sizes they are seen at n=93, 139, and 199 in the positive mass spectrum and at n=91 and 137 in the negative. A slight decrease was observed around n=200 in the mass spectrum of negative clusters, which, it is supposed, corresponds to the variation at n=197. In the range of larger cluster sizes, the clear-cut variation is not observed because of the back ground noise caused from the metastable decay of cluster ions, which decompose in flight in the mass spectrometer.

The mass spectra of gold clusters were obtained up to the cluster size n=150, and complete resolution was achieved up to n=70 for the positive clusters and n=65 for the negative clusters. The size distributions of gold clusters are shown in Fig. 2 on a logarithmic scale. Two types of anomalies are again observed. The discontinuous variation is observed at n=3, 9, 19, 21, 35, 59, 93, and 139 in the positive mass spectrum and at n=17, 33, 57, 91, and 137 in the negative mass spectrum.

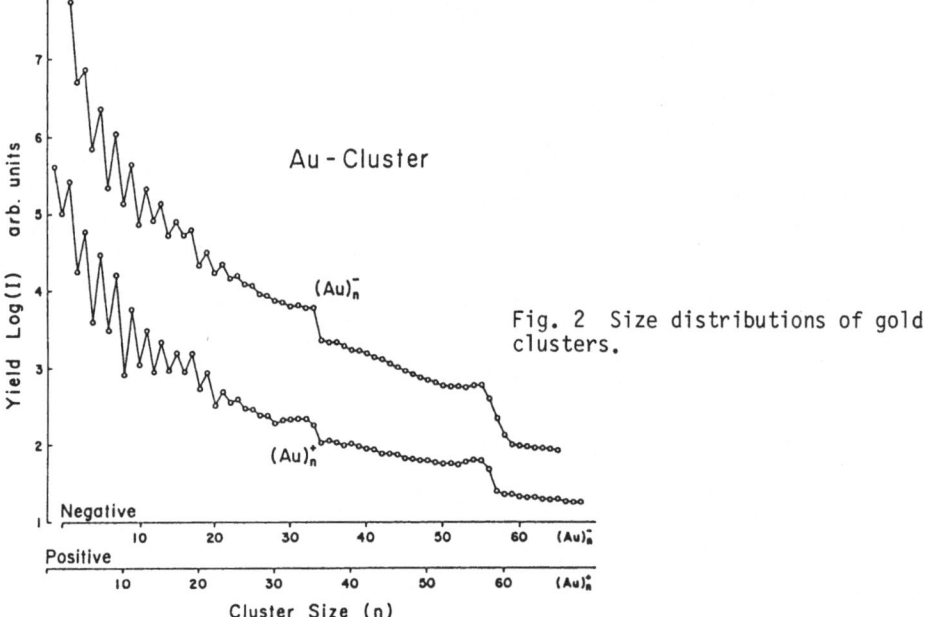

Fig. 2 Size distributions of gold clusters.

The mass spectra of copper clusters were obtained up to n=150, and resolution of adjacent clusters was achieved up to n=80 in the positive clusters and n=65 in the negative clusters. The intensities of the monomer and dimer of the negative copper clusters were not determined because of high background noise caused by an organic compound (trichloroacetic acid=TCA) used in previous experiments to increase the ionization efficiency of peptides. The back ground noise decreases exponentially with increasing mass number, therefore it is not serious for the silver and gold clusters having greater atomic masses. Figure 3 shows the size distribution of copper clusters. The discontinuous variation is observed at n=3, 9, 21, 35, 41, 59, 93, and 139 in the positive spectrum and n=7, 19, 33, 39, 57, 91, and 137 in the negative.

Table 1 shows magic numbers n_p and n_n of positive and negative clusters respectively. The magic numbers of the positive clusters n_p are defined as

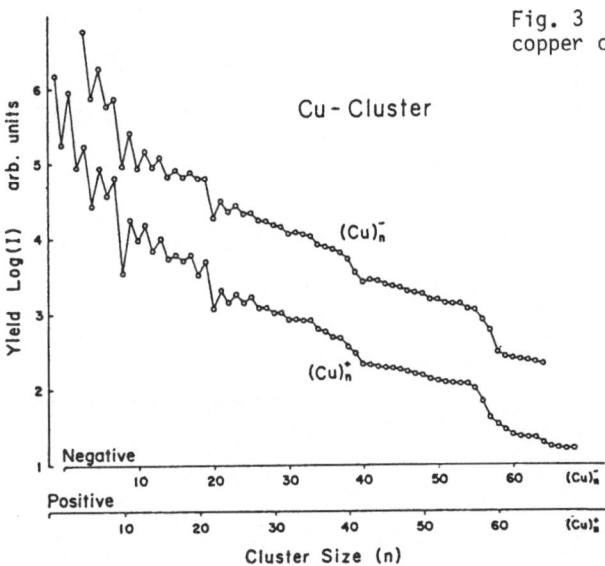

Fig. 3 Size distributions of copper clusters.

Table 1

n	configurations	copper		silver		gold	
		$(Cu)_n^+$	$(Cu)_n^-$	$(Ag)_n^+$	$(Ag)_n^-$	$(Au)_n^+$	$(Au)_n^-$
2	1s	o		o	o	o	x
8	1p	o	o	o	o	o	x
18	1d	Δ	Δ	Δ	Δ	Δ	o
20	2s	o	o	o	o	o	x
34	1f	o	o	o	o	o	o
40	2p	o	o	o	o	x	x
58	1g	o	o	o	o	o	o
92	2d, 1h, 3s	o	o	o	o	o	o
138	2f, 1i, 3p	o	Δ	o	o	o	Δ
198	2g, 1j, 3d, 4s			o	Δ		

$n_p = n+1$ and those of the negative clusters as $n_n = n-1$. The symbol "o" represents an observation of a clear discontinuous variation, the symbol "Δ" represents the observation of a weak variation and "x" represent absence of the variation.

The results obtained in this experiment can be explained on the bases of the one-electron shell model described by Knight et al.[1] in the case of sodium clusters. They observed a shell-closing effect at n=2, 8, 20, 40, 58, and 92. The even-odd alternation was also observed, where even-n clusters are generally larger than odd-n clusters. The magic numbers of sodium clusters related directly to neutral clusters, because the total number of s-valence electrons in a neutral n-mer is n. The shell-closing effect occurs at the (n+1)-mer in the positive clusters and at the (n-1)-mer in the negative clusters, as the total number of s-valence electrons of a positive n-mer is n-1 and that of the negative is n+1. No shell closing effect at n=2 and 8 is seen for the negative gold clusters and effect at n=18 is weak in all cases except for the negative gold cluster. No shell-closing effect at n=40 is observed for gold clusters. The shell-closing effect seems to be more parameter sensitive at n=2, 8, 18, and 40 than at other numbers. A weak shell-closing effect is observed at n=138 for the negative copper and gold clusters, and at n=198 for negative silver clusters. It is not clear whether the effect is really weak or it merely looks weak because the negative ion current at larger cluster sizes is very weak.

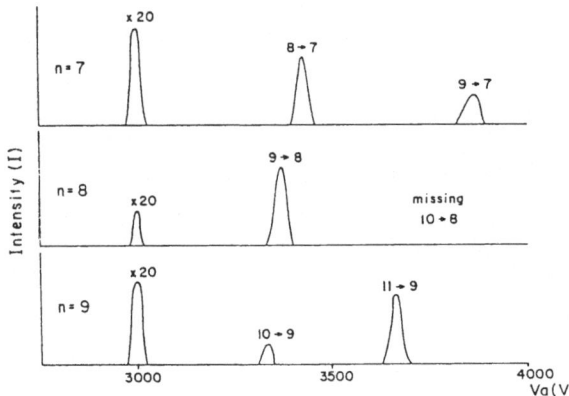

Fig. 4 Intensity behaviour of metastable decay of silver clusters.

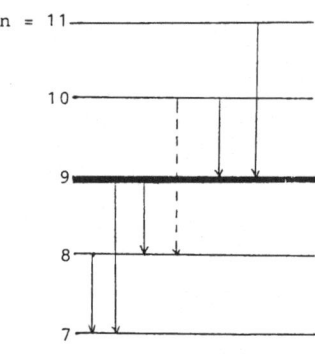

Fig. 5 Decay pattern of 11-, 10-, 9-, and 8-mers of silver clusters.

The metastable decay of 7-, 8-, and 9-mers of the silver clusters was investigated. Figure 4 shows the intensities due to unimolecular decomposition. The sign 8→7 means unimolecular decay from the 8-mer to 7-mer. Peaks corresponding to the decays from 8- to 7-, 9- to 7-, 9- to 8-, 10- to 9-, and 10- to 8-mer were observed, however, a peak corresponding to decay from 10- to 8-mer was not observed. Figure 5 shows this decay pattern in analogy to nuclear decay. From Figs. 4 and 5, it is concluded that the 9-mer ion is more stable than others and constitutes an anchor point for the metastable decay chains of larger n-mers. The same reasoning is assumed to be applicable to other clusters with shell-closing numbers.

The intensity patterns of Zn and Cd clusters have been measured. In these cases, each atom has two free electrons, therefore the total number of valence electrons is always odd. For any singly charged cluster it is not possible for the number of electrons to be equal to the shell-closing number. If the valence electrons adopt a shell structure, as in the cases of noble and alkali metals, it might be expected that some shell-closing effect would be observed at around cluster sizes equal to half of the shell-closing numbers of noble metals. Figure 6 shows the size distribution of the positive and negative clusters of zinc. The trace of the negative clusters is shifted by one cluster size towards the right-hand side. The ion intensities up to n=6 of the positive clusters and up to n=24 of the negative cluster were not determined because of high back ground noise. The traces of two mass spectra are quite similar and high ion intensities are observed at the cluster sizes n=10, 18, 20, 28, 30, 32, 35, 40, 41, 46, 47, 54, 57, 60, and 69 in the positive mass spectrum and at n=27, 29, 31, 34, 40, 45, 46, 52, 54, 56, 60, 61, and 68 in the negative mass spectrum. Similar mass distributions were obtained for Cd clusters. Table 2 shows the cluster sizes, n,

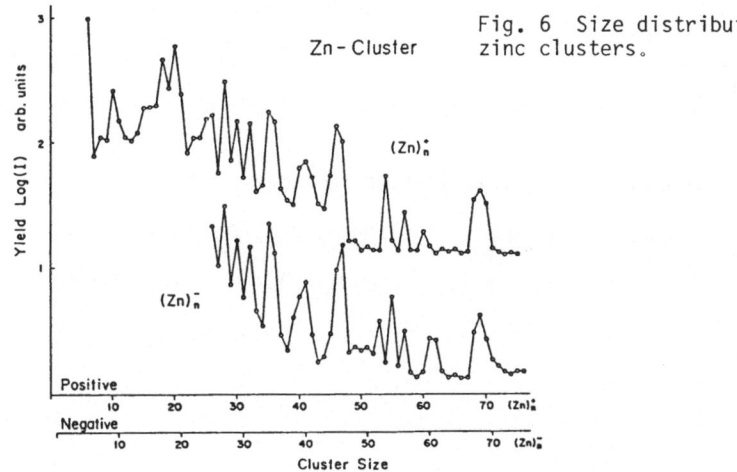

Fig. 6 Size distributions of zinc clusters.

at which the high ion intensities occur, together with the total numbers of free electrons, n_e, and the theoretical shell-closing numbers, n_t. The numbers of the electrons are calculated from the equation $n_e = 2n-1$ (positive clusters and $n_e = 2n+1$ (negative clusters) and the theoretical shell-closing numbers are those of the finite square well potential with round edges. Cluster sizes of almost all pronounced peaks in the negative mass spectrum are shifted by one cluster size from pronounced peaks in the positive mass spectrum, however, the total numbers of free electrons of them are the same in the two cases. The pronounced peaks at n=10, 18, 20, 30(29), 35(34), 46(45), 54(52), 57(56), 69(68) are supposed to correspond to shell closing at n=20, 34, 40, 58, 70, 92, 106, 112, and 138, where the figures in parentheses mean the cluster size of the negative. No peaks corresponding to shell-closing number at n=68 could be observed.

Table 2

n positive	n negative	n_e	n_t	$n_e - n_t$
10	?	19	20	-1
18	?	35	34	1
20	?	39	40	-1
28	27	55	--	-
30	29	59	58	1
32	31	63	--	-
--	--	--	68	-
35	34	69	68	1
40	--	79	--	-
41	40	81	--	-
46	45	91	92	-1
--	52	105	106	-1
54	54	107, 109	106	1, 3
57	56	113	112	1
60	60	120, 121	--	-
--	61	123	--	-
69	68	137	138	-1

From the above facts the stability of zinc and cadmium clusters is assumed to be explained by some modification of the electronic shell model and the numbers of free electrons play an important role in determining stabilities, as in the case of noble and alkali metals.

References

1) W. D. Knight, K. Clemenger, W. A. de Heer. W. A. Saunders, M. Y. Chou and M. L. Cohen: Phys. Rev. Lett., 52, 2141 (1984)
2) I. Katakuse, T. Ichihara, H. Nakabushi, T. Matsuo and H. Matsuda: J. Mass Spectrosc. Jpn., 31, 111 (1983)
3) H. Matsuda: Atomic Masses Fundam. Const., 5, 185 (1976)

Stability of Alkali Metal Clusters

Y. Ishii and S. Sugano

The Institute for Solid State Physics, The University of Tokyo, Roppongi, Tokyo 106, Japan

1. Introduction

In the recent experiments of mass spectroscopy for various kinds of small clusters, the pronounced abundance has been observed at certain numbers of atoms forming a cluster. These special numbers of atoms are usually called "magic numbers" of a cluster. The magic number is considered to be determined by relative stability of clusters of different sizes. KNIGHT et al. [1] have assigned the magic number of sodium clusters as the number of valence electrons giving a closed shell configuration of quantum states in the spherical square well or its modification called Wood-Saxon potential. This model is quite similar to the shell model in nuclear physics.

The physical reason why the spherical square well can be used as the effective potential for valence electrons may be stated as follows: In metal clusters we have valence electrons and core ions. Since the core ions are less mobile, they exert a static field on the valence electrons. This static field reflecting the spatial configuration of ions is neither spherical nor homogeneous. For alkali metals, however, a large part of the static field is screened by the valence electrons and the nearly-free-electron model describes bulk properties very well. Therefore one may consider that the valence electrons in the cluster move independently in a smooth effective field which lacks a great part of the information concerning the geometrical structure of the cluster. It is the main purpose of this paper to examine this statement from a microscopic point of view.

2. Spherical Jellium Model[2-4]

For alkali metals, the nearly-free-electron model, or equivalently the jellium model, is a good approximation for calculation of bulk properties. In this section, we investigate the spherical jellium model, which is an extension of the nearly-free-electron model to a finite system. The effects of spatial configuration of metal ions in a cluster is examined in the following section.

In the spherical jellium model, the positive charge density is distributed uniformly in a sphere with a radius R_0 determined by

$$\int d^3r \, \rho_+(\vec{r}) = N_a, \tag{1}$$

where $\rho_+(\vec{r})$ is the positive charge density and N_a the number of atoms forming a cluster. If the positive charge density is assumed to be equal to that in a bulk solid, R_0 is obtained as

$$R_0 = N_a^{1/3} \, r_s, \tag{2}$$

17

where r_s is the Wigner-Seitz radius for a bulk solid. Then $\rho_+(r)$ is given by

$$\rho_+(\vec{r}) = [(4\pi/3)r_s^3]^{-1} \, \Theta(R_0 - r),$$ (3)

with the unit step function $\Theta(x)$ and $|\vec{r}|=r$. The electrostatic potential due to the positive charge, which is not constant because of a sudden cutoff of the positive charge density at R_0, is calculated as

$$V_+(\vec{r}) = -(N_a/2R_0) \, [3 - (r/R_0)^2], \qquad r < R_0$$

$$= - N_a/r. \qquad\qquad\qquad r > R_0 \qquad (4)$$

The electronic structure is calculated by using the spin-polarized local-density-functional approximation(LDF)[6]. The total energy is obtained as a function of the number of atoms, N_a, the number of electrons, N_e, and the spin polarization, M, as $E_{total}(N_a;N_e,M)$. The ground state is found by changing the spin polarization for fixed N_a and N_e. The calculation is performed for a neutral system, $N_a=N_e=N$, and r_s is taken to be 3.93.a.u. corresponding to sodium.

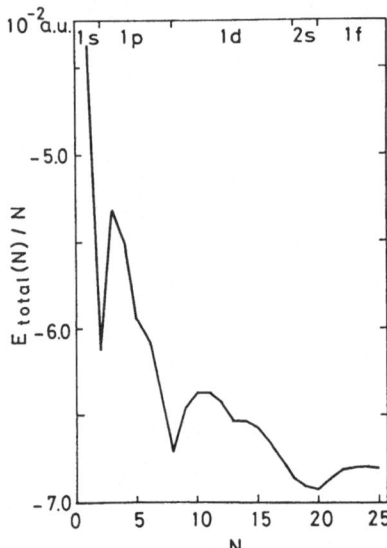

Fig.1. The total energy per atom for a jellium sphere with r_s=3.93 a.u.

The total energy per atom is shown in Fig.1. The relative stability shows downward cusps at N=2, 5, 8, 13, 18, 19 and 20. Here, N=2, 8, 18, and 20 are, respectively, the shell-closing numbers of the 1s, 1p, 1d, and 2s orbitals while N=5, 13 and 19 are, respectively, the numbers giving the half-filled 1s, 1d and 2s orbitals with the high spin configuration. The stabilization at the half-filled shell numbers is caused by the exchange energy gain expected for high spin configurations. It should be noted here that the high degeneracy of single-electron levels is essential in stabilizing the high spin state. If the nonspherical perturbation due to core potentials of metal ions lifts the degeneracy of single-electron levels, the orbital energy cost might overcome the exchange energy gain. This is discussed in the following section.

It is important to consider why a level scheme similar to that for the spherical square well is obtained for the spherical jellium model. This coincidence means that the effective potential for interacting electrons in a jellium sphere has similar radial dependence to the square well potential, which is constant for $r < R_0$ and zero for $r > R_0$. If $r_s \ll R_0$, then the electrostatic field is screened completely and the effective field arises from the exchange-correlation one. Since the exchange-correlation potential is determined locally by the electron density, which is homogeneous in a jellium sphere in this case, an electron is regarded as moving in a field homogeneous everywhere in a jellium sphere. If the screening is not complete, the electron density is not homogeneous due to the finite size effects undergoing Friedel oscillation. Let us suppose that we have a region where the electron density is large compared with the positive background, the electrostatic potential of this region becomes positive and large but the exchange-correlation is negative and large to compensate the inhomogeneity of the potential. Therefore the effective potential for interacting electrons in a jellium sphere is fairly homogeneous in the sphere.

From the total energies for the spin-polarized states, we can estimate the interaction energy between electrons[4]. Here we confine our discussion to the open 1d-shell configuration, which corresponds to the N=9-18 cases. Here we assume that the total-energy difference between the different spin-polarized states comes entirely from the d-d interaction energy as

$$E_{int} = \Delta\varepsilon_{exch}\, n_\uparrow\, n_\downarrow, \tag{5}$$

where n_σ is the number of d electrons of spin σ. The present calculation gives

$$\Delta\varepsilon_{exch} = (3.5 \pm 0.5) \times 10^{-3} \text{ a.u.} \tag{6}$$

3. Effects of Non-Spherical Potential[4]

For calculating the cohesive properties of alkali metals with a great success, the jellium model has been adopted as a first approximation and then the core potentials of metal ions have been taken into account in terms of model pseudopotentials in a perturbative way[5,7]. Applying a similar treatment to the spherical jellium model, we discuss the effects of the non-spherical potential, which reflects the spatial configuration of metal ions in a cluster.

The non-spherical potential is constructed by superposing an appropriate model potential at $\vec{R}_i$ as

$$V_{ion}(\vec{r}) = \sum_i v(\vec{r} - \vec{R}_i), \tag{7}$$

where $\vec{R}_i$ is the positional vector of the sodium atom. In the present calculation the jellium potential for $N_a = 1$,

$$v_{JEL}(\vec{r}) = -(1/2r_s)\,[3 - (r/r_s)^2], \qquad r < r_s$$

$$= -1/r, \qquad r > r_s \tag{8}$$

19

with $r_s=3.93$ a.u. and the model pseudopotential proposed by ASHCROFT and LANGRETH[7],

$$v_{AL}(\vec{r}) = 0.0 \qquad\qquad\qquad r < r_c$$

$$= -1/r, \qquad\qquad\qquad r > r_c \qquad\qquad (9)$$

with $r_c=1.67$ a.u. are used as the model potentials.

Since the unperturbed system is the spherical jellium, which is solved for the potential $V_+(\vec{r})$, we calculate the level splitting by treating

$$\Delta V(\vec{r}) = V_{ion}(\vec{r}) - V_+(\vec{r}), \qquad\qquad\qquad (10)$$

as a perturbation. For the $N_a=15$ cluster, the geometry of body-centered-cubic(bcc) is possible. The level splitting in the bcc geometry is shown in Fig.2. Comparing with the results of the linear combination of atomic orbitals[8] for a $N_a=15$ lithium cluster, we consider that the present perturbative treatment gives the quantitatively satisfactory evaluation of the effects of the non-spherical potential due to the spatial configuration of metal ions in a cluster.

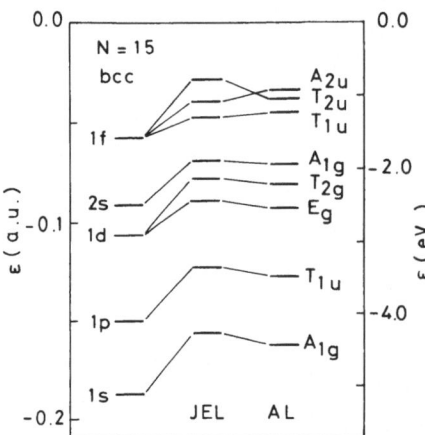

Fig.2. Level splitting in the bcc geometry for a $N_a=15$ cluster

For the $N_a=13$ cluster, several geometries for the spatial configuration of metal ions are possible: face-centered-cubic(fcc), hexagonal-close-packed(hcp) and icosahedral(icos) geometries (Fig.3). In Figs.4, the level scheme for each geometry is shown. In any case, the level splitting due to the non-spherical perturbation is smaller than the separation of levels of the original jellium orbitals except for that between the 1d and 2s orbitals. Furthermore the occupation of the molecular orbitals is unchanged in the presence of the non-spherical perturbation: the occupied jellium orbital is still occupied. These results are due to a good screening of the core ion potentials by valence electrons, that gives relatively small non-spherical perturbation.

The competition between the exchange energy gain and the orbital energy cost can be examined more precisely for the $N_a=13$ cluster[4]. Here we

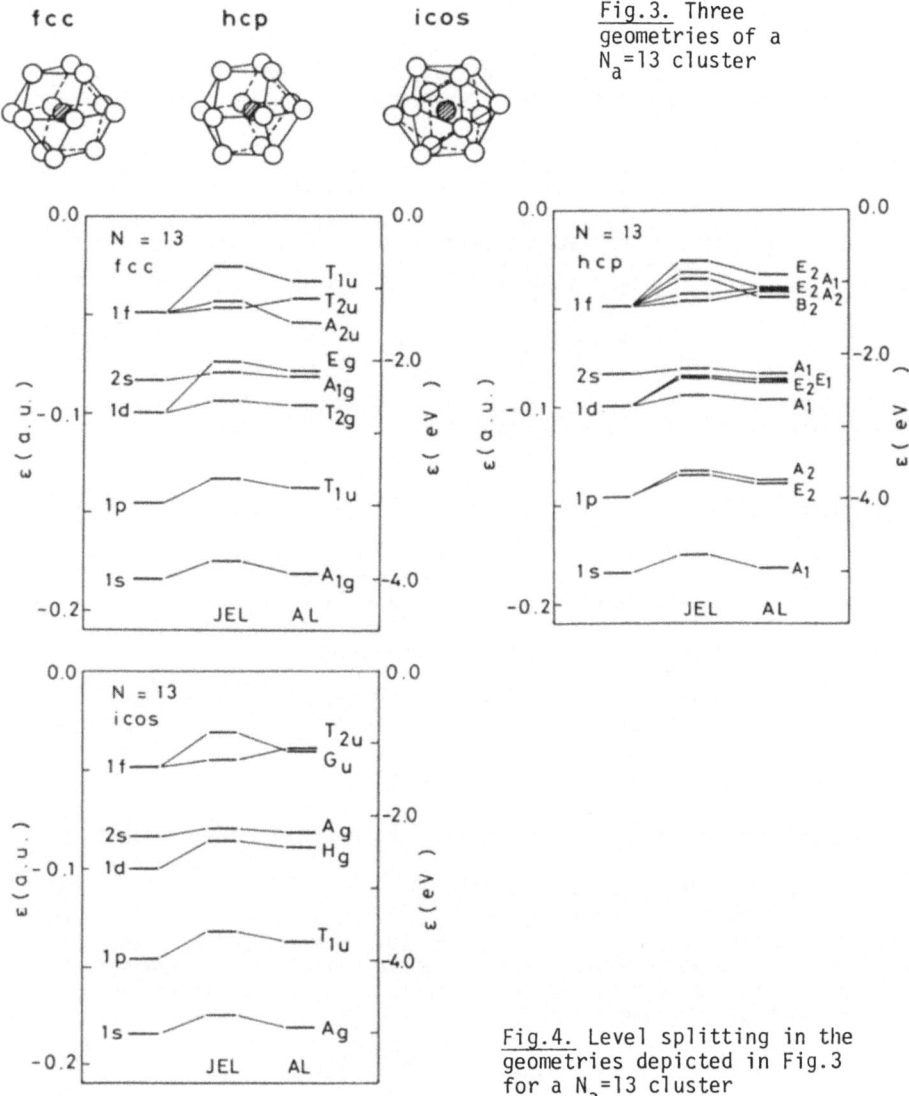

fcc   **hcp**   **icos**

Fig.3. Three geometries of a $N_a=13$ cluster

Fig.4. Level splitting in the geometries depicted in Fig.3 for a $N_a=13$ cluster

confine ourselves to the fcc cluster. Ignoring the A_{1g} orbital from 2s, we calculate the energy difference between the low-spin and high-spin states with M=1 and 5, respectively, which are obtained by accommodating five electrons in the triply degenerate T_{2g} and the doubly degenerate E_g orbitals in appropriate ways;

$$E_{total}(13; 13, 1) - E_{total}(13; 13, 5) = 6\Delta\varepsilon_{exch} - 2\Delta_{fcc}, \qquad (11)$$

where Δ_{fcc} is the splitting of the d level. Using eq.(6) and the calculated value of $\Delta_{fcc}=2.0\times10^{-2}$ a.u., we find the energy difference in eq.(11) is negative;

$$E_{total}(13; 13, 1) < E_{total}(13; 13, 5).$$

Thus it is concluded that the orbital energy cost overcomes the exchange energy gain resulting in the low-spin ground state.

It should be noted here that our spin-polarized state is that averaged over all the multiplets with a given spin. For example, the states with the highest spin are those averaged over ^3F and ^3P for d^2, ^4F and ^4P for d^3. Therefore, our spin-polarized states have to be further stabilized by the multiplet splitting. The detailed examination[4] has revealed that the stabilization energy due to the multiplet formation of all the highest spin states for $d^2(N=10)$, $d^3(N=11)$, $d^7(N=15)$ and $d^8(N=16)$ is given as

$$\Delta\varepsilon_{mult} = 0.45 \, \Delta\varepsilon_{exch} = (1.6 \pm 0.2) \times 10^{-3} \text{ a.u..} \tag{12}$$

In the present paper, we have not discussed this stabilization in detail, as it is smaller than the exchange stabilization.

4. Concluding Remarks

The analysis assuming that the valence electrons move in a spherical and homogeneous field confined in a sphere of radius R_0 in eq.(2) works well for explaining the magic numbers of alkali metal clusters. We have discussed the physical origin of a "spherical" and "homogeneous" field from a microscopic point of view. However, the N-dependence of the ionization potential does not seem to be consistent with the assumption of the spherical and homogeneous effective field. To clarify this point, we first note that the ionization potential is defined by

$$I_N = E_{total}(N; N, 0) - E_{total}(N; N-1, 0),$$

while the relative stability is considered to be reflected in the quantity,

$$\Delta E = E_{total}(N; N, 0) - E_{total}(N-1; N-1, 0).$$

The anomalies of the mass spectra at N=2, 8, 20,$\cdots$ are identified to those of ΔE. If the absence of the magic number anomalies is assumed for I_N as found in the measurement of the ionization potential, one is led to the conclusion that the quantity,

$$\Delta E - I_N = E_{total}(N; N-1, 0) - E_{total}(N-1; N-1, 0), \tag{13}$$

has the magic number anomalies in contradiction to expectation: one expects that the effective potential for valence electrons is a smooth function of the number of atoms as far as the spherical and homogeneous effective field is assumed. At this point we emphasize the necessity of further experimental and theoretical studies to understand physics of the magic numbers more clearly. In particular, studies of dynamical processes such as fragmentation and coagulation of clusters are strongly desired.

1. W. D. Knight, K. Clemenger, W. A. de Heer, W. A. Saunders, G. Y. Chou and M. L. Cohen: Phys. Rev. Lett. 52, 2141 (1984).
2. M. Y. Chou, A. Cleland and M. L. Cohen: Solid State Commun. 52, 654 (1984).

    M. Y. Chou and M. L. Cohen: Phys. Lett. 113A, 420 (1986).
3. B. K. Rao, P. Jena and M. Manninen: Phys. Rev. B32, 477 (1985).
    B. K. Rao and P. Jena: Phys. Rev. B32, 2058 (1985).
4. Y. Ishii, S. Ohnishi and S. Sugano: Phys. Rev. B33, 5271 (1986).
5. V. Heine and D. Weaire: In Solid State Physics, Vol.24, (Academic Press, New York 1970), p.250.
6. P. Hohenberg and W. Kohn: Phys. Rev. 136, B864 (1964).
    W. Kohn and L. J. Sham: Phys. Rev. 140, A1133 (1965).
7. N. W. Ashcroft and D. C. Langreth: Phys. Rev. 155, 682 (1967).
    N. W. Ashcroft: J. Phys. C1, 232 (1968).
8. F. R. Redfern, R. C. Chaney and R. G. Rudolf: Phys. Rev. B32, 5023 (1985).
9. A. Herrmann, E. Schumacher and L. Woste: J. Chem. Phys. 68, 2327 (1978).
    W. A. Saunders, K. Clemenger, W. A. de Heer and W. D. Knight: Phys. Rev. B32, 1366 (1985).

Part II

Electronic Structure

Electron Correlations and Chemical Bonding in Aggregates – a Simplified Approach

J. Friedel

Université Paris-Sud, Laboratoire de Physique des Solides, F-91405 Orsay Cedex, France

One discusses the consequences and validity of a simple Hubbard approach to electron correlations in aggregates.

1. Introduction

Electron correlations are treated to great accuracy by configuration interactions techniques in small molecules. This technique however, requires computations of impossible length in large aggregates or macroscopic solids. One has then to fall back on approximate treatments.

The standard technique today treats the electrons as independent and delocalised in the aggregate or the solid ; correlations only appear in a local correction of the effective self-consistent potential, computed as for a gas of free electrons with the same density : this idea, introduced by Slater, was developed by Kohn and Sham.

Such an approach cannot possibly treat the Verwey-Mott transition where the electron repulsions become large enough with respect to their kinetic energy to localise them. To treat such cases, one is led to simplify the description of the electronic properties by making atomic averages. One thus describes the wave functions $|\psi\rangle$ using linearised combinations of atomic orbitals $|a\rangle$:

$$|\psi\rangle \cong \sum_a \alpha_a |a\rangle \qquad (1)$$

and uses interatomic transfer integrals to describe the kinetic energy

$$t = \langle a | V_a | b \rangle \qquad (2)$$

where V_a is the potential energy for the $|a\rangle$ orbital, in the field of its inner ion ; b is usually limited to be nearest neighbour to a, owing to the long-range exponential decrease of $|a>$ and $|b>$. One can then use a hierarchy of intra and interatomic Coulomb integrals (b neighbour to a) :

$$U = <a(1) \ a(2) \ |V_{12}|a(1) \ a(2)> \quad (3)$$

$$u = <a(1) \ b(2) \ |V_{12}|a(1) \ b(2)> \quad (4)$$

$$j = <a(1) \ b(2) \ |V_{12}|b(1) \ a(2)> \quad (5)$$

$$v = <a(1) \ a(2) \ |V_{12}|a(1) \ b(2)> \quad (5)'$$

The <u>simplest</u> such model, where one treats the atomic orbitals as nearly orthogonal ($S = <a|b> \cong 0$) and neglects all u, j, v terms with respect to the intraatomic repulsions U, was introduced by Slater[1] in the Hartree Fock approximation ; it was first treated in a variational way by Hubbard[2], and by Gutzwiller[3], and also exactly by Kanamori[4] in the limit of few carriers. In the Hartree Fock approximation, the independent electrons can be considered as distributed randomly over all atomic spinorbitals so as to respect locally the Pauli principle[5]. The effect of positive U is qualitatively to reduce the charge fluctuations produced on each atom by this disorder.

In the limit of weak U's, a development in $|U/t|$ of the total energy is valid[6]. The term linear in U is the Hartree Fock correction due to electron interactions, computed for independent delocalised electrons[7]. The term in U^2 is the first correction from independent behaviour due to electron correlations. More precisely, it is due to a reduction of the atomic charge fluctuations, owing to <u>correlations of the interatomic jumps of the electrons</u>[6]. It is thus less local than the Slater Kohn Sham term.

The purpose of this paper is to show that this perturbation limit can help to systematise correlation effects on the atomic structure of aggregates of elements. We discuss in detail the quantitative validity of such an approach.

2. Dimers versus closepacked macroscopic aggregates.

In the Hubbard approximation recalled above, the orbitals $|a>$ are 1s states for H aggregates.

The energy of a H_2 <u>dimer</u> reads, in its stable singlet state[8],

$$E_2 \cong 2E_0 + R + \frac{1}{2}\left(U - \sqrt{U^2 + 16\,t^2}\right) \tag{6}$$

where $2E_0$ is the energy of two separate neutral atoms and R their electrostatic repulsion. In the limit $|U/t|$ small, this reads $(t < 0)$:

$$E_2 \cong 2E_0 + R + 2t + \frac{U}{2} + \frac{U^2}{16t} + O(U^3/t^2) \tag{7}$$

For order of magnitude estimates, one can assume

$$
\begin{aligned}
t &\cong t_0 \exp(-pr) \\
R &\cong R_0 \exp(-qr)
\end{aligned} \tag{8}
$$

with

$$2 < q/p < 3 \tag{9}$$

where r is the distance between protons. At equilibrium,

$$E_2 \cong 2E_0 + 2\left(1 - \frac{p}{q}\right)t_2 + \frac{U}{2} + \left(1 + \frac{p}{q}\right)\frac{U^2}{16t_2} \tag{10}$$

with

$$t_2 = to \exp(-p\,r_2)$$

and

$$(q-p)r_2 \cong \ln\frac{q\,R_0}{2p|t_0|} + \frac{U^2}{32t_2^2} \tag{11}$$

For a macroscopic closepacked aggregate H_N, the energy per atom can be written in a similar way[8] :

$$\frac{1}{N}E_N \cong E_0 + \frac{n}{2}R - \frac{w}{4} + \frac{U}{4} - \frac{U^2}{16\,w} + O(U^3/w^2) \tag{12}$$

where n is the number of nearest neighbours per atom. The 1s band width w is approximately measured, in terms of its second moment, by[9]

$$w \cong \sqrt{12\,n\,|t|} \tag{13}$$

Minimising with respect to r gives

$$\frac{1}{N}E_N \cong E_0 - \left(1 - \frac{p}{q}\right)\frac{w}{4} + \frac{U}{4} - \left(1 + \frac{p}{q}\right)\frac{U^2}{w} \tag{14}$$

with

$$w = \sqrt{12\,n}\,|t_N| = \sqrt{12\,n}\,|t_0|\exp(-pr_N) \tag{15}$$

and

$$(q-p)r_N \cong \ln\sqrt{\frac{n}{3}}\,\frac{q}{p}\,\frac{R_0}{|t_0|} + \frac{U^2}{4w^2} \tag{16}$$

Similar formulae would hold for the valence s electrons of alkali metals, and the treatment can, in principle, be extended to polyvalent elements.

Comparing (12) to (7) shows that the band term in w stabilises a closepacked aggregate with maximum number n of neighbours, but that the correlation term in U^2 stabilises, if large enough, a collection of dimers ; the repulsive terms in R tip the balance in favour of dimers, and the more so if their spatial variation is less abrupt (q small, more comparable to p).

There is therefore a critical value of $|U/t_2|$ below which a closepacked aggregate is more stable than a collection of dimers, and above which dimers are more stable[8]. This value depends on a somewhat critical way of the value of q/p (Table I). For any reasonable value of q/p as given in (10), it is anyway well below the crossover value $|U/t| = 4$ above which, according to (6), a development in $|U/t|$ is no longer valid and must be replaced by a development in $|t/U|$.

TABLE I - Critical values of $|U/t_2|$ when comparing a closepacked aggregate and a collection of dimers.

q/p	2	2.5	3
$|U_c/t_2|$	O	2.1	3.5

The fact that H produces dimers while alkali metals favour closepacked aggregates strongly suggests that, in a Hubbard model, one must take $U > U_c$ for H and $U < U_c \neq O$ for alkali metals. For later estimates, we shall take

$$q/p = 2.5 \qquad (17)$$

for which $|U_c/t_2|$ is slightly larger than 2.

More generally, the fact that only H, O, N and the halogens favour dimers and all the other elements prefer macroscopic aggregates can be seen as indicating that only those elements, with large electronegativities, have large values of

U and thus large correlation effects. We shall come back to this in more detail.

Note that, in this order of magnitude estimate, we have neglected the small dispersion forces which stabilise the molecular crystals made of dimers.

3. Trimers of monovalent elements.

H_3 and H_3^+ aggregates have been studied in connection with astrophysics[10]. Trimers of alkali and noble metals (Cu, Ag, Au) have been studied in jets[11]. They have given rise to a large corpus of refined computations of their electronic structures[12][13].

Owing to these theoretical and experimental studies, such trimers are therefore fairly well known ; and one can ask whether the simple Hubbard approach mentioned above can be fitted with their behaviour.

The points I want to discuss are that H_3 is unstable and that stable H_3^+ form an equilateral triangle, while neutral trimers of alkalis form isosceles triangles with a summit angle θ between 60 and 90°. This difference in behaviour between H and alkalis can again be related to a larger value of $|U/t_2|$ for H. (We neglect again here the small dispersion forces which bind very weakly H to H_2).

For H_3 and H_3^+, the simplest Hubbard model again only uses 1s atomic orbitals. A straightforward if somewhat lengthy computation leads, for their energies E_3 and E_3^+ to the following expressions[14]

$$E_3 = 3E_o + 3R + 3t + \frac{2}{3}U + \frac{U^2}{18t} + O(U^3/t^2) \qquad (18)$$

$$E'_3 = 3E_o + 2R + 2\sqrt{2}t + \frac{5}{8}U + \frac{U^2}{17,3t} + O(U^3/t^2) \qquad (19)$$

$$E''_3 = E_2 + E_o = 3E_o + R + 2t + \frac{U}{2} + \frac{U^2}{16t} + O(U^3/t^2) \qquad (20)$$

$$E_3^+ = 2E_o + 3R + 4t + \frac{1}{3}U + O(U^2/t) \qquad (21)$$

$$E_3^{+'} = 2E_o + 2R + 2\sqrt{2}t + \frac{3}{8}U + O(U^2/t) \qquad (22)$$

$$E_3^{+''} = E_2 = 2E_o + R + 2t + \frac{U}{2} + O(U^2/t) \qquad (23)$$

In these expressions, E_3 and E_3^+ refer to an equilateral triangle ; E'_3 and $E_3^{+'}$ to a linear chain ; and E''_3 and $E_3^{+''}$ to a dimer

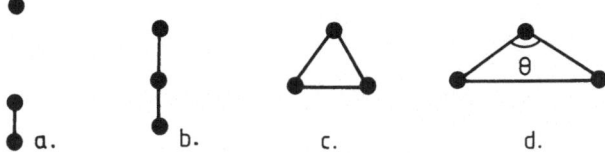

a.   b.   c.   d.

Figure 1. Various configurations for a trimer : a - dimer + atom or ion ; b - linear chain ; c - equilateral triangle ; d - isosceles triangle.

plus an atom or an ion (figure 1 a, b, c). The differences in the 'band' terms in t arise from differences in the spectral distribution of delocalised one-electron states (figure 2).

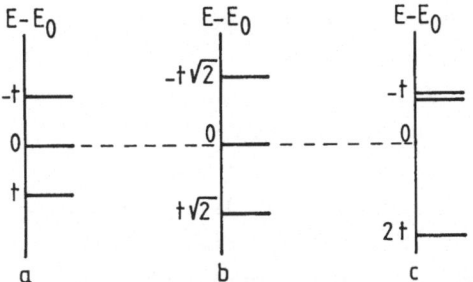

Figure 2. Spectral distribution of one-electron delocalised states : a- dimer + atom ; b- linear chain ; c- equilateral triangle.

For H^+_3, with two electrons, they clearly stabilise the <u>equilateral triangle</u>. This is also true for the term linear in U_0.

      This is also true for H_3, but by a much smaller amount. Here, not only the terms in U^2 but also the terms in U favour the dimer, as do the repulsive terms. One expects therefore the dimer to be preferred if $|U/t_2|$ is large enough.

$$E_3 = 3E_0 + 3(1 - \frac{p}{q})t_3 + \frac{2U}{3} + (1 + \frac{p}{q})\frac{U^2}{18t_3} + .. \qquad (24)$$

$$E'_3 = 3E_0 + 2\sqrt{2}(1 - \frac{p}{q})t'_3 + \frac{5U}{8} + (1 + \frac{p}{q})\frac{U^2}{17,3t'_3} + .. \qquad (25)$$

$$E^+_3 = 2E_0 + 4(1 - \frac{p}{q})t^+_3 + \frac{3U}{8} + .. \qquad (26)$$

$$E^{+'}_3 = 2E_0 + 2\sqrt{2}(1 - \frac{p}{q})t^{+'}_3 + \frac{U}{3} + .. \qquad (27)$$

31

with

$$(q-p)r_3 = \ell n \; \frac{g}{p} \; \frac{R_o}{|t_o|} + \frac{U^2}{54t^2_3} + .. \tag{28}$$

$$(q-p)r'_3 = \ell n \; \frac{q}{\sqrt{2}p} \; \frac{R_o}{|t_o|} + \frac{U^2}{49t'^2_3} + .. \tag{29}$$

$$(q-p)r^+_3 = \ell n \; \frac{3q}{4p} \; \frac{R_o}{|t_o|} + .. \tag{30}$$

$$(q-p)r^{+'}_3 = \ell n \; \frac{q}{\sqrt{2}p} \; \frac{R_o}{|t_o|} + .. \tag{31}$$

One deduces from these relations that, for $q/p=2.5$, there is again a critical value of $|U/t_2|$ of order 2. For $|U/t_2|<2$, the equilateral triangle is more stable than the linear chain or the dimer. For $|U/t_2|> 2$, the dimer is the most stable. Account has been taken that for the critical value $|U/t_2|= 2$, the selfconsistent solutions of (11), (28) and (29) give[14] $|t_3/t_2|= 0.6$ and $t'_3/t_2 = 0.75$. This result extends to trimers the result for closepacked aggregates, but for a somewhat different physical reason, as the contribution of the terms in U depends now on the configuration considered. Thus the instability of H_3 versus a decomposition into a dimer and an atom suggests again that $|U/t_2|> 2$ for hydrogen ; it is coherent with the fact that a closepacked macroscopic aggregate of H atoms is less stable than a collection of dimers.

The preceding treatment extends directly to <u>trimers of alkali metals or noble metals</u>, replacing again the 1s orbitals by valency orbitals. Coherent with the larger stability of macroscopic closepacked aggregates for these metals, thus $|U/t_2|< 2$; one expects an equilateral triangle to be more stable than a linear chain or a dimer.

However, inspection of figure 2c shows that for the equilateral triangle, the one-electron state of energy E_o-t, occupied by a single electron, is doubly degenerate. One then expects a <u>Jahn Teller distortion</u> of neutral trimers from the equilateral form. A development to second order in $\theta - \frac{\pi}{6}$ where θ is the summit angle of an isosceles triangle (figure 1 d) and up to the term in U, thus valid for $|U/t_2| << 2$, shows[14] that the stable state corresponds to $\frac{\pi}{2} > \theta > \frac{\pi}{3}$. This is in agreement with most computations and the few experiments made so far. The lowering of energy by Jahn Teller distortion is not large, so does not alter significantly the discussion.

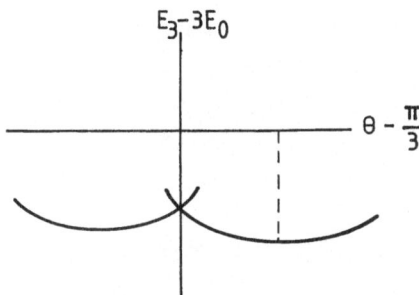

E_3-3E_0

$\theta - \dfrac{\pi}{3}$

Figure 3. Jahn Teller distortion from the equilateral form $(\theta = \dfrac{\pi}{3})$.

4. Validity of the Hubbard model.

One can try and apply this model quantitatively.

For H_2, one computes[8] $U = 17$ eV. From the experimental value of the cohesive energy[9] $2E_o-E_2 = 4.4.$ eV, one obtains $|U/t_2| \cong 2$ for $q/p = 2.5$. This justifies the perturbation scheme used and seems to give a somewhat low but reasonable value of $|U/t_2|$.

One should however be aware of three types of complications, as shown in more complete LCAO computations[17].

a) First, as the interatomic Coulomb interaction u is retained in the definition of R, it would be more coherent to retain it with respect to U. One finds then that, in (6), (7), (10), one must replace U by U - u, a self-evident result [16]. For $r_2 = 1.4$ a_o, the experimental distance for H_2, u = 19 eV. Strictly speaking, all correlation effects disappear in H_2 in this limit.

b) One can then argue that interatomic configuration interactions modify the effective values of U and u.

One knows indeed that for two 1s electrons localised on different atoms, virtual excitations (mainly 1s → 2p) compensate u by a van der Waals attraction[18] that become of the same order as u for the very small equilibrium distance r_2 of H_2. For alkali dimers where r_2 is much larger, u becomes a smaller fraction of U and the van der Waals term correlatively much more reduced ; but u can be somewhat reduced by the polarisation of the inner ions. In larger aggregates, a further small reduction of u is expected from the screening effect of other electrons[16].

33

Correlatively, for two electrons on the same atom, virtual excitations (mainly $1s \rightarrow 2s$, thus different from those responsible for van der Waals interactions) lower U to an effective[8][19] value U_e which takes into account interatomic electron correlations. U_e can be deduced experimentally from properties of free atoms, as the energy involved in the reaction changing two neutral atoms A into a pair of ions :

$$2A \rightarrow A^+ + A^- - U_e$$

Thus

$$U_e = I - A$$

where I is the first ionisation potential and A the electron affinity. Figure 4 shows that U_e varies rather regularly in the periodic table, in a way similar to Pauling's[20] electronegativity E. This is not surprising, as $E \propto I + A$ by definition, and $I >> A$.

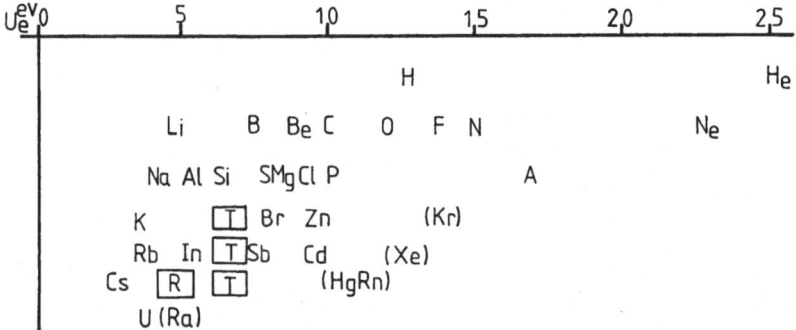

Figure 4. Variation of $U_e = I - A$ in the periodic table. Values of A taken from[26]. Elements in parenthesis : value of A unknown ; T transitional elements ; R rare earths elements.

Following this reasoning, one expects the Hubbard model to hold with an effective value of order of U_e. Indeed this explains qualitatively that elements with large electronegativity tend to dimerise, while those with small electronegativity tend to build closepacked aggregates, and especially explains the difference of behaviour between H and alkali metals. It also suggests that divalent metals, with abnormally large values U_e, should have stronger correlation effects[21]. Finally it is generally coherent with the fact that the observed alternations in stability of metallic clusters, depending on the parity of their number of electrons[27], are not washed out by correlations[28].

A strict Hubbard model where U is replaced by $U_e =$ I-A however does not fit numerically : values of t thus deduced from the measured cohesive energies[22] of dimers would give $|U/t| < 2$ for H and $|U/t| > 2$ for alkalis ! In the same way, the value of U to be used in a Hubbard model for transition element[6][23][24] is near to 2 eV, much smaller than U_e as given figure 4. It is clear that some smaller but significative contribution from an effective value of u should be included, as well as, for H_2, a van der Waals correction.

c) But deeper complications[16] come from <u>the role of transfer integrals on interaction configurations</u>. First the role of t is to cut down the inter and intra atomic correlations corrections[14] on u and U ; in the limit of large t, a factor $\frac{1}{4}$ develops for a dimer from the fact that the pair of electrons spends quarter of its time on a given atomic configuration. Second hybridisation due to virtual intraatomic excitations increases the effective value of t. Third virtual interatomic excitations become possible. The two last corrections decrease the ratios $|U/t|$ and $|u/t|$, thus work against the first one; and it is not clear what exact role they play in specific cases.

In conclusion, it seems that a simple Hubbard model contains the essentials to describe at least qualitatively the role of correlations on chemical bonding in aggregates as well as solids. This has the great advantage of simplicity and generality. However, the exact value to take for the characteristic intraatomic energy U seems, except for H, systematically smaller than the simple atomic estimate $U_e \simeq I - A$. The case of H_2, because of the small bond length, is an extreme and somewhat non-characteristic case of these complications[25].

References.

1 - Slater J.C. Phys. Rev. <u>49</u>, 537 (1936).
2 - Hubbard J. J. Proc. Roy. Soc. A276, 238 (1963).
3 - Gutzwiller M.C., Phys. Rev. <u>134</u>, 4923 (1964).
4 - Kanamori J., J. Appl. Phys. <u>36</u>, 939 (1965).
5 - Friedel J., J. Physique Rad. <u>16</u>, 829 (1955).
6 - Friedel J. and Sayers C.M., J. Physique <u>38</u>, 697 (1977) ;
    Treglia G., Ducastelle F. and Spanjaard D., J. Physique <u>41</u>,
    281 (1980).

7 - Allan G., Sayers C.M. and Friedel J., J. Physique Lett.
   <u>41</u>, 287 (1980).
8 - Friedel J. in Physics and Chemistry of electrons and ions
   in Condensed Matter, J.V. Acrivos, N.F. Mott and Yoffee
   A.D. ed. NATO ASI Series C Math and Phys. Sci. <u>130</u> (1984)
   Deidel Publ. Dordrecht Holland.
9 - Friedel J. in J.M. Ziman ed. The Physics of Metals I,
   Cambridge University Press, Cambridge (1961).
10 - Carney G.D. Can. J. Phys. <u>62</u>, 187 (1984).
11 - Knight W.D. Helv. Phys. Acta, <u>56</u>, 521 (1983).
12 - Hirschfelder J.O., Eyring H. and Rosen N., J. Chem. Phys.
   <u>4</u>, 121 (1936) cf also Tennyson J. Chem. Phys. Lett. <u>86</u>,
   181 (1982).
13 - Flad J., Igel-Mann G., Preuss H. and Stoll H., Surf. Sci.
   <u>156</u>, 399 (1985) ; Martins J.L., Buttet J. and Car R.,
   Berl. Bun. Phys. Chem. <u>88</u>, 239 (1984) ; Andreoni W. and
   Martins J.L., Surf. Sci. <u>156</u> (1985).
14 - Friedel J. to be published in J. Physique.
15 - Kittel C., Introduction to Solid State Physics 3^d Ed. John
   Wiley New York (1966).
16 - Friedel J. and Noguera C. Intern. J. Quantum Chem. <u>23</u>,1209
   (1983).
17 - Rao B.X. and Jena P. Phys. Rev. B<u>32</u>, 2058 (1985).
18 - Schiff L.I., Quantum Mechanics, McGraw Hill New York (1955).
19 - Friedel J., Physica <u>109</u> and <u>110</u> B, 1421 (1982).
20 - Pauling L. The nature of the chemical bond, Cornell Univ.
   Press, New York (1960).
21 - Durand G., Dandey J.P. and Malrieu J.P., J. Physique
   (in press)
22 - Herzberg G. Molecular Spectra and Molecular Structure I.
   Spectra of diatonic Molecules 2^{d}Ed.van Reinhold Princeton N.J.(1950)
23 - Treglia G., Ducastelle F. and Spanjaard D., Phys. Rev.
   B<u>22</u>, 6472 (1980).
24 - Bourdin J.P., Desjonquères M.C., Spanjaard D. and Friedel
   J., Surf. Sci. <u>137</u>, L345 (1985).
25 - cf Laforgue A. and Varandas A.J.C. Compte Rendus (Paris)
   <u>302</u>, 395 (1986).
26 - Hotep H. and Lineberger H.C., J. Phys. Chem. Ref. data
   <u>4</u>, 2 (1975).
27 - Joyes P., J. Phys. Chem. Sol. <u>32</u>, 1269 (1971).
28 - Joyes P., Surf. Sci. <u>106</u>, 272 (1981).

Phenomenological and Microscopic Theories of Structure

D.G. Pettifor

Department of Mathematics, Imperial College of Science and Technology, London SW7 2BZ, UK

1. Introduction

The NEC Corporation are to be congratulated for their foresight in setting up these biennial symposia on "The Fundamental Approach to New Material Phases". This is indeed a timely step, for during the past few years a meaningful dialogue has started between fundamental theorists, materials scientists, and engineers who are interested in designing materials (see, for example [1]). This dialogue has arisen partly because quantum theorists are now beginning to predict from first principles the cohesion and relative structural stability of equilibrium and metastable phases. The advent of fast computers and the development of efficient computer codes have demonstrated that the intuitive and very simple Hartree-type equations of the Local Density Functional (LDF) approximation [2] are usually sufficient [3] to calculate the ground state properties of molecules and solids within a few percent (see, for example, [4], [5], and [6]). The present symposium has numerous papers which assess the accuracy of the LDF approximation in predicting the properties of microclusters.

However, rather than focussing on a particular system whose properties will be computed as accurately as possible within density functional theory, I will address the more global problem of ordering and explaining the trends in the structural stability of binary systems. The designers of a new material phase are faced with a multitude of choice, since they have at their disposal some one hundred elements that they can mix together in endless permutations. Even if they were to design the simplest possible binary system, a solid with equal proportions of two constituent atoms, they would still face some 5000 (or 100 x 99 ÷ 2) different systems displaying hundreds of types of structure. In section 2 I will show that the large empirical data base on the equilibrium structures of binary systems can be ordered with the aid of two-dimensional structure maps. In section 3 the microscopic origin of the observed structural domains of the pd-bonded AB compounds will be examined within the Tight-Binding Bond (TBB) model. This allows the influence of atomic size, electronegativity difference, and the average number of valence electrons per atom to be studied explicitly from first principles. In section 4 I will conclude by stressing the applicability of the TBB model to the computation of cohesion and interatomic forces in microclusters.

2. Phenomenological Structure Maps

The structural stability of binary systems is known [7] to be determined by numerous factors such as the electronegativity difference ΔX between the constituent atoms, the atomic size difference ΔY and the average number of valence electrons per atom $\bar{Z}$. These factors have recently been used by VILLARS [8] for constructing three-dimensional structure maps in which any given compound with a particular stoichiometry is characterized by some point (ΔX, ΔY, $\bar{Z}$). He considered, for example, the equiatomic AB compounds and found that 20 different structure types such as NaCl or CsCl usually separated into different regions within his three-dimensional space. Unfortunately, three-dimensional structure maps are not easy to work with, sixteen separate two-dimensional plots being required to display the data for the equiatomic AB compounds. More seriously, the fifth most important structure type, NiAs, could not be separated and was omitted. Classical concepts such as the electronegativity difference ΔX, the atomic size difference ΔY [9], the average number of valence electrons per atom $\bar{Z}$ can not embrace the quantum mechanical bonding behaviour of valence s, p and d orbitals with their different angular lobes.

On the other hand, rather than trying to find a set of <u>physical</u> co-ordinates which will produce a separation of the structure types within some n-dimensional space, we can look instead for a single <u>phenomenological</u> co-ordinate which will lead to good structural separation of the empirical data in two dimensions [10]. This may be achieved by running a one-dimensional string through the two-dimensional periodic table as shown in table 1 [11]. Pulling the ends of the string apart orders all the elements along a one-dimensional axis, their sequential order being given by what I have called the Mendeleev number $\mathcal{M}$. This simple procedure is found to provide excellent structural separation [12] of all binary compounds with a

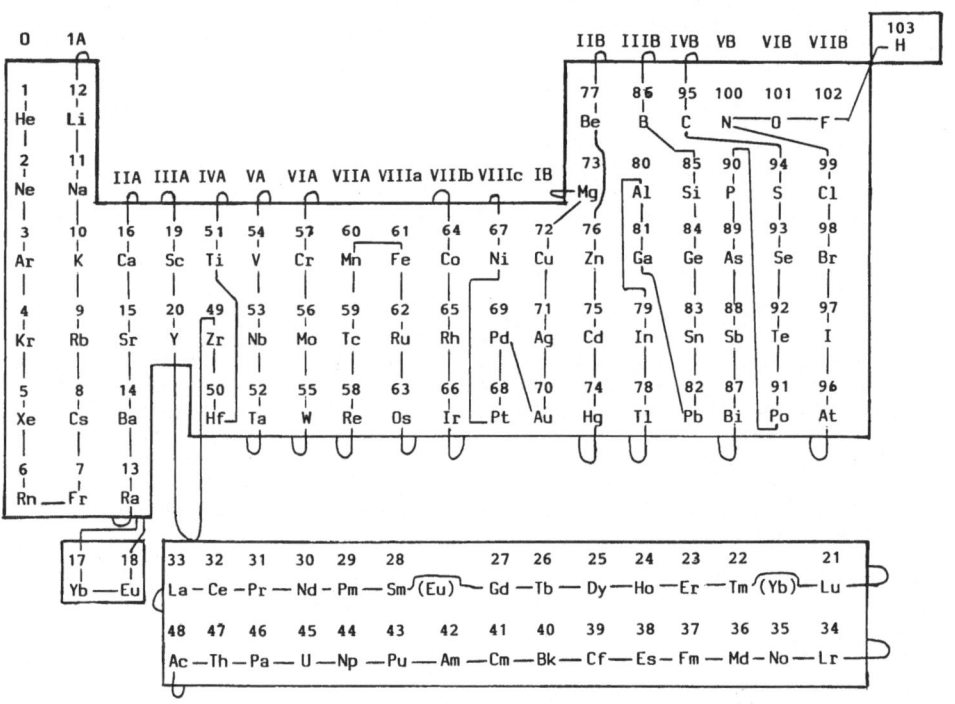

Table 1: The string running through this modified Periodic Table puts all the elements in sequential order, given by the Mendeleev number. Note that the group IIA elements beryllium and magnesium have been grouped with group IIB and that the divalent rare earths have been separated from the trivalent.

given stoichiometry A_mB_n within a single two-dimensional structure map $(\mathcal{M}_A, \mathcal{M}_B)$. The following stoichiometries have been plotted and discussed in [13]: AB, AB_2, AB_3, AB_4, AB_5, AB_6, AB_{11}, AB_{12}, AB_{13}, A_2B_3, A_2B_5, A_2B_{17}, A_3B_4, A_3B_5, A_3B_7, A_4B_5, and A_6B_{23}.

These structure maps are important for both pedagogical and predictive purposes. They will help the materials designer to interpolate more reliably the known data. For example, in the search for new alloys capable of withstanding the high temperatures in jet engines, designers are currently examining ordered Ni_3Al alloys which have the Cu_3Au type of structure. Ordered compounds with this structure usually get stronger as they get hotter [1]. The AB_3 structure maps of binary and pseudo-binary systems might suggest possible candidates for the next generation of high-temperature alloys.

The Mendeleev number may also be used to display the domains corresponding to molecules with different shapes or structure. For example, fig. 1 shows the structure map for sp-bonded AB_2 trimers using the data compiled by ANDREONI et al. [14]. We see that good structural separation is obtained between bent and linear molecules, only Al_2O and Li_2O being in the wrong domain. The plots based on a certain choice of physical co-ordinates [14] require two different maps corresponding to N = 16 and N < 16 respectively where N is the number of valence electrons per molecule. Moreover, the good separation claimed would be removed if the molecules LiSb, Li_2Sb, Na_2S, CaS_2, SrS , CdS_2 and Cd_2S had their assumed linear structures confirmed by experiment.

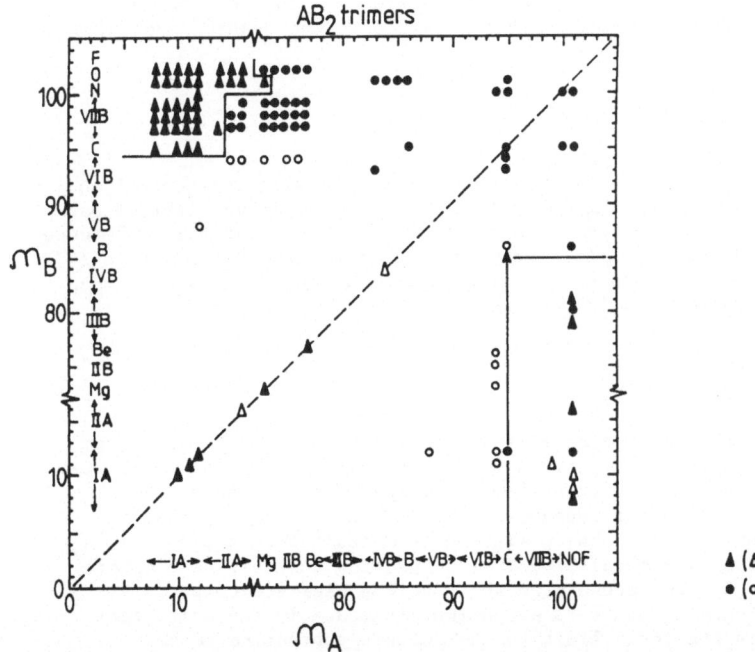

Figure 1: The structure map for the sp-bonded AB_2 trimers. The full triangles and circles correspond to bent and linear molecules respectively whose shape is well established from experiment or self-consistent-field (SCF) configuration-interaction (CI) calculations. The open symbols correspond to ambiguous evidence or non-CI calculations [14].

3. Microscopic Origin of Structural Stability

The structural separation achieved by the Mendeleev number does not give it any fundamental microscopic significance. It simply reflects the fact that elements in the same or neighbouring groups of the periodic table usually behave in a chemically similar fashion. The origin of the structural domains must be sought within a quantum mechanical framework.

Moreover, although the microscopic LDF theory would probably predict the correct crystal structure for any particular compound, the calculated total energies by themselves would not provide a physical explanation for the trends observed across a given structure map. The energy must be broken down in order to relate with one's physical intuition regarding the importance of factors such as relative atomic size, electronegativity difference, average electron per atom ratio or the angular momentum character of the valence electrons.

The equilibrium energy difference theorem [13] allows direct contact to be made between the microscopic calculations and our physical intuition. Suppose the binding energy has been written as the sum of two terms, namely

$$U = U^{rep} + U^{bond} \tag{1}$$

where the labels rep and bond imply that a division has been made so that around the equilibrium volume the first contribution is repulsive and the second attractive. For example, in the ionic limit the bond contribution would be the Madelung electrostatic energy, whereas in the covalent or metallic limit for Tight-Binding systems it would be the quantum mechanical bonding energy

$$U^{bond}_{qm} = \sum_i \int^{\varepsilon_F} (\varepsilon - \varepsilon_i) \, n_i(\varepsilon) \, d\varepsilon \tag{2}$$

where $n_i(\varepsilon)$ is the local density of states associated with the orbital on atom i with energy ε_i (assuming for simplicity of notation one orbital per site) and ε_F is the Fermi energy. Then, it can be shown that the difference in energy between two equilibrium structures is given to first order in $\Delta U/U$ by the difference in the bond energy alone, provided the lattices have been fixed to show the same repulsive energy i.e.

$$\Delta U^{(1)} = \left[\Delta U^{bond}\right]_{\Delta Urep=0} \tag{3}$$

The importance of this theorem is that it allows the structural energy difference to be interpreted within a two-stage process. In the first step the volumes of the different structures are prepared to guarantee the same repulsive energy. This stage depends only on the nature of the repulsive interaction and reflects the atomic sizes of the constituents [15]. It generalizes the usual classical procedure of packing together hard spheres until they touch. In the second step the bond energies are compared at these prepared volumes in order to see which structure is the most stable. This corresponds in the ionic limit to the customary practice of comparing the electrostatic Madelung energies (see, for example, fig. 3.08 of [16]). Away from the extreme ionic limit this stage depends on the nature of the quantum mechanical bonding between the atoms, and is controlled by the angular momentum character of the bonding orbitals, the electronegativity difference, and the band filling or average electron per atom ratio.

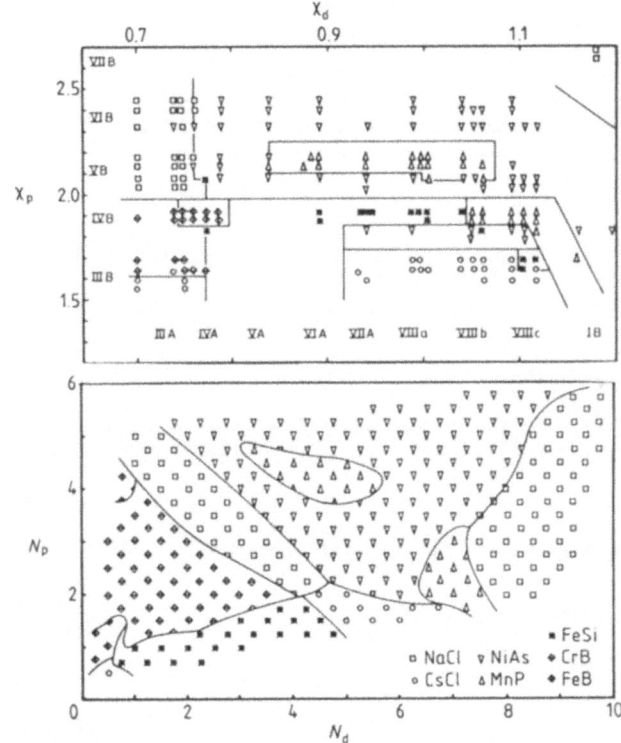

Figure 2: Upper panel: The structure map (χ_p, χ_d) for 169 pd-bonded AB compounds, where χ_p and χ_d are the values of the chemical scale for the A and B elements respectively. The earlier χ-scale orders the elements in a similar fashion to the present Mendeleev number. Each transition-metal group comprises columns corresponding to 3d, 4d, and 5d elements; each IIIB to VIIB metalloid group comprises rows corresponding to 3p, 4p, 5p and 6p elements. The 2p elements are not included in the present figure.
Lower panel: The theoretical structure map (N_p, N_d), where N_p and N_d are the number of p and d valence electrons respectively on the CsCl lattice.

This theorem has been used to examine the microscopic origin of the structural domains of the pd-bonded AB compounds which are shown in the upper panel of fig. 2 [15]. The total binding energy, eq. (1), is represented by the Tight-Binding Bond (TBB) model in which the repulsive energy is given by the sum of central pair potentials and the bond energy by eq. (2) under the assumption that local charge neutrality is maintained. This TBB model can be justified within density functional theory by approximating the exact ground state charge density which enters the Schrödinger equation by the sum of overlapping atomic charge densities [4, 7, 18]. The error involved is second order in the difference between the overlapping free atom and the exact ground state charge densities [4, 17]. The errors are less than 15% for the equilibrium separation and vibrational frequency of diatomic molecules such as C_2, N_2, and Cu_2 [4]. By identifying the second contribution in eq. (1) with the bond energy rather than the band energy which enters the density functional scheme, the first contribution is just the sum of the Coulomb, exchange and correlation interactions between individual pairs of atoms [18].

We have taken the simplest possible TBB model for the pd-bonded systems [15]. The valence s electrons are neglected, so that we need only consider the bonding between the valence p and d states on the metalloid and transition element sites respectively. Their hopping or bond integrals are given by the canonical TB parameters that result from the Atomic Sphere Approximation of ANDERSEN [19]. They are given explicitly [20] by

$$dd(\sigma,\pi,\delta) = (-6,4,-1) \, (r_d/R)^5$$

$$pp(\sigma,\pi) = (2,-1)(r_p/R)^3 \tag{4}$$

$$pd(\sigma,\pi) = (-3,3^{1/2}) \, (r_p^3 r_d^5)^{1/2}/R^4$$

where R is the internuclear separation.

The repulsive pair potentials were chosen to fall off with distance as the square of the corresponding bond integrals, i.e. $\phi_{dd} = c_d^2/R^{10}$, $\phi_{pp} = c_p^2/R^6$ and $\phi_{pd} = (\phi_{pp}\phi_{dd})^{1/2}$.

The repulsive energy per formula unit may then be written

$$U^{rep} = (c_p c_d/V_{AB}^{8/3}) \, (\alpha_{pd} + 1/2 \, \mathcal{R}^{-1} \, \alpha_{dd} + 1/2 \, \mathcal{R} \, \alpha_{pp}) \tag{5}$$

where

$$\alpha_{\ell\ell'} = \mathcal{N}^{-1} \sum (V_{AB}^{1/3}/R)^{2(\ell+\ell'+1)} \tag{6}$$

with $\alpha_{dd} = \alpha_{22}$, $\alpha_{pp} = \alpha_{11}$ and $\alpha_{pd} = \alpha_{12}$ and the sum in (6) extending over the relevant dd, pp or pd interactions on the lattice. V_{AB} is the volume per formula unit. The coefficients $\alpha_{\ell\ell'}$ depend only on structure and not on volume. The <u>relative size factor</u> $\mathcal{R}$ is defined by

$$\mathcal{R} = (c_p/c_d) \, V_{AB}^{2/3} \tag{7}$$

where the volume dependence enters as the result of the different distance behaviour of the p and d potentials ϕ_{pp} and ϕ_{dd} respectively. It is a measure of the relative size of the p and d atoms as can be seen as follows. Consider a pair of transition metal atoms a distance $2R_d = V_{AB}^{1/3}$ apart. Then a pair of metalloid atoms will show the same repulsive energy when they are separated by the distance $2R_p$ such that

$$(R_p/R_d)^3 = \mathcal{R} . \tag{8}$$

Thus, $\mathcal{R}$ generalizes the concept of the relative size of hard spheres to the case where the mutual interaction varies smoothly rather than discontinuously.

The relative preparation volumes, which are required by the equilibrium energy difference theorem, may be obtained from (5) by setting $\Delta U^{rep} = 0$. The resultant first-order change in volume ΔV is given by

$$\frac{\Delta V}{V} = \frac{6 \, \Delta\alpha_{pd} + 3 \, \mathcal{R}^{-1} \, \Delta\alpha_{dd} + 3 \, \mathcal{R} \, \Delta\alpha_{pp}}{16 \, \alpha_{pd} + 10 \, \mathcal{R}^{-1} \, \alpha_{dd} + 6 \, \mathcal{R} \, \alpha_{pp}} \tag{9}$$

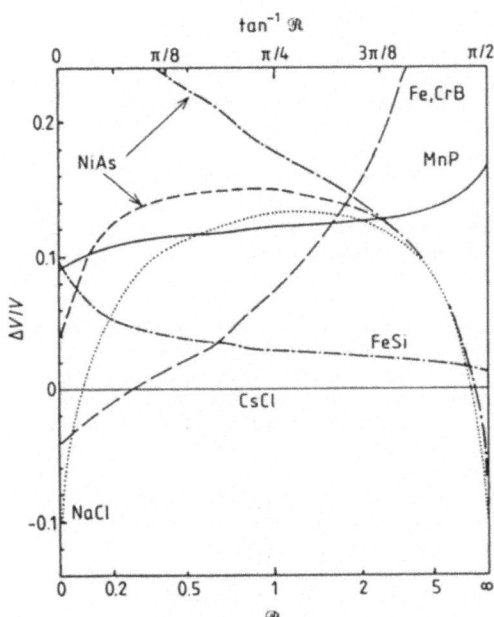

Figure 3: The fractional change in prepared volume ΔV/V with respect to the CsCl lattice versus the relative size factor $\mathcal{R}$. The upper and lower NiAs curves correspond to c/a = 1.39 and $(8/3)^{1/2}$ respectively.

Figure 4: The structural energy curves as a function of band filling N and atomic energy level separation ε_{pd}. The value of ε_{pd} in eV is given at the top of each panel. The predicted structural trend for a given choice of ε_{pd} is shown by the symbols along the bottom of the corresponding panel.

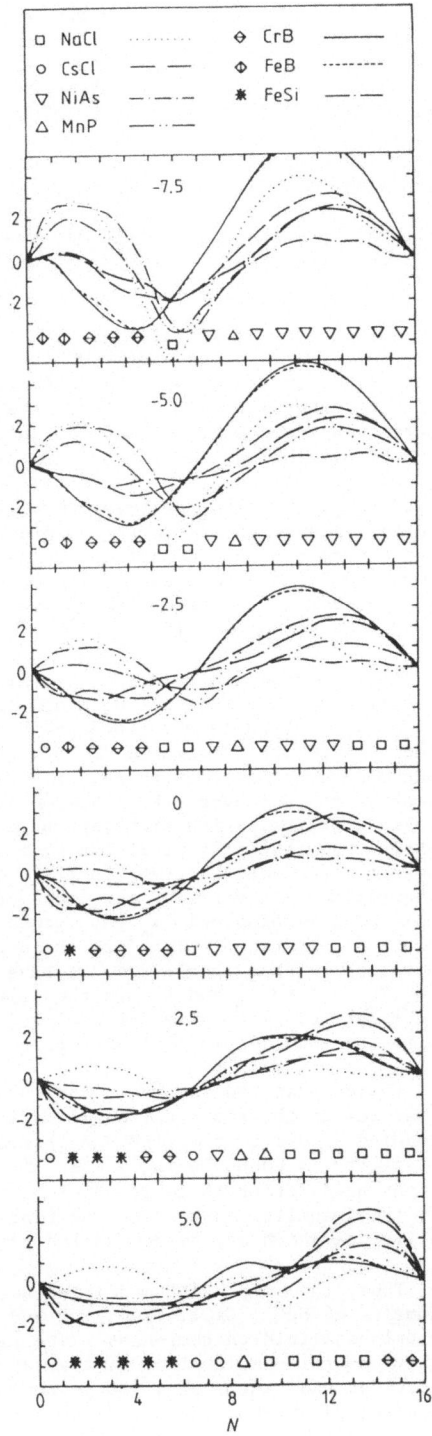

where the $\Delta\alpha_{\ell\ell'}$ are the corresponding changes in the repulsive coefficients. Thus, the fractional change in volume is a function only of the relative size factor $\mathcal{R}$ so that a universal curve may be plotted for each structure with a given set of internal co-ordinates. Fig. 3 shows these curves with the CsCl lattice as reference. As expected the NaCl lattice has the smallest volume at either end of the $\mathcal{R}$-scale, because as c_p or c_d tends to zero the repulsion is dominated by one or other of the close-packed fcc sublattices. On the other hand, in the middle of the scale where the nearest-neighbour (NN) pd repulsion dominates, the volume of the NaCl lattice with six NNs is about 13% larger than the CsCl with eight NNs. The packing of hard spheres would have predicted the much larger volume difference of 30%.

The structural stability of the pd-bonded AB compounds may now be studied by comparing the bond energy of the different lattices at the volumes determined by the relative size factor $\mathcal{R}$. Since we are interested in the global features of fig. 2 rather than predicting the structure of a particular compound, we have chosen for simplicity that value of $\mathcal{R}$ implicit in the assumption that the volumes are prepared so that there is no change in the second moment μ_2 of the TB density of states from one lattice to another i.e. $\Delta\mu_2 = 0$. This corresponds to the choice of $\mathcal{R} \simeq 0.8$ when hopping integrals appropriate to the 4d and 5p elements are taken. The fact that, apart from the borides and NiAs, the fractional volume changes are not too sensitive to the particular choice of $\mathcal{R}$ for $0.3 < \mathcal{R} < 3$ is responsible for the success of the theory in treating the global trends in fig. 2 with a single choice of $\mathcal{R}$ [15].

Fig. 4 shows the relative stabilities of the seven different lattices as a function of the band filling N and the p and d atomic energy level separation $\varepsilon_{pd} = \varepsilon_p - \varepsilon_d$ which were obtained from the computed TB density of states. We see at once the importance of both the average number of valence electrons per atom and the electronegativity difference (strictly ε_{pd}) on structural stability. Moreover, the shape of these curves would be very different for valence orbitals with different angular momentum character. The resultant structural stability predicted by this TBB model is displayed explicitly beneath each set of curves in fig. 5. Together the points constitute a theoretical ε_{pd} versus N structure map which may be compared with BURDETT's [21] phenomenological (ε_{pd},N) structure map. However, because the ε_{pd} that enter the TB calculations are those appropriate to the binary compound and not the isolated free atoms, we have chosen the rotated frame of axes (N_p,N_d), where N_p and N_d are the number of p and d valence electrons on the CsCl lattice for a given choice of (ε_{pd},N). This allows direct comparison with the phenomenological structure map in the upper panel of fig. 2.

We see that the theory in the lower panel predicts the broad topological features of the empirical data: NaCl in the top left-hand corner adjoins NiAs running across to the right and borides running down to the bottom. MnP stability is found in the middle of the NiAs domain and towards the bottom right-hand corner where it adjoins CsCl to the bottom left. The main failure is the inability to predict the FeSi stability of the transition metal silicides which may be due to the neglect of the valence s electrons.

Thus, the simple TBB model can account qualitatively for the observed domains of NaCl, CsCl, NiAs, MnP and boride stability amongst the transition-metal-metalloid AB compounds. The classical electrostatic Madelung energy plays no role in metallic systems as the atoms may be taken to be perfectly screened and, therefore, charge neutral. The structural stability is determined by the quantum mechanical bond energy (2), which may be expressed

44

directly in terms of the contributions from individual bonds by writing

$$U_{qm}^{bond} = \mathcal{N}^{-1} \sum_{\substack{i,j \\ i \neq j}} h_{ij} \left[- 1/\pi \ \text{Im} \int^{\varepsilon_F} G_{ji}(\varepsilon) \ d\varepsilon \right] \tag{10}$$

where h_{ij} is the bond integral between neighbouring atoms i and j, G_{ij} is the inter-site Green function, and the expression inside the large brackets is the bond order [22]. Very recently MAJEWSKI and VOGL [23] have found that the structural stability of semi-conducting and insulating sp-bonded octet compounds is also driven by changes in the quantum mechanical bond energy rather than the classical Madelung energy.

4. Conclusion

Phenomenological and microscopic approaches to structural stability perform very different but complementary roles [10]. Phenomenological structure maps are useful when there already exists a large empirical data base which may be used for interpolating the structures of systems not yet studied. On the other hand, approximate quantum mechanical schemes allow the factors determining structural stability to be investigated, but they obviously cannot produce structure maps as accurate as the experimental data itself. However, besides their pedagogical value, microscopic schemes are becoming increasingly important for predicting metastable phases and the structure of grain boundaries, microclusters etc. because these are not readily accessible to experiment and hence amenable to phenomenological interpolation. In this last regard the TBB model is ideally placed since it not only incorporates the quantum mechanics correctly for tightly-bound systems, but it also is sufficiently simple for use in atmoistic relaxation packages, the interatomic forces following directly from (10). The next few years will see an explosion of predicted structures which are relaxed using quantum mechanical forces.

References

1. High Temperature Alloys: Theory and Design, ed. by J. O. Stiegler, Bethesda Conference Proceedings (The Metallurgical Soc. of AIME, Warrendale 1984).
2. P. Hohenberg and W. Kohn: Phys. Rev. 136, B864 (1964); W. Kohn and L. J. Sham: Phys. Rev. 140 , A1133 (1965).
3. Non-locality of exchange and correlation can sometimes be important as, for example, in Be$_2$ [4].
4. J. Harris: Phys. Rev. B31, 1770 (1985).
5. A. R. Williams, C. D. Gelatt, and V. L. Moruzzi: Phys. Rev. Lett. 44, 429 (1980).
6. M. L. Cohen: Phys. Rep. 110, 293 (1984).
7. W. B. Pearson: The Crystal Chemistry and Physics of Metals and Alloys (Wiley, New York 1972).
8. P. Villars: J. Less Common Met. 92, 215 (1983); 99, 33 (1984).
9. The quantum mechanical pseudopotential radii used by Villars in [8] essentially measure the size of the free atom. This decreases monotonically across a given period unlike the bulk atomic volume which is determined by competition between the size of the free atom and the strength of its bonding to its neighbours (see §6 of D. G. Pettifor: in Sol. St. Phys, ed. by H. Ehrenreich and D. Turnbull, (Ac. Press, Florida, to be published, Spring 1987)).
10. D. G. Pettifor and R. Podloucky: Phys. Rev. Lett. 55, 261 (1985).

11. D. G. Pettifor, New Scientist 110, No. 1510, 48 (1986).
    Note that there are some differences in the relative ordering of the
    4d and 5d transition elements provided by the string compared to the
    earlier table 1 of [13]. This does not affect the structural
    separation. The string values for the Mendeleev-number will be used
    consistently in future.
12. Other authors such as A. Zunger in Phys. Rev. Lett. 47, 1086 (1981).
    and P. Villars and K. Girgis in Z. Metallk. 73, 445 (1982) have used
    the periodic table to present in matrix form structural or other
    empirical data. However, Zunger and Villars were both discouraged by
    the poor structural separation they obtained so that they favoured
    the use of physical co-ordinates in their search for structural
    ordering. The way the string runs through the periodic table is of
    crucial importance for successful structural separation; as too is
    the realization that two-dimensional ordering of all the data for a
    given stoichiometry can only be achieved with a single
    phenomenological co-ordinate.
13. D. G. Pettifor: J. Less. Common Met. 114, 7 (1985);
    J. Phys. C 19, 285 (1986).
14. W. Andreoni, G. Galli, and M. Tosi: Phys. Rev. Lett. 55, 1734
    (1985).
15. D. G. Pettifor and R. Podloucky: Phys. Rev. Lett. 53, 1080 (1984);
    J. Phys. C 19, 315 (1986).
16. R. C. Evans: An Introduction to Crystal Chemistry, 2nd ed. (C.U.P.,
    Cambridge, 1966).
17. W. M. C. Foulkes: (to be published).
18. M. W. Finnis, D. G. Pettifor, A. P. Sutton, and Y. Ohta (to be
    published).
19. O. K. Andersen: Phys. Rev. B21, 3060 (1975).
20. D. G. Pettifor: J. Phys. F 7, 613 (1977);
    O. K. Andersen, W. Klose, and M. Nohl: Phys. Rev. B17, 1209 (1978).
21. J. K. Burdett: J. Solid State Chem. 45, 399 (1982).
22. M. W. Finnis and D. G. Pettifor: in The Recursion Method and its
    Applications, ed. D. G. Pettifor and D. L. Weaire, Springer Ser.
    Solid-State Sci., Vol. 58 (Springer, Berlin, Heidelberg 1985) p.
    120.
23. J. Majewski and P. Vogl: Phys. Rev. Lett. (to be published).

Electronic Structure of Microclusters

P. Jena, S.N. Khanna, and B.K. Rao

Physics Department, Virginia Commonwealth University,
Richmond, VA 23284, USA

The evolution of the structural and electronic properties of small atomic clusters, as atoms come together, has been studied as a function of the size of clusters using ab initio self-consistent field methods based on a linear combination of atomic orbitals. The physics of these atomic clusters has been investigated in terms of their equilibrium geometries, relative stabilities, ionization potentials, preferred spin configuration and its relationship to geometry, structural changes when subjected to ionizing radiation, fragmentation channels and impurity contamination.

1. INTRODUCTION

Recent experiments[1] on naked and matrix-isolated metal atom clusters have revealed many interesting properties that are unique to this new kind of matter. For example, there is strong evidence[2] that micro-clusters consisting of a few atoms are structurally very different from what can be derived by assuming that they form miniature solids. The electronic properties such as electron charge and spin distribution, "band-gap", electron affinity, ionization potentials, and reactivity with gas molecules are strongly dependent on the size and constituents of the cluster. These studies have prompted a great deal of research activity in recent years because of the possibility of leading to some fundamental changes in our current understanding of how solid state properties evolve. There is also the possibility of controlling the size of the clusters to make bulk materials with specific properties. Since the cluster properties are size specific, it is essential to determine the structure of clusters. No experimental techniques exist at this time to determine the geometrical structure of microclusters. Thus, such determination must begin from theoretical studies.

In this paper, we present studies based upon ab initio methods where the equilibrium geometries of microclusters in neutral and ionized phases have been obtained. We also discuss a variety of electronic properties to illustrate its nature of evolution as clusters grow.

2. THEORETICAL PROCEDURE

Our theoretical procedure has been presented elsewhere[3,4] in detail. We only provide some of its saliant features for the sake of completeness and easy reading. Our method is based upon self-consistent field-linear combination of atomic orbitals-molecular orbital method (SCF-LCAO-MO). We begin with the atoms in a cluster placed at random locations. The molecular orbital is constructed by taking a linear combination of atomic orbitals centered at each atomic site. The coefficients of this combination are obtained variationally by solving the SCF equation which treats kinetic energy exactly, electrostatic and exchange energy within the Har-

tree Fock approximation, and correlation contribution in the configuration
interaction scheme involving all pair-excitations of core and valence electrons. To enhance the accuracy of the calculated total energy, we represent the atomic orbitals in terms of Gaussian orbitals. The reader is referred to our original paper[3] for details regarding the basis set and numerical procedure.

From the computed energy, the forces at atomic sites are calculated
by the gradient technique. The atoms are moved to a new adjacent location
along the direction of the force and the SCF procedure is repeated to
again find the total energy and the gradient forces. The above two steps
are repeated until the forces at every site vanish. To insure that this
configuration does not represent a local minimum in the potential energy
surface, the above calculations are repeated starting with new random distributions until one reaches the global minimum in the structure of the
cluster. Since, in the beginning, it is not clear as to what the spin
multiplicity of the ground state should be, calculations are carried out
for various possible spin states of the clusters. In the following section, we present the equilibrium geometries of some neutral and ionized
clusters and study the evolution of their electronic properties.

3. RESULTS

In Fig. 1, we present the equilibrium geometries and spin multiplicity for the ground state of Li_n, Be_n, Mg_n and C_n clusters. It is immediately apparent that the electronic structure plays an important role in

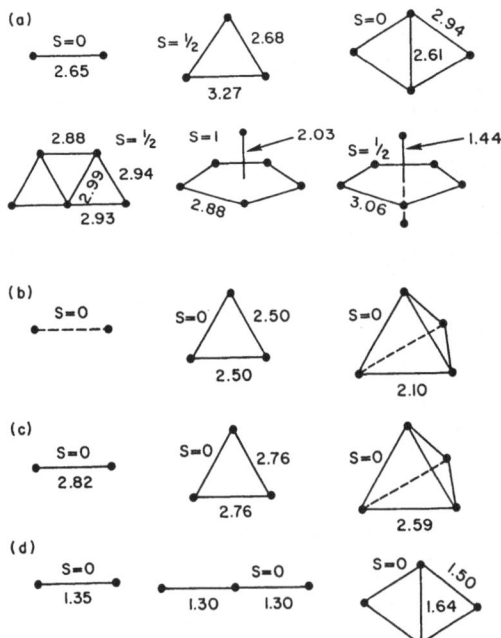

Fig. 1   Optimized geometries and spin multiplicity of neutral (a) Li, (b)
         Be, (c) Mg, and (d) C clusters. The bond lengths are given in
         Angstroms. Be_2 is shown with dashes as, in the present calcula-
         tions, it is not bound.

determining the equilibrium geometry. For example, while Li_3 is in the shape of an isosceles triangle, Be_3 and Mg_3 are equilateral triangles and C_3 is a linear chain. Li_n clusters do not assume three-dimensional structure until there are at least six atoms in the cluster, whereas Be_n and Mg_n clusters have tetrahedral geometry for n=4. Be and Mg atoms have closed electronic shells[5]. Consequently, their structures are determined by the principle of hard sphere packing. A comparison of bond lengths between equal-sized Be and Mg clusters indicate that the bonding in Mg is weaker than that in Be.

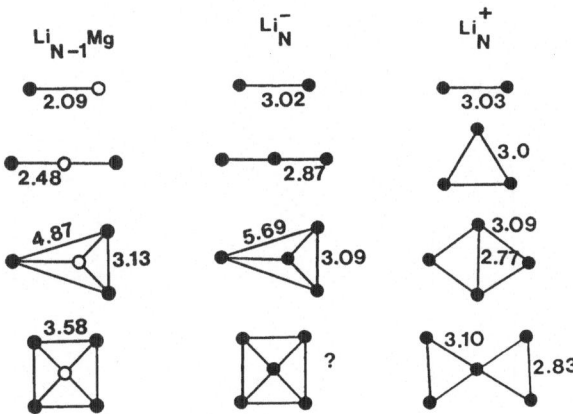

Fig.2   Optimized geometries of $Li_{N-1}Mg$, Li_N^-, and Li_N^+ clusters. All bond lengths are in Angstroms. Values for Li_5^- bond length are not completely optimized.

In Fig. 2, the equilibrium geometries of Li_nMg, Li_n^+, and Li_n^- clusters are given. A comparison between the geometries of Li_n^+ and Li_n (in Fig. 1) reveal that the positively charged clusters, in general, have undergone an expansion. The energy gained in this structural relaxation upon ionization is typically of the order of 0.1 eV. For a quantitative comparison, we present in Table 1 the vertical and adiabatic potentials of Li_n^+ clusters. For vertical ionization, the energy corresponds to a Li_n^+ cluster whose geometry is unchanged from the neutral phase. The adiabatic potential, on the other hand corresponds to one in which the cluster is allowed to relax to a new equilibrium configuration after ionization. Table 1 clearly shows that with the exception of Li_5, the energy gained by structural relaxation is about 0.1 eV.

The equilibrium geometries of Li_n^- in Fig. 2 are very different from those of Li_n^+. Interestingly, the geometries of Li_n^- and Li_nMg have much more similarity with each other. We should note that for a given value of n, the number of electrons in Li_n^- and Li_nMg are the same. Thus, the successive filling of molecular orbital energy levels are expected to proceed in the same manner. This further confirms our assertion from Fig. 1 that the electronic structure of a cluster is primarily responsible for its equilibrium structure.

The relative stability of the clusters can be studied by analyzing the energetics. In Fig. 3 and 4 we plot the energy necessary to add an atom to an existing cluster,

Table 1. Comparison between the vertical and
adiabatic ionization potentials of
Li_n^+ clusters.

N	V.I.P. (eV)	A.I.P (eV)
1	4.860	4.860
2	4.439	4.327
3	3.194	3.085
4	3.787	3.785
5	3.407	3.047

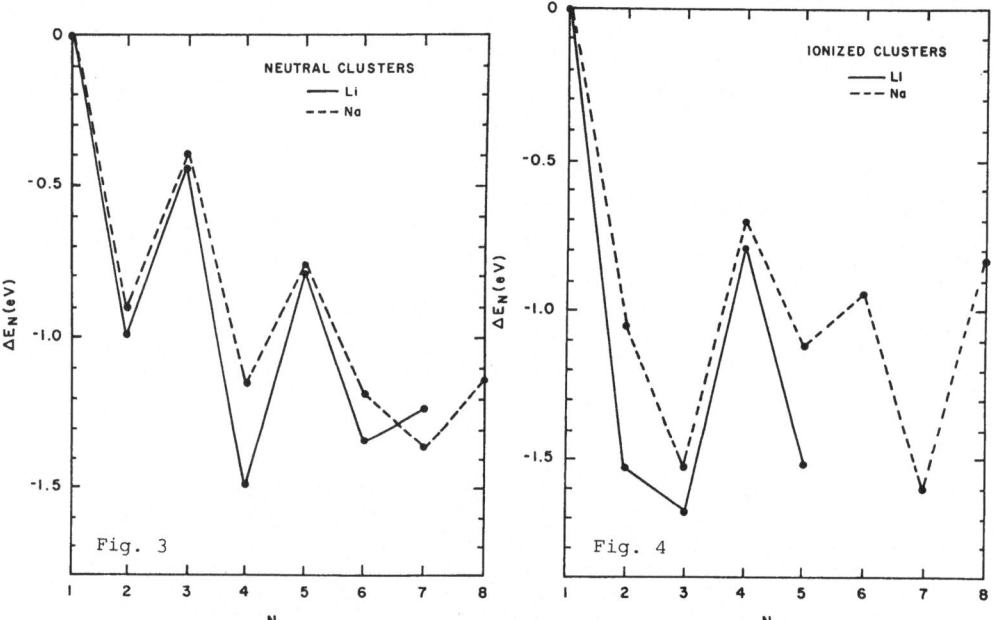

Fig.3   Plot of ΔE_N versus N for Li_N and Na_N clusters. Data for Na_N
clusters are from ref. 8.

Fig. 4   Plot of ΔE_N^+ versus N for Li_N^+ and Na_N^+ clusters. Data for Na_N^+
clusters are from ref. 8.

$$\Delta E_n = E_n - E_{n-1} - E_1 \qquad (1)$$

(normalized to a free atom) for neutral and positively ionized clusters
of alkali atoms. Note that for neutral clusters, the minima occur at even
numbers of n whereas for charged clusters they occur at odd-values. Thus,
based upon the energy considerations alone, one can conclude that even-neu-
tral (odd-charged) clusters are more stable than odd-neutral (even-charged)
clusters. This is in complete agreement with the experimental findings
in alkali[5,6] as well as noble metal[7] clusters.

From the geometries of clusters in Fig. 1 and 2, it is clear that these do not represent a miniature solid, ie. one cannot, at first glance, say that these micro-clusters are fragments of the bulk solid. It is, however, quite certain that at some stage clusters must resemble the bulk phase. When does it start? In the following, we discuss this issue for simple metals. The analysis should be regarded as "empirical" since no fundamental justifications for its validity has been given. It is our hope that our interpretation on the evolution of extended structures will fuel further thoughts and encourage attempts to study different systems with varying crystalline structures.

We first discuss the structural relationship of larger clusters to smaller ones. The two isomeric states of Li_3, as mentioned previously, have bond angles of 72° and 53°. The spin-singlet ground state Li_4 structure in Fig. 1(a) can be pictured as composed of two triangular structures of bond angle of 53°. In the case of Li_6, the atom which is off the plane makes five equal triangles with the sides of the pentagon as the base. The apex angle of this triangle is again 53.9°! These comparisons illustrate that the triangles can not only be used as building blocks, but also the bond angle of the equilibrium trimer is preserved in larger structures. There are, however, exceptions to the later finding. The various bond angles of the ground state Li_5 structure are approximately 60°. For the Li_7 structure, the atoms off the plane on either side of the pentagonal base make angles of 62°. While these two structures do not preserve the trimer bond angle, they can be seen to form portions of an icosahedric phase. Thus, the apex angles of the triangles constituting all the above clusters are either 53° or 60° approximately. To understand this alternation, we look at the energies of Li_3 as a function of apex angle, θ. The energy difference between the triangles with apex angles of 60° and 53° is only 0.07 eV - an energy small enough to be achieved by a minor structural distortion. Thus, one could imagine that structural evolution of small clusters passes through an _intermediate_ icosahedric phase. In the bulk, as will be discussed in the following, the characteristic bond angle for the bcc phase assumes a value of 71° which is very close to the bond angle (72°) of the other isomer of Li_3 cluster.

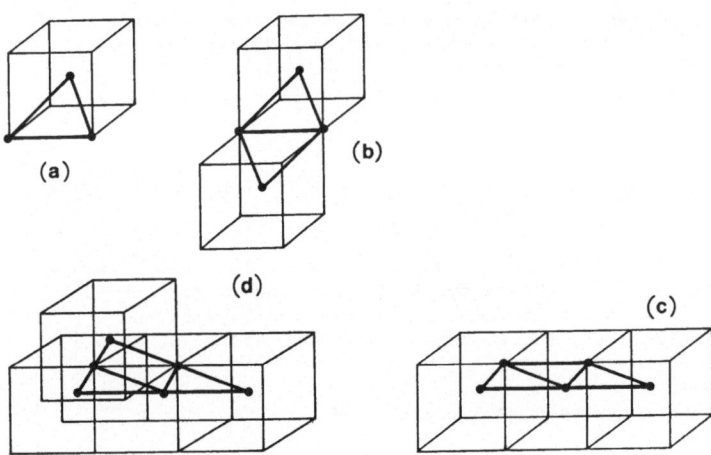

Fig. 5 Crystal growth for bcc lattices.

In Fig. 5 we show the smallest triangle consisting of three atoms that can be constructed from a bcc lattice. It is easy to see that the bond angle of this triangular structure is 71° which is very close to the equilibrium bond angle (72°) of one of the isomers of the alkali trimers. Thus, with a knowledge of the optimum geometry of alkali trimers, one could conclude that they would condense to form a bcc lattice. To visualize how the lattice evolves, we show in Fig. 5 (b, c. d) the atoms in the bcc network that can be joined to resemble the optimum geometries of alkali clusters discussed previously in Fig. 1. The clusters having pentagonal symmetry cannot be considered as a fragment of any conventional lattice structure. However, there may be other isomeric states of a given cluster that have energies not too far above the pentagonal ground state. For example, Li_6 and Na_6 have been found[8] to have isomeric structures similar to that shown in Fig. 5 (d). Thus, as clusters grow in size, their structures can be visualized as Jahn-Teller distortions of what would have been a fragment of the bulk.

In order to probe the relationship of equilibrium trimer bond angle to crystalline structure further, we return to Be_3 and Mg_3 trimers which crystallize in the hcp structure. The smallest triangle in the hcp structure is equilateral and has a bond angle of 60°. For both Be_3 and Mg_3 clusters, the equilibrium structures are equilateral triangles. Even though the trimer bond angle seems to be preserved, it is not possible to distinguish between the hcp and fcc structure at the trimer state, since the smallest triangle in the fcc structure is also equilateral. The energy differences between these structures are very small and it is hard to distinguish between their relative stabilities. For Al_3 cluster, the bond angle has been obtained by Upton[9] to be 60°. This could be identified with the fcc structure. Although the observation of the similarities be-between the bond angles of the trimer and extended structures is empirical, it may have an underlying physical significance. Note that for a trimer, two and three-body forces are operative. The analysis, therefore, suggests that the total energy can be approximately expressed in terms of two- and three-body terms. Such an expression can be used to study the equilibrium geometries of clusters larger than those considered here, since the present method is prohibitively difficult. Such an investigation is under way.

This work was supported in part by grants from the Army Research Office (DAAG 29-85-K-0244), Thomas F. Jeffress and Kate Miller Jeffress Trust, and the National Science Foundation.

REFERENCES

1. T. D. Mark and A. W. Castleman, Jr., Adv. at. and Mol. Phy. 20, 65 (1985). See the references therein.

2. J. Farges, M. F. Deferaudy, B. Raoult, and P. Torchet, J. Chem. Phys. 78, 5067 (1983); J. Farges, M. F. Deferaudy, B. Raoult, and P. Torchet in the "Proceedings of the International Symposium on the Physics and Chemistry of Small Clusters", Ed. P. Jena, B. K. Rao, and S. N. Khanna (in press).

3. B. K. Rao and P. Jena, Phys. Rev. B 32, 2058 (1985).

4. W. J. Hehre, L. Radom, P. v. R. Schleyer, and J. A. Pople, "Ab initio molecular orbital theory", John Wiley, N.Y. (1986).

52

5. W. D. Knight, K. Clemenger, W. A. deHeer, W. A. Saunders, M. Y. Chou, and M. L. Cohen, Phys. Rev. Lett. 52, 2141 (1984).

6. K. Kimoto, I. Nishida, H. Takahishi, and H. Kato, Jpn. J. Appl. Phys. 19, 1821 (1980).

7. C. Solliard and M. Flueli, Surf. Sci. 156, 487 (1985).

8. J. L. Martins, J. Buttet, and R. Car, Phys. Rev. B 31, 1804 (1985).

9. T. H. Upton, Phys. Rev. Lett. 56, 2168 (1986).

Electronic Structure of Small Clusters

K.H. Bennemann and G. Pastor

Freie Universität Berlin, Institut für Theoretische Physik, Arnimallee 14, D-1000 Berlin 33, Germany

Various properties of small clusters have been studied using a Tight-Binding type electronic theory.

1. Introduction

First, we determined the change in the electronic structure and bond-character as a function of cluster size. Results are presented for the cohesive-energy (ionization potential) and s,p electronic density of states of Hg_n-clusters. In particular, we calculated the change of the gap between s- and p-states as a function of cluster size.

Secondly, we calculated for transition-metals the ionization-energy and cohesive-energy as a function of cluster size. Numerical results obtained for Fe_n are compared with recent experimental results by Kaldor et al. and exhibit a non-monotonic behaviour as a function of the number of cluster atoms. It is shown that most likely small iron clusters have b.c.c. like atomic structure. Using recent theories by Hubbard, Hasegawa et al. we calculated also magnetic properties of Fe_n.

Thirdly, we studied the stability of small clusters like $(Cs_nCl)^+$ etc. and explain that clusters with even number of valence electrons are more stable than those with odd number. Intra-atomic correlations reduce the oscillations in the cluster stability with respect to number of valence electrons.

Finally, results are presented for the stability of multiply charged clusters. To explain experimental results for Pb_n^{m+} we must assume that the screening of the repulsive Coulomb interaction changes as a function of cluster size.

2. Theory and Results

In the following we discuss the determination of various cluster properties. All results presented were obtained by using a simplified tight-binding type theory for the binding electrons of the cluster. Thus, the size-dependence of the atomic- and electronic structure of clusters M_n has been calculated. The electronic Green's functions and the resultant density of states are calculated using the continued fraction method. For small clusters M_n, $n < 8$, we assume atomic structures as shown in Fig. 4 (some of them obtained as most stable in previous calculations). For larger clusters we assume either bcc-like or fcc-like structures. Cluster relaxations have been neglected for simplicity. It seems possible that these are important, in particular for smaller clusters. Also, our calculations lack self-consistency regarding electronic charge transfer and screening of Coulomb interactions. Nevertheless, we expect that our results indicate properly the physical properties of small clusters.

(a) Size-Dependence of the Bond-Character of Small Clusters

In contrast to van-der-Waals type bonding systems metal atom clusters change their binding character as a function of cluster size. For example, for Hg_n clusters we expect a change from van-der-Waals type bonds to metallic bonds as the cluster size increases. Such behavior is also expected for other clusters consisting of atoms with s, p valence electrons and results from the closing of the energy-gap Δ between the s- and p-states due to band-broadening and s,p hybridization. This is illustrated in Fig. 1. It is of interest to find out whether such a non-metal-metal-like transition occurs smoothly or abruptly. Therefore, we calculated (1) $\Delta(n)$, (2) $N(\varepsilon,n)$, and (3) the cohesive-energy $E_{coh}(n)$ using the Hamiltonian

$$H = \sum_i \varepsilon_i \, n_i + \sum_{i,j} t_{ij} \, c_i^+ \, c_j \qquad (1)$$

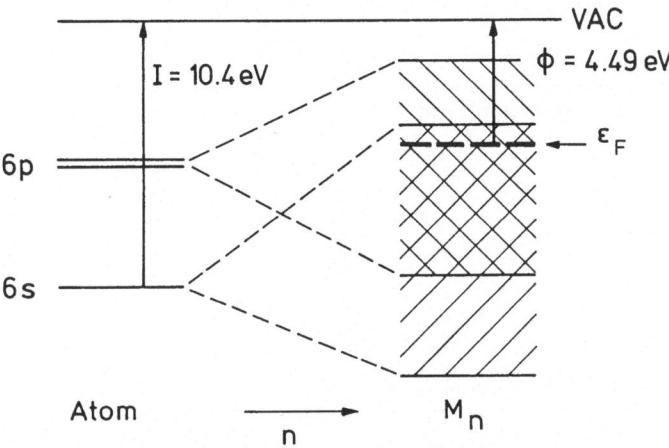

Fig. 1: Illustration of the closing of the atomic s,p-gap Δ for increasing cluster size for Hg_n

Here, ε_i denotes the original energy-levels of the s,p electrons which hop via the hopping integrals t_{ij} from site i to j, and c_i^+, c_j are the usual electronic creation - and annihilation operators. The hopping integral for n.n and n.n.n atoms are determined in such a way that proper band-structure results for bulk Hg are obtained. The t_{ij} cause shifts of ε_i and band-broadening. Results are shown in Fig. 2. For simplicity we assumed (for n > 8) a f.c.c.-like cluster structure. Clearly, our results indicate a smooth transition from non-metallic to metallic bonding. Further studies indicate that this will be also the case if cluster relaxation or other atomic cluster structure are assumed. Since s, p electrons have similar screening capacity it is likely that an explicit treatment of Coulomb-interactions will not change this conclusion. Also, we find from our calculations that the ionization potential, being 10.43 eV for Hg-atom and 4.49 eV (work-function) for metallic Hg, changes smoothly reaching bulk value for n ≃ 40 corresponding to Z_{eff} = 7.[1]

55

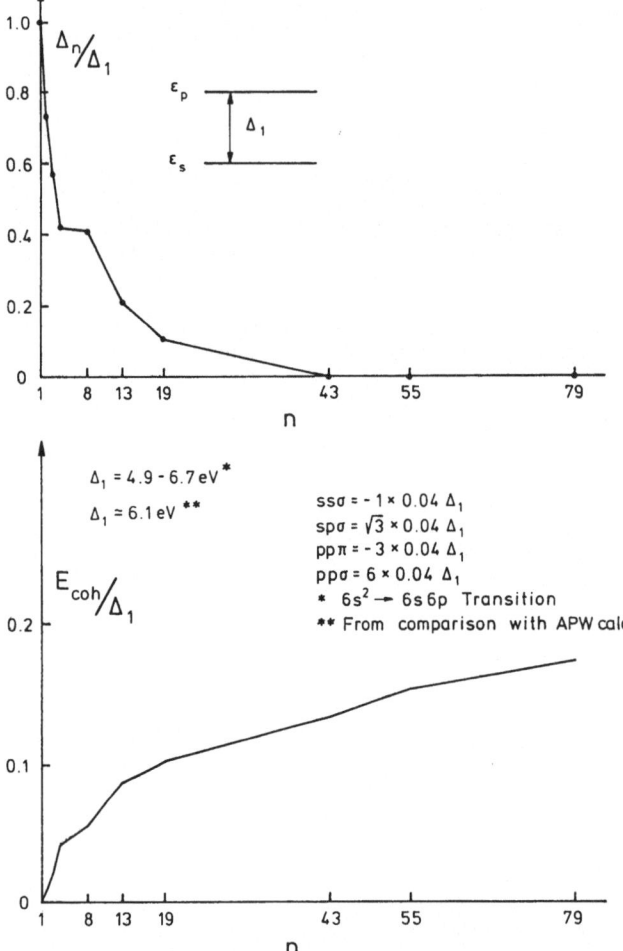

Fig. 2 (a) Results for the s,p-gap Δ and the cohesive-energy E_{coh} as a function of cluster size for Hg_n clusters

(b) Ionization- and Cohesive-Energy of Metallic Clusters

The electronic structure of small clusters should be most clearly reflected in the cluster size dependence of the ionization-potential and cohesive-energy.[2] Therefore, we have calculated these properties for Fe_n for which detailed experimental results are available. The calculations were performed assuming both fcc-like and bcc-like small clusters (n > 8) and by using the Hubbard-Hamiltonian, which extends Eq.(1) by adding intra-atomic Coulomb-interactions. The parameters invoked like ε_i, t_{ij} and Coulomb integral U_{ii} were taken such that proper bulk properties of Fe are obtained. Then, the ionization-threshold energy is given by

$$I_n = (E_n^{i-1} - E_n^i) + \Delta I_n , \qquad (2)$$

where $E_n^i = \sum_\ell \int d\varepsilon^{\varepsilon_{max}} \varepsilon \, N_\ell(\varepsilon)$ is the total electronic energy of a cluster of

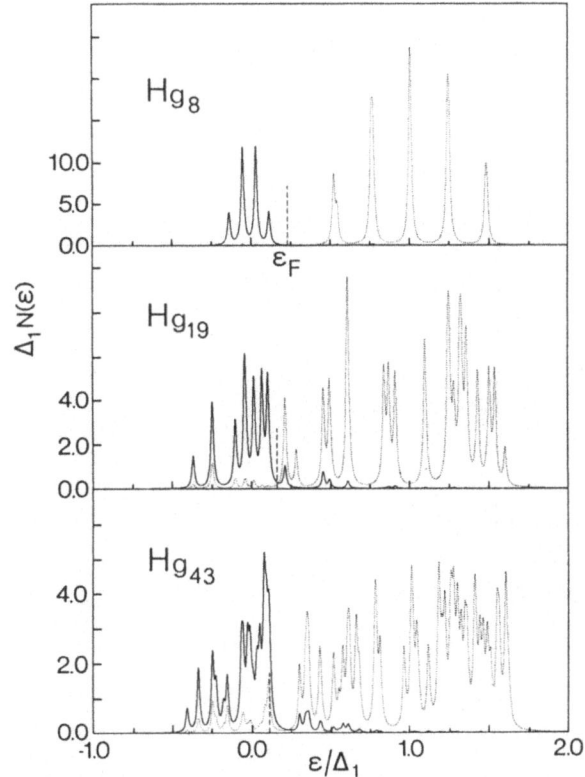

Fig. 2 (b) Results for the electronic density of states $N(\varepsilon)$ as a function of cluster size for Hg_n clusters

n-atoms and i-electrons, and ΔI_n the many-electron contribution due to po-larization, etc. ε_{max} refers to the energy of the highest occupied state. Using a classical approximation ΔI_n is given by $\Delta I_n = const + (3/8)e^2/R$, with $R \simeq (\frac{3V}{4\pi}n)^{1/3}$ being the radius of the cluster (V = Volume). The cohesi-ve-energy per cluster atom is given by

$$E_{coh}(n) = \frac{1}{n} \sum_\ell E_{coh}(\ell), \quad E_{coh}(\ell) = \int^{\varepsilon_{max}} d\varepsilon (\varepsilon_\ell - \varepsilon) \; N_\ell(\varepsilon) - E_R \; . \quad (3)$$

Here, E_R denotes the (Born-Mayer) repulsive energy, particulary important for determining cluster relaxation, etc.

Results for I_n and E_{coh} are shown in Figs. 3,4 and were obtained by using the Hubbard Hamiltonian, the continued fraction technique (2. moment approximation) for determining the electronic-Green's function, and $\varepsilon_\ell = \varepsilon_\ell^0 + \Delta\varepsilon_\ell(n)$.[3] For simplicity, the level shifts $\Delta\varepsilon_\ell$ are given by $\Delta\varepsilon_\ell = - a z_\ell$, z_ℓ is the coordination number and the constant a was taken to be 0.2 (eV) as may be estimated from screened Coulomb potentials for the nuclear charges. The structure in I_n as a function of cluster size compares well with recent experimental results by Kaldor et al.[2] and is physically understandable in terms of changes of $\Delta\varepsilon_\ell$ and band-width. These occur since

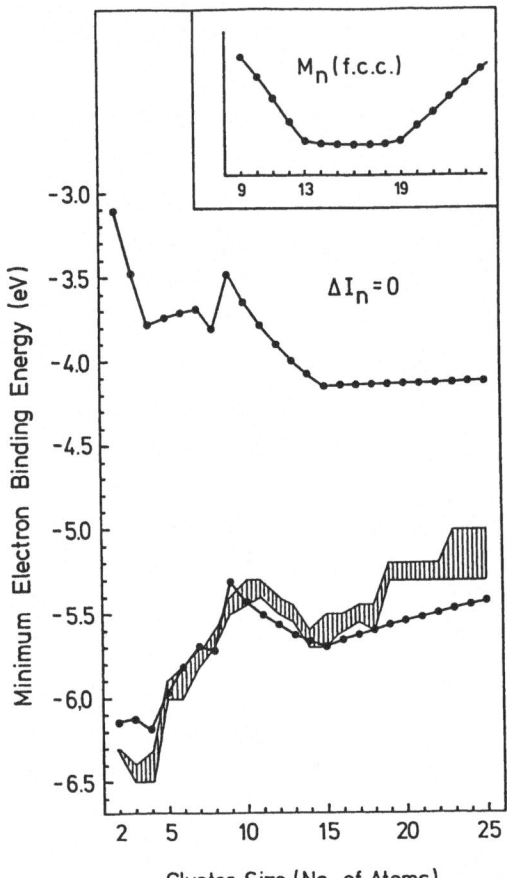

Fig. 3: The minimum electron binding energy as a function of cluster size for b.c.c.-like Fe_n. Experimental results are indicated by the hatched region. The inset Fig. shows results for f.c.c.-like clusters.

the effective coordination number varies in a characteristic way as the cluster grows. From comparison with experiment we are tempted to conclude that small Fe_n-clusters have bcc-like structures.[3] Since ε_{max} correlates also with the catalytic activity one expects I_n and the catalytic activity to exhibit a parallel size dependence.

Our calculations are not limited to Fe_n, but may be applied to other metal clusters as well. A particularly interesting application would be the determination of alloy formation as a function of cluster size, for example, for CsAu alloy clusters.

Applying our theory to the magnetic properties we find that the magnetic moment per atom is somewhat larger in small bcc-like Fe_n-clusters than in bulk Fe, and also that fcc-like small Fe_n-clusters are magnetic.[4] For studying the magnetic order it is very important to include spin-correlations, in particular for cluster sizes which are of the same order as the range of the short-range spin correlations. Calculations are in progress.

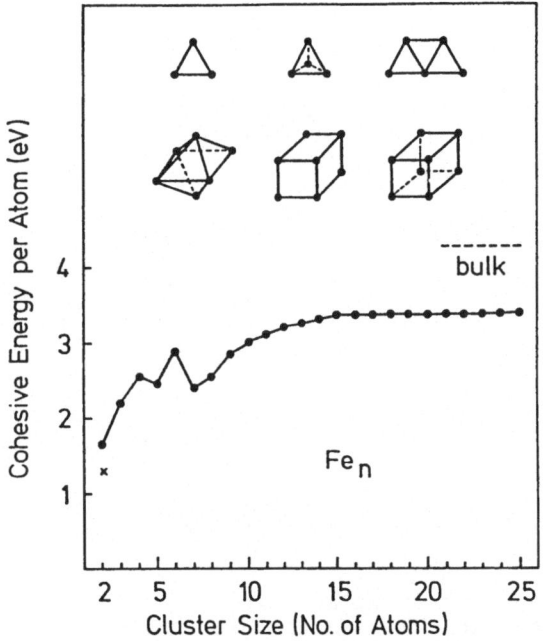

Fig. 4: Cohesive energy for bcc-like Fe_n-clusters. For $n \leq 6$ we use
structures as indicated in the Fig.

The figure axes: Cohesive Energy per Atom (eV) vs Cluster Size (No. of Atoms), with label Fe_n and "bulk".

(c) Stability of Clusters as a Function of Cluster Size

Of course, $E_{coh}(n)$ (and I_n) reflect also the stability of clusters and determine the size distribution of a cluster ensemble (in a beam, etc.) and
the cluster decay due to Coulomb charging. Interesting systems for stability
studies are Hg_n, note Hg_2^+ is stronger bonded than Hg_2, and heterogeneous
clusters like $(Cs_nCl)^{m+}$. For these one observes that clusters with even
number of valence electrons are more stable than those with odd number.
Using the Hubbard Hamiltonian, which yields

$$E = \sum_{\ell} \int d\varepsilon (\varepsilon - \varepsilon_\ell) \, N_\ell(\varepsilon) + E_c \, , \qquad (4)$$

with Coulomb-energy

$$E_c = \{ - \frac{1}{4} (N_v/2N)^2 \, U/W + (N_v/2N)^2 \, (1 - \frac{N_v}{2N})^2 \, (\frac{U}{W})^2 \} \, ,$$

one obtains for the incremental energy $\Delta E = (E_{n+1} - E_n)$ the results shown
in Fig. 5.[5] E_c was derived[6] for a rectangular density of states of width
W and the number of valence electrons of a cluster of n-atoms of kind 1,2 is
given by $N_v = z_1N_1 + z_2N_2$. N_1 is the number of atoms of kind 1 with z_1 valence
electrons. To simulate $(Cs_nCl)^+$ for example. We take into account
$\Delta = \varepsilon_{Cs} - \varepsilon_{Cl} \neq 0$, one s-orbital for each Cs-atom and 3p-orbitals for the
chlorine atom. Binding is dominated by the first term in Eq.(4), but the
even-odd effects in $\Delta E(n)$, e.g. the change in energy when one atom (Cs) is
added, are reduced due to E_c as physically expected. In Cs_nCl-clusters the
Cs-Cl bonds have partial ionic character, this depends on $\Delta = \varepsilon_{Cs} - \varepsilon_{Cl}$. For
For simplicity we assumed that the clusters have b.c.c. like structure. The

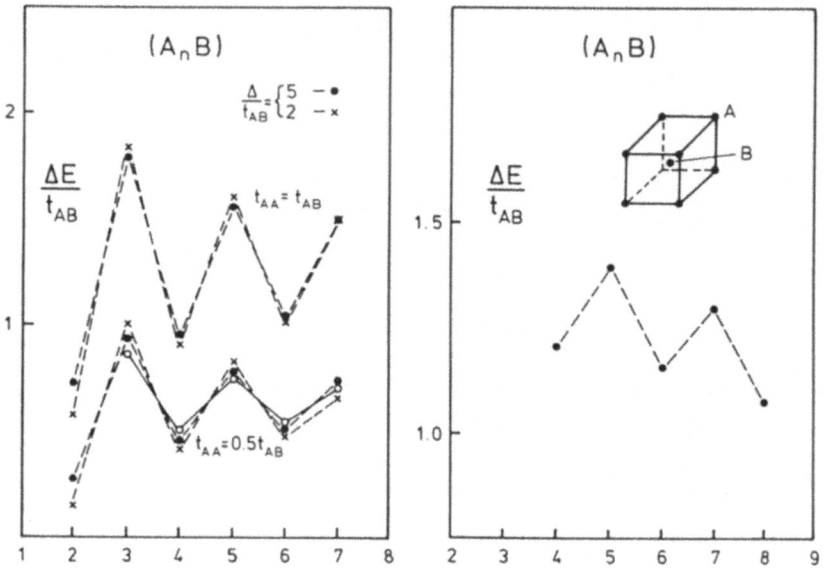

Fig. 5: The oscillating behaviour of the incremental energy
$\delta E = E_{n+1} - E_n$. Results refer to linear and bcc-like clusters.

relative stability decreases as the number of valence orbitals N_O of the
cluster increases. Therefore, for Cs_nCl clusters the even-odd structure in
ΔE decreases as the number of Cs-atoms increases. Moreover, we find that the
relative stability of the clusters decreases at a faster rate when the
number of Cl-atoms increases, since each Cl-atom adds 3p-orbitals to the
cluster.[5] This explains the experimental observation that Cs_nCl_m-clusters
with several Cl-atoms are less stable.

In summary, the even-odd effects can be understood from the fact, that
whenever a bond is formed and occupied by two electrons, the energy is
lowered. Consequently, clusters with even number of electrons are energetically favoured.[5] Finally, note we have not explicitly taken into account
electrostatic energy which may be important for aggregates with small
number of atoms, larger charges and strongly ionic bonds. For clusters
Cs_nCl^+, etc. we expect that for larger n only Cs-atoms n.n to Cl form
ionic like Cs-Cl bonds, but Cs-atoms further away from Cl will form metallic
like Cs-Cs bonds.

(d) Coulomb-Explosion of Charged Clusters

The decay of positively charged up clusters, $M_n^{(m+)}$, reflects sensitively the
atomic- and electronic-structure, the size dependence of electronic screening, of bond character, etc. In the following we discuss in particular the
stability of $Pb_n^{(m+)}$ and transition-metal clusters for which experimental results are available.[7] Assuming that a cluster decays according to
$M_n^{(m+)} \rightarrow M_{n-1}^{(m-1)+} + M^+$, due to large bond-energy, then a Coulomb explosion occurs
if

$$\delta = E_{n-1} - E_n$$

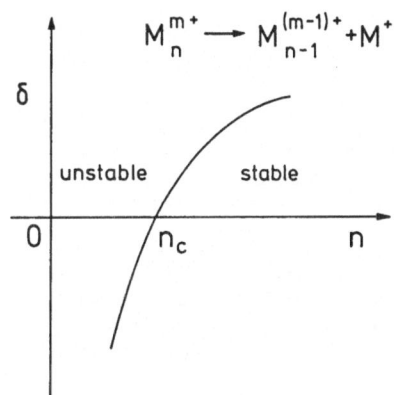

$$M_n^{m+} \longrightarrow M_{n-1}^{(m-1)+} + M^+$$

Fig. 6: Illustration of the cluster stability against Coulomb explosion. $E_{n,i(f)}$ refers to initial and final total energy.

$$\delta = E_{n-1,f} - E_{n,i}$$

becomes negative. However, note, more correctly one should also take into account a possible energy-barrier E between the initial state and the final state, since for $kT < \delta E$ a charged up cluster will not decay even if its final state, $(n,m+) \rightarrow (n-1,(m-1)+)$ is energetically favourable. The energy barrier δE may arise from polarization effects, for example. In calculating δ one must include increase of binding due to electron loss, s. Hg_n Hg_n^+, etc., and odd-even effects, which were discussed above. Furthermore, for small clusters M_n, $n \lesssim 7$, one must allow for cluster relaxation to linear clusters in order to minimize the repulsive Coulomb energy in double or multiply charged up clusters. For the repulsive Coulomb interaction we assume for simplicity

$$V(r) \sim (e^{-\lambda r}/r)$$

and avoid the difficult question about the localisation of the electrical charges.

Results obtained are given in table 1. It is remarkable that we obtain stable clusters M_n^{++}, for $n \lesssim 10$, only if they are linear, and that spherical like Pb_n^{++} clusters are stable if $n \gtrsim 30$ (if we assume Thomas-Fermi

Table 1: Stability of linear M_n^{2+} clusters. The sign (+) indicates stable, and (-) refers to unstable clusters. Linear clusters Pb_n^{2+} are stable for $7 \lesssim n \lesssim 12$. Approximately, linear small clusters M_n with Coulomb energy $E_c \lesssim 1.3\,E_2$ are stable.

M_n^{2+}	Stability
Ni	+
Sn	+
Hg	+
Au	+
Ag	-
Cu	-
Pb	-

like screening, but somewhat poorer than in bulk Pb) in good agreement with experiment.[7] However, experimental results for Pb_n^{+++}, Pb_n^{4+} can only be explained easily if we assume that the screening increases as the cluster size increases from $n = 30$ to $n = 60 \div 70$ (when the screening length reaches bulk value). Assuming a size independent screening-length our simple theory would yield that Pb_n^{+++} are stable if $n < 100$ in sharp contrast to experiment. For Hg_n^{++} we estimate that they are stable for $n \gtrsim 4$. Further experimental and theoretical studies are necessary to understand the Coulomb explosion of small clusters.

In summary, we have studied various properties of small metallic-like clusters to learn more about the atomic- and electronic-structure from which all cluster properties follow. One of our goals was to demonstrate that a tight-binding type electronic theory is able to determine reliably cluster properties.

References

1. C. Bréchignac, M. Broyer, Ph. Cahuzac, G. Delacretaz, P. Labastie and L. Wöste, Chem. Phys. Lett. 118, 174 (1985);
   G. Pastor, P. Stampfli, and K.H. Bennemann (results on Hg_n), to be published
2. R.L. Whetten, D.M. Cox, D.J. Trevor, and A. Kaldor, Phys. Rev. Lett. 54, 1494 (1985)
3. G. Pastor, J. Dorântes-Davila, and K.H. Bennemann, subm. P.R. Lett.
4. G. Pastor, J. Dorântes-Davila, and K.H. Bennemann, preprint
5. S. Ghatak and K.H. Bennemann, J. of Chem. Phys.
6. C.M. Sayers and J. Friedel, J. de Physique 38, 697 (1978);
   P. Joyes, Surf. Sci. 106, 272 (1981); Phys. Rev. B26, 6307 (1982)
7. K. Sattler, J. Mühlbach, O. Echt, P. Pfau, and E. Recknagel, Phys. Rev. Lett. 42, 160 (1981); T. Jentsch, W. Drachgel, and J. Block, Chem. Phys. Lett. 93, 144 (1982); S. Mukherjee and K.H. Bennemann, Surf. Sci. 156, 580 (1985)

Electronic Structure, Cohesion, and Effective Interatomic Potentials in Small Transition Metal Particles

D.E. Ellis and H.P. Cheng

Department of Physics and Astronomy and Materials Research Center,
Northwestern University, Evanston, IL 60201, USA

1. INTRODUCTION

The electronic properties of small metal particles are of interest for a
variety of scientific and technical reasons. These reasons range from the
development of molecular beam techniques for the production of well-
characterized N-atom particle data to efforts to understand and control
clustering and segregation in alloys, and to the exploitation of chemical
reactivity and selectivity essential for catalytic applications. In recent
years, it has become possible to apply self-consistent field models in the
framework of local density theory to provide a description of electronic
energy levels and spectroscopic properties of systems of a reasonable size,
say N < 100. In addition, sufficiently precise numerical methods are now
available to permit the calculation of atomic binding energies and to
determine the relative stability of different cluster geometries. This
opens the door to fruitful interactions between "first-principles" theory
and the semiempirical models used in dynamical and thermodynamical
simulations, and to interpretations of current data on physically
interesting systems.

In this report we describe calculations on the electronic structure,
cohesion, and effective interatomic potentials in small transition metal
particles. For background and computational details, we refer to the
literature [1-6]. By providing a sampling of recent results obtained by a
particular method, the discrete variational (DV-Xα) scheme, we hope to show
the capabilities of local density (LD) theory at its present stage of
development, and to stimulate other workers to explore the many application
areas which now are open to study.

2. TRANSITION METAL DIMERS

The transition metal (TM) dimers are sufficiently small so that they can
be studied by a variety of theoretical techniques. Hartree-Fock,
configuration interaction (CI), generalized valence bond (GVB), local
density, and a variety of approximate and semiempirical approaches have
been used. [7-11] It is interesting that the complex dense multiplet level
structure of the dimers causes great difficulty for the ab initio "classical" methods of quantum theory, while the gross averaging of exchange
and correlation interactions made in LD theory leads to useful and accurate
binding energies and bond lengths. Since the LD model arises from many-
body theory, it is not surprising that macroscopic bulk metal systems are
rather well described, but it is surprising that the model works so well
at the dimer level. Of course, if one is interested in properties of a
particular multiplet level, it is necessary to formulate a proper wavefunction. Here too, the LD orbitals can be utilized in construction of
a conventional quantum mechanical description.

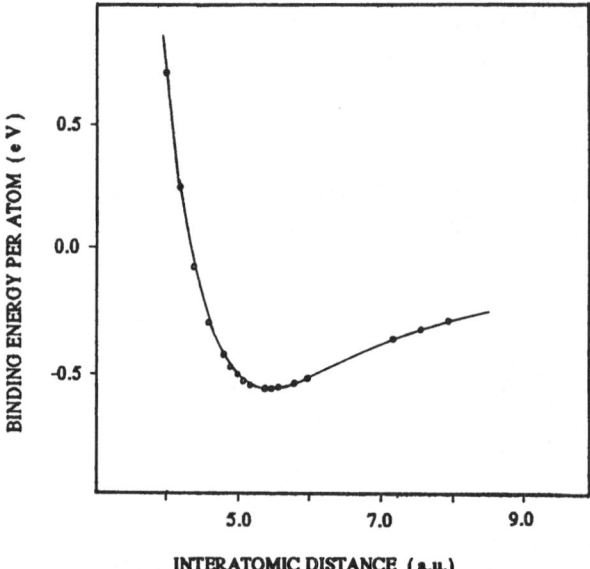

Fig. 1. Nonrelativistic binding energy versus bond length for Pt dimer

In Fig. 1 we show the binding energy curve for Pt_2 as obtained in a nonrelativistic LD calculation, using a simple Kohn-Sham-Slater exchange-correlation potential with $\alpha=0.7$. This value is known to be near-optimal for properties of small molecules, and can be thought of as the exchange-only value (2/3) plus a small additional binding to simulate the correlation effects. The self-consistent-field (SCF) ground state was calculated as a function of internuclear separation using the DV-Xα scheme with numerical basis functions. The basis set was generated as solutions of the free Pt atom, embedded in a potential well to induce additional bound states. [1] The total energy is written as:

$$E_t = \sum_\sigma \left\{ \sum_i n_i \, \epsilon_{i\sigma} - \frac{1}{2} \int \rho_\sigma(\vec{r}) \, V_C^e(\vec{r}) + \int \rho_\sigma(\vec{r}) \left[E_{XC}(\vec{r}) - V_{XC}(\vec{r}) \right] \right\} + \frac{1}{2} \sum_{\nu\mu}' \frac{Z_\nu Z_\mu}{R_{\nu\mu}} \tag{1}$$

and the binding energy is determined as:

$$E_b = E_{tot}(SCF) - E_{tot}(Ref) \tag{2}$$

Here SCF refers to the one-electron energies and density of the self-consistent-field solution for the system of interest, and Ref. denotes the corresponding quantities in some chosen reference state (here dissociated atoms). The implementation of Eq. (2) by numerical integrations, such that a detailed cancellation of errors occurs, is of great practical immportance in obtaining results of sufficient precision to be of use.

A glance at Fig. 1 will reveal that the nonrelativistic (NR) binding is probably much too weak and that the predicted equilibrium bond length is much too long. This result is consistent with the findings of ZIEGLER et al./7/ for the gold dimer, using a DV-Xα scheme with analytic Slater-type basis functions. When we repeat the calculations in a relativistic model by solving the Dirac equation with four-component atomic basis functions considerable changes are observed. The predicted binding energy nearly doubles, to 2.4 eV, and the equilibrium point moves in to ~ 4 a.u.[12] These results are again consistent with the iterative perturbation results for relativistic effects obtained for Au_2 by the Amsterdam group. [7] Thus, it is clear (and probably not surprising) that quantitative models of stability and bonding in the 5d series will need to include relativistic effects explicitly.

Nevertheless, in the following section we present some nonrelativistic (NR) results for Pt_N particles with N = 2, 3, 4 to provide the basis for comparisons with computationally more demanding Dirac-model results. There at least exists the possibility that the relativistic effects become less pronounced in the larger particles, and in interactions with light ligands which we wish to explore.

3.   COHESION AND EFFECTIVE INTERATOMIC POTENTIALS

There exists an infinite number of ways to decompose the molecular binding energy E_b of an N-body particle into interactions between its constituent atoms. The most interesting schemes appear as a sum of two- body interactions plus some many-body corrections:

$$E_b = \sum_{i<j} E_{ij} + F \qquad (3)$$

Here too, the two-body interaction E_{ij} is essentially determined by some "reasonable" conventions or criteria. We have been interested in developing some reasonable criteria for the extraction of interaction potentials from theoretical calculations on small particles. In general, we would like to represent E_b to reasonable accuracy for the actual N-atom particle on which the data are generated and also to be able to extrapolate the derived interactions to larger and more complex systems such as alloys and surfaces.

Among the various published schemes, we find the "embedded atom", approach derived from LD theory and made practical by application to thermodynamic data, to be most interesting.[13] Here the function $F = F(\rho,Z)$ can be simply parametrized either by reference to calculated E_b or experimental data and inserted directly into thermodynamic models and dynamical simulations. We will see that this scheme works reasonably well in describing the binding of Pt_N in a variety of geometrical configurations.

In Fig. 2, are given the effective pair potentials, E_{ij}, for platinum dimers, trimers, and tetramers in the NR model. The equilateral triangle and regular tetrahedron are used here to obtain a single structural parameter R_{ij} and an equivalent set of pair-wise interactions. We see that the two-body potential, obtained by optimizing the parameters of the form:

$$E_{ij}(R) = E_0 -e^{\beta(R-R_0)} + \frac{C}{R^\alpha} \qquad (4)$$

in fitting E_b, is remarkably similar in all three cases. The fits produce errors in reproducing the binding energy and equilibrium distance which are well below errors due to computational considerations. These considerations, including basis truncation and numerical

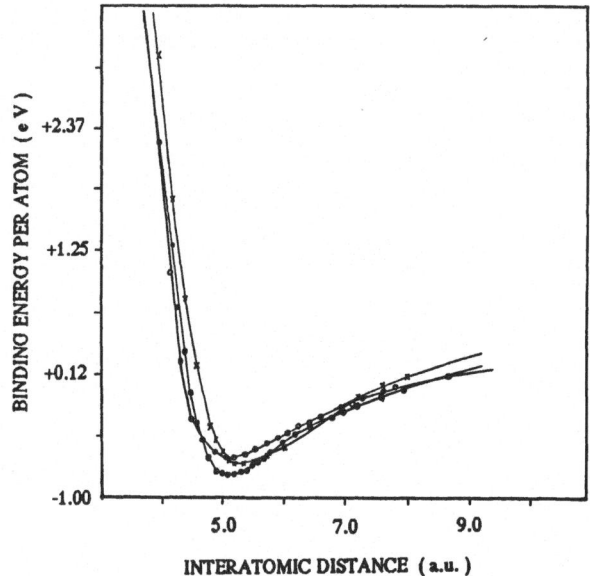

Fig. 2. Effective pair potentials versus atomic separation for Pt_N,
N = 2, 3, 4

integration error, amount to an uncertainty of perhaps ± 0.5 eV in
binding and ± 0.1 a.u. in bond length. Of course, we must recall the
intrinsic NR model error is very much larger !

The utility of the "embedding interaction" $F(\rho,Z)$ becomes evident
when we consider distortions of the trimer and tetramer away from the

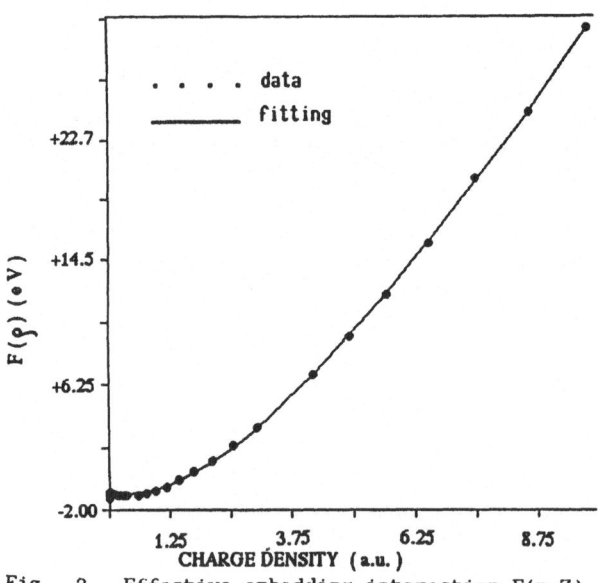

Fig. 3. Effective embedding interaction $F(\rho,Z)$ for Pt particles

highly symmetric configuration. By comparing theoretical E_b for distorted particles with the sum of two-body interactions optimized for the symmetric case, we obtain a "plot" of the effective F versus density, which is shown in Fig. 3. This figure is qualitatively similar to results obtained from fits to thermodynamic data on alloys. Thus we believe that it would be interesting to insert theoretical F-functions into models capable of simulating the structure and dynamics of both particles and bulk systems.

<u>4. LIGATED CLUSTERS</u>

Inorganic chemists have synthesized and characterized a large number of metal cluster compounds, of which the best known are probably the carbonyls $M_n(CO)_m$. Since many of these compounds have stable crystal structures, a great deal is known about their properties, and useful analogies have been drawn between these "covered metal particles" and chemisorption systems. The interaction between the metal core and its ligands, and the modification of electronic properties under substitution of ligands, is a lively subject of both experimental and theoretical research.[3]

A schematic diagram of the cobalt compound $Co_4(CO)_{12}$ and the substituted "tripod" compound $Co_4(CO)_9(PH_2)_3CH$ is given in Fig. 4. The dodecacarbonyl is a relatively stable diamagnetic compound, with a very simple energy level structure shown in Fig. 5.[3] The parent metal

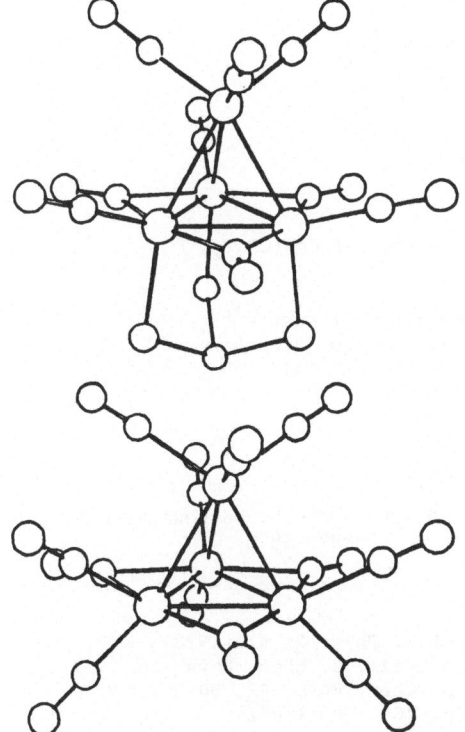

Fig. 4. Nuclear geometry of $Co_4(CO)_{12}$ and $Co_4(CO)_9[(PH_2)_3CH]$

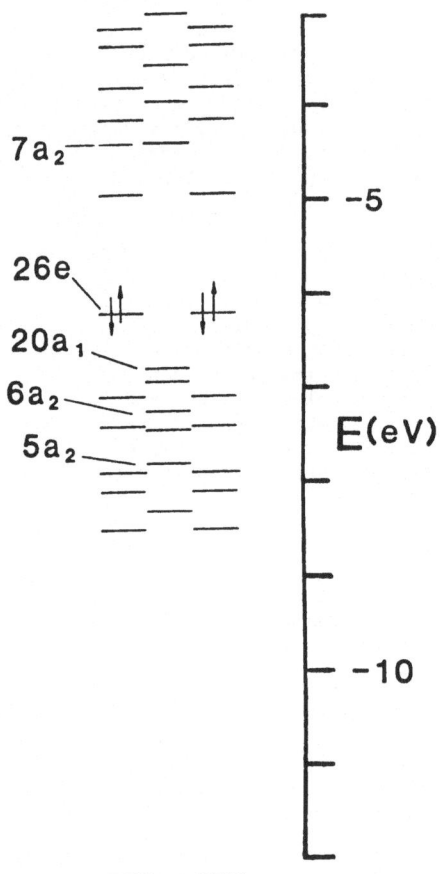

Fig. 5. Ground state one-electron energy levels of $Co_4(CO)_{12}$

tetramer Co_4 is predicted to be ferromagnetic, consistent with experimental observations on Fe_4 [14], and we find that the quenching of magnetism is a local, essentially atomic phenomenon [14]. That is, by adding ligands one at a time, we can observe the reduction and eventual disappearance of magnetic moment at each metal site.

ACKNOWLEDGEMENTS

This work was supported in part by the NSF through the Northwestern University Materials Research Center, Grant No. DMR82-16972.

REFERENCES

1. E.J. Baerends, D.E. Ellis and P. Ros, Chem. Phys. 2, 41 (1973); A. Rosén, D.E. Ellis, H. Adachi and F.W. Averill, J. Chem. Phys. 65, 3629 (1976); A. Rosén and D.E. Ellis, J. Chem. Phys. 62, 3039 (1975); B. Delley and D.E. Ellis, J. Chem. Phys. 76, 1949 (1982).
2. B. Delley, D.E. Ellis, A.J. Freeman, E.J. Baerends, and D. Post, Phys. Rev. B27, 2132 (1983).

3. G.F. Holland and D.E. Ellis, and W.C. Trogler, J. Am. Chem. Soc. 108, 1884 (1986); B. Delley, M.C. Manning, D.E. Ellis, J. Berkowitz, and W.C. Trogler, Inorg. Chem. 21, 2247 (1982).

4. R.A. Johnson and W.D. Wilson, in "Interatomic Potential and Simulation of Lattice Defects", ed. P.C. Gehlen, J.R. Beeler, and R.I. Jaffee, (Plenum, New York 1971)

5. A.E. Carlsson, C.D. Gellatt, and H. Ehrenreich, Philos. Mag. A41, 241 (1980).

6. M.M. Goodgame and W.A. Goddard III, Phys. Rev. Lett. 48, 135 (1982); T.H. Upton and W.A. Goddard III, CRC Crit. Rev. Solid State Mat. Sci. 10, 261 (1981).

7. T. Ziegler, J.G. Snijders, and E.J. Baerends, J. Chem. Phys. 74, 1271 (1981); J.G. Snijders and E.J. Baerends, in Electron Distribution and The Chemical Bond, ed. by M.B. Hall, P. Coppens (Plenum NY, 1982) p.111

8. J. Harris and R.O. Jones, J. Chem. Phys. 70, 830 (1979).

9. G.S. Painter and F.W. Averill, Phys. Rev. 28, 5536 (1983).

10. D. Delley, A.J. Freeman and D.E. Ellis, Phys. Rev. Lett. 50, 488 (1983).

11. T.H. Upton and W.A. Goddard, III, J. Am. Chem. Soc. 100, 5659 (1978); I. Shim, J.P. Dahl and H. Johansen, Int. J. Quant. Chem. 15, 311 (1979).

12. H.P. Cheng and D.E. Ellis, (manuscript in preparation).

13. M.S. Daw and M.I. Baskes, Phys. Rev. B29, 6433 (1984).

14. G.F. Holland, D.E. Ellis and W.C. Trogler, J. Chem. Phys. 83, 3507 (1985).

Force Calculation in Density Functional and Muffin-Tin Orbital Framework

T. Fujiwara[1] and Y. Ishii[2]

[1]Department of Applied Physics, University of Tokyo, Tokyo 113, Japan
[2]Institute for Solid State Physics, University of Tokyo, Tokyo 106, Japan

We present the calculation of electronic structure of heavy metal cluster Cu_2. The total energy and force are calculated in the framework of the local density functional theory and the linear muffin-tin orbital method. Each contribution of force, i.e. the Hellmann-Feynman force and the base-derivative term (and their subparts), is investigated separately. The Hellmann-Feynman force seems to determine the internuclear distance, which fairly agrees with the experimental one. The base derivative term, which is introduced to compensate an error due to the incompleteness of the basis set, causes a large disagreement in the present scheme.

1. Introduction

One of the recent topics in the electronic structure calculation is the stable structure determination by calculating the force. Most of the work uses the basis set of linear combination of atomic orbitals (LCAO) expanded in terms of the Gaussian orbitals. The motivation for this choice is owing to the existence of the exact integration formula. On the other hand, we usually use the numerically solved "exactly" orthogonal basis set in the solid state case and can obtain remarkably good results of not only the one-electron properties but also the cohesive properties of the bulk system by the local density functional (LDF) formalism /1/. It may be highly desirable at the present stage to extend the general force calculation to the cases in the ill-condensed systems such as the atomic clusters, adsorbed surfaces, grain boundaries and amorphous systems. Among the several calculating methods for the bulk electronic properties, the linear muffin-tin orbital (LMTO) method /2/ is the most efficient and accurate enough. Gunnarsson, Harris and Jones /3/ presented the application of LMTO method for atomic clusters such as diatomic molecules. The LMTO method assures that the valence orbitals are orthogonal to the core orbitals because of their boundary condition and we can use the frozen core approximation. Harris, Jones and Müller /4/ presented the force formula within the frozen core approximation and calculated the binding energy and the force for H_2, Li_2 and Be_2. Recently, Averill and Painter /5/ presented the force calculation for Cu_2 by using LCAO-Gaussian orbitals and claimed the large effect of the core polarization. We believe that the frozen core approximation is still good.

In the present paper, we calculate the binding energy and the force for Cu_2 dimer in the framework of the LDF and LMTO with the frozen core approximation, including the base derivative term which is introduced to compensate an error of the incompleteness of the basis set /6/. We analyse several contributions for the force formula separately.

2. Definition of cells and muffin-tin orbitals

We follow the formulation by Gunnarsson, Harris and Jones /3/ and Harris, Jones and Müller /4/. The whole diatomic system is divided into three cells; two convex cells and one concave cell. The convex cell is the appropriate region around each atom. The concave cell is the outer region of the cluster (i.e. vacuum region). Each cell is subdivided into a muffin-tin sphere and a non-muffin-tin region. In a convex cell, the muffin-tin sphere is the largest sphere with the center at the nucleus. In the concave cell, the muffin-tin sphere is defined as the outer spherical region whose center locates at the midpoint between two nuclei.

We define the muffin-tin orbitals (MTO) χ_L^i inside the cell i as

$$\chi_L^i(\kappa,\mathbf{r}_i) = \Phi_L^i(G,\mathbf{r}_i)$$

$$= Y_L(\hat{\mathbf{r}}_i) \begin{cases} \Phi_{\ell i}(D(G_\ell^i),r_i): \ \mathbf{r}_i \text{ in the muffin-tin sphere} \\ \\ N_\ell^i(G)G_\ell^i(\kappa,r_i): \ \text{otherwise} \end{cases} \tag{1}$$

where $\mathbf{r}_i=\mathbf{r}-\mathbf{R}_i$ with $\mathbf{R}_i$ the center position of the i-th cell and an abbreviated notation $L=(\ell,m)$ for the angular momentum. The factor $N_\ell^i(G)$ ensures the continuity and smoothness of the MTO at the boundary of the muffin-tin sphere. The function $G_\ell^i(\kappa,r_i)$ is defined as follows,

$$G_\ell^i(\kappa,r_i) = \begin{cases} K_\ell(\kappa,r_i), & i:\text{convex} \\ \\ J_\ell(\kappa,r_i), & i:\text{concave} \end{cases} \tag{2}$$

The spherical Bessel functions K_ℓ and J_ℓ are defined in Ref.3. The function $\Phi_{\ell i}(D,r_i)$ is the linear combination of $\phi(\epsilon_{\ell i},r_i)$ and $\dot{\phi}(\epsilon_{\ell i},r_i)$ as

$$\Phi_{\ell i}(D,r_i) = \phi_{\ell i}(\epsilon_{\ell i},r_i) + \omega_{\ell i}(D)\dot{\phi}_{\ell i}(\epsilon_{\ell i},r_i) , \tag{3}$$

where ϕ is the solution of the radial Schrödinger equation with an "energy" $\epsilon_{\ell i}$ inside the muffin-tin sphere and $\dot{\phi}$ is its energy derivative function. The coefficient $\omega_{\ell i}(D)$ is determined so that its logarithmic derivative equals to D. The spherical Bessel function contains a parameter κ^2 which corresponds to the kinetic energy in the intermediate region. The Bessel function $G_L^i=G_\ell^i Y_L$ inside the i-cell can be expanded in terms of another set of Bessel function F_ℓ^j inside the j-th cell (j≠i).

$$F_\ell^j(\kappa, r_j) = \begin{cases} J_\ell(\kappa, r_j), & j:\text{convex} \\ \\ K_\ell(\kappa, r_j), & j:\text{concave} \end{cases} \tag{4}$$

which are the non-singular solution in the j-th cell.

In the following, the physical quantities such as the charge density, Coulomb potential, exchange-correlation potential etc. are treated separately in the above-mentioned different regions. In the muffin-tin region, the spherical component of each quantity is calculated on the radial logarithmic mesh (200 points in the convex cell and 40 points in the concave cell) and the non-spherical component is calculated on the radial linear mesh (15 points). All components in non-MT region are calculated on the radial linear mesh with 14 points.

The charge density is expanded in partial waves and each term can be calculated in terms of MTOs very easily. The total electronic Coulomb potential in a cell is the sum of that of the electronic charge inside its own cell and that of the electronic charge outside the cell. The evaluation of the latter term must be carefully treated. We use the exchange-correlation potential fomula of Ceperley and Alder /7/ which is now believed as the most accurate LDF formula especially in the high density limit.

3. Total energy and force

With the frozen core approximation, the total energy can be written down as the sum of the core (E_c) and valence (E_v) contributions. The latter contribution is

$$\begin{aligned}
E_v = &\sum_n f_n E_n \\
&- \sum_i \int_{i\text{-cell}} d\mathbf{r}_i\, n_v(\mathbf{r}_i)[(1/2)\phi_v(\mathbf{r}_i) + V^{XC}(\mathbf{r}_i) - \varepsilon^{XC}(\mathbf{r}_i)] \\
&+ \sum_i \int_{i\text{-cell}} d\mathbf{r}_i\, n_c(\mathbf{r}_i)[\varepsilon^{XC}(n(\mathbf{r}_i)) - \varepsilon^{XC}((n_c(\mathbf{r}_i))] \\
&+ (1/2) \sum_{i \neq j} \int d\mathbf{r}_i \int d\mathbf{r}_j'\, Z_i^*(\mathbf{r}_i)\, Z_j^*(\mathbf{r}_j')/|\mathbf{r}-\mathbf{r}'| \tag{5}
\end{aligned}$$

where $Z_j^*(\mathbf{r}_j) = Z_j \delta(\mathbf{r}_j) - n_c(\mathbf{r}_j)$ is the effective core charge density. The

first term is the sum of the one-electron orbital energies, and the second term for the double counting electron-electron interaction energy and the part of the exchange-correlation energy. The third term is the core electron exchange-correlation energy. The last term is the electrostatic energy between effective core charges.

The expression for the force is also given in the framework of the LDF. They consist of two parts, the first part is the Hellmann-Feynman force F^{H-F} and the second part is the base derivative term F^{BD}, which is introduced to compensate an error due to the incompleteness of the basis set /6/.

$$F = -\frac{dE}{d\lambda} = F^{H-F} + F^{BD} \tag{6}$$

$$F^{H-F} = -\frac{\partial}{\partial\lambda}(1/2)\sum_{i\neq j}\iint \frac{Z_i^*(\mathbf{r}_i)Z_j^*(\mathbf{r}_j')}{|\mathbf{r}-\mathbf{r}'|}d\mathbf{r}_i d\mathbf{r}_j'$$

$$-\int n_v(\mathbf{r})\frac{\partial}{\partial\lambda}\sum_i \frac{-Z_i}{|\mathbf{r}-\mathbf{R}_i|}d\mathbf{r}$$

$$-\int \{\phi_v(\mathbf{r}) + V^{xc}(n(\mathbf{r}))\}\frac{\partial}{\partial\lambda}n_c(\mathbf{r})\,d\mathbf{r} \tag{7}$$

$$F^{BD} = -\sum_n f_n \langle \frac{\partial}{\partial\lambda}\phi_n | H - E_n | \phi_n \rangle + c.c. \tag{8}$$

The first term of F^{H-F} is the force between effective core charges and is really well expressed as that between point charges. The second term is the electrostatic force between the nucleus and the valence electrons. The third term is the force between the core electrons and the valence electrons. These two terms, the second and the third, must be carefully evaluated because the behaviour is rather singular near the nucleus and the anomalous parts are due to the non-spherical charge density or the non-spherical Coulomb potential. We will present the value of the individual parts separately.

4. Numerical results

We present the results of two cases with the internuclear distance R=4.1Bohr and 4.2Bohr. The one-electron orbital energy is in order of $E(\sigma_g) < E(\pi_u) < E(\sigma_g) < E(\delta_g) < E(\delta_u) < E(\sigma_u) < E(\pi_g)$. The highest occupied state is π_g and nearly degenerate with σ_u state, which is different from the results by Painter and Averill /5/. Their highest occupied state is σ_u and the state σ_g locates between δ_g and δ_u. The numerical result of the valence contributions to the total energy and the force is shown in Table 1. The corresponding valence contribution to the total energy in an isolated Cu atom is 102.647Ryd/atom. From the calculated total energy, the stable internuclear distance may be between these two values (the experimental value is 4.2 Bohr) and the binding energy is about 2.8eV (the experimental value is 2.03eV). Here, we can see that the contribution of the non-spherical potential in the muffin-tin region is very small but not negligible.

The result of the force is more subtle. The Hellmann-Feynman force between effective core charges contributes a rather large part but can be expressed as the force between point charges at the position of the nucleus. Therefore, the most important part is the Hellmann-Feynman force between core and valence electrons and that between nucleus and valence electrons. From these results, we can see the contribution of non-spherical charge distribution in the muffin-tin region is the most important. The sum of the Hellmann-Feynman force suggests the equilibrium position may be larger than R=4.2Bohr.

The base derivative term F^{BD} can not compensate the error of the Hellmann-Feynman force and seems to increase the disagreement of the stable internuclear distance. In the LMTO formalism the terms mainly contributing to the total energy and the force are different partial wave components. On the other hand, in the Gaussian base formalism, the force formula with F^{BD} may correspond to the equation of the rigorous numerical derivatives of the total energy expression.

Table 1 Valence contribution E_v and force.

internuclear distance (Bohr)	:	4.1	4.2
Total Energy (Ryd)	:	-205.499	-205.492
(1) sum of one electron orbital energy	:	-10.620	-10.707
(2) a convex (atomic) cell			
1. valence electrons in MT.	:	-113.474	-113.261
2. core electrons in MT.	:	-4.192	-4.188
3. valence electrons in MT.(non-sph)	:	-0.140	-0.137
4. valence electrons in non-MT.	:	-5.341	-5.059
5. core electrons in non-MT.	:	-0.008	-0.006
(3) a concave cell			
1. valence electrons in MT.	:	-4.562	-4.211
2. valence electrons in MT.(non-sph)	:	-0.176	-0.170
3. valence electrons in non-MT.	:	-2.863	-2.724
(4) between nuclei and core electrons	:	59.027	57.622
Force (Ryd/Bohr)	:	-0.264	-0.220
(1) F^{H-F} between effective cores	:	-14.400	-13.718
(2) F^{H-F} between nuclei and electrons			
1.Nucl.(1)-valence electrons in MT(1)	:	2.401	2.383
2.Nucl.(1)-valence electrons in non-MT(1):		1.715	1.585
3.Nucl.(1)-valence electrons in 2nd cell :		35.587	34.032
4.nucl.(1)-valence elect. in concave cell:		-0.169	-0.166
(3) F^{H-F} acting on core electrons	:	-25.289	-24.287
(4) base derivative force	:	-0.111	-0.148

5. Conclusions

We have calculated the total energy and the force in Cu dimers in the framework of LDF and LMTO. The total energy is reasonable if the integral for the spherical component is carefully treated. The force calculation for heavy metal dimer seems to be a still hard test for LMTO because the integration of the non-spherical partial wave components can not be evaluated with the same accuracy as that of the spherical one.

In the framework of the LMTO, the base derivative term does not show any tendency of compensating the error of the Hellmann-Feynman force.

References

1. P.Hohenberg and W.Kohn, Phys.Rev.136, B864 (1964);
   W.Kohn and L.J.Sham, Phys.Rev.140, A1133 (1965).
2. O.K.Andersen, Phys.Rev.B12, 3060 (1975).
3. O.Gunnarsson, J.Harris and R.O.Jones, Phys.Rev.B15, 3027 (1977).
4. J.Harris, R.O.Jones and J.E.Muller, J.Chem.Phys.75, 3904 (1981).
5. F.W.Averill and G.S.Painter, Phys.Rev.B32, 2141 (1985).
6. C.Satoko, Phys.Rev.B30, 1754 (1984).
7. J.P.Perdew and A.Zunger, Phys.Rev.B23, 5048 (1981).

The Electronic Structure of Transition Metal Clusters

H. Tatewaki[1]*, M. Tomonari*[1]*, T. Nakamura*[1]*, and E. Miyoshi*[2]

[1]Research Institute for Catalysis, Hokkaido University, Sapporo 060, Japan
[2]Fukuoka Dental College, Fukuoka, Japan

1. Introduction

The investigations on metallic clusters increase day by day.
Experimentalists as well as theorists are interested to know
whether the small metallic clusters have similar properties to
those of bulk metals. The review article on this field was
first given by OZIN [1]. In this report, the electronic structure of Cu, Ni, and Zn clusters will be discussed. Using X_α-
SCF method, MESSMER et al.[2] have shown that small transition
metal clusters such as Cu_8 have a density of states (DOS)
parallel to that of corresponding metals: the d band is embedded in the s band or the top of the d band and that of the s
band consist of the Fermi level. On the other hand, using ab
initio SCF calculations and resulting orbital energies, BASCH
et al.[3] and BACHMANN et al. [4] have given completely different
pictures: the d band is far from the the s band for Ni clusters
[3] as well as Cu clusters [4]. The difference between the two
is schematically shown in Fig. 1. We, therefore, have determined to investigate the electronic structure of the transition
metal clusters with the aid of the ab initio SCF and configuration interaction (CI) calculations. Whether the small transition metal clusters have the characteristics of the bulk or

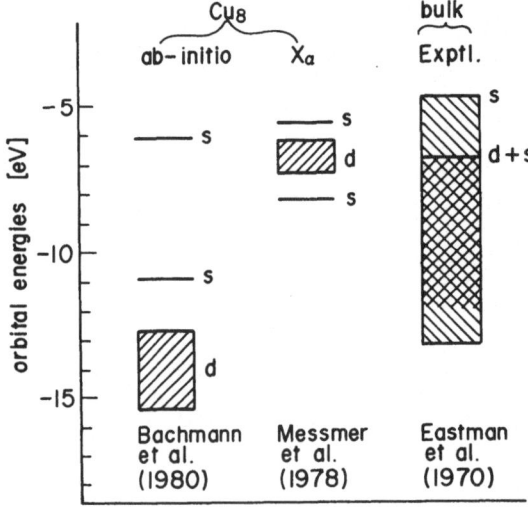

Figure 1

Energy level distri-
butions by X_α and ab
initio SCF calcula-
tions

not is tested by
1) the position of the Fermi level,
2) the position of the d band relative to the Fermi level,
3) the s and d band widths, and
4) the magnitude of the electron affinity.

The electron affinity of the metal is equal to the work function. In this report, 2) and 3) will be mainly discussed. The object transition metal clusters are Cu_{1-6}, Zn_{1-6}, and Ni_{1-6}.

2. Method of Calculations.

Except for Ni, Ni_2, Zn, Zn_2, and Zn_3 where CI calculations were performed, we used ab initio SCF calculations proposed by ROOT-HAAN and BAGUS [5]. The programs used were 'JAMOL' and 'COMICAL' for SCF and CI calculations, respectively. The former is written by KASHIWAGI et al.[6] and the latter TANAKA et al.[7]. In 'COMI-CAL' the first order interacting space [8],[9] has been used to span the CI space. The basis sets were newly developed contracted Gaussian type functions (CGTF's) by the method of TATEWAKI and HUZINAGA [10]. They have a form of $(43321/421/311)+P_1+P_2$ or similar ones to this. The slashes in the parentheses specify the symmetries, s, p, and d. The numbers 4, 3, 3, 2, and 1 before the first slash indicate that five CGTF's are employed for the s symmetry and the respective CGTF's are expanded with four, three, three, two, and one primitive GTF's, and so on. The P_1 and P_2 are polarization functions for 4s electrons. The characteristic of these CGTF's is their ability to describe the valence region of the atoms with rather short expansions.

3. Geometries of Clusters.

The linear, triangular(D_{3h}), square planar(D_{4h}), square pyramid(C_{4v}), and bisquare pyramid(O_h) are assumed for M_{1-6} (M=Cu, Zn, and Ni) clusters. The bond distances for the respective clusters are the nearest neighbor of the respective metals; 4.83, 5.04, and 4.71 a.u. for Cu_n, Zn_n, and Ni_n.

4. Band Structure from Koopmans' Theorem

Band structure given by the Koopmans' theorem for Cu_n is shown in Table 1. The band width is obtained by the difference of the highest occupied and lowest occupied orbital energies. The Fermi level is defined as the the highest occupied orbital energy. The top of the d band relative to the Fermi level is also given in the last column of the table. The d band width increases from Cu_4 to CU_6. The increment in the width which arises from the energy lowering in the lowest orbital, is 0.2-0.3 eV. The width for Cu_6 is 1.9 eV. BACHMANN et al. have reported that the width is 2.85 eV for Cu_4 [4]. The half-width of the d band is 2.9-3.2 eV for the bulk [11]. More atoms are required to get the experimental bulk value. We, however, feel that the calculated result is fairly close to that of bulk. The s band width shows large increase from Cu_4 to Cu_5 and change is small from Cu_5 to Cu_6. The width for Cu_6 is 4.7 eV,

Table 1 Band structure for Cu_n by Koopmans' theorem (eV)

	d band			s band			top of d band
	LO	HO	width	LO	HO	width	
Cu_1	-12.9	-12.9	0.0	-6.3	-6.3	0.0	-6.6
Cu_2	-13.4	-12.3	1.1	-6.1	-6.1	0.0	-6.3
Cu_3	-14.1	-12.7	1.4	-7.4	-4.2	3.2	-8.5
Cu_4	-14.0	-12.7	1.3	-7.8	-5.3	2.5	-7.4
Cu_5	-14.7	-13.0	1.7	-9.2	-4.8	4.4	-8.2
Cu_6	-15.0	-13.1	1.9	-10.1	-5.4	4.7	-7.6

which is 2.5 times larger than that of the d band. The absolute
values of the highest occupied orbital energies for the s band
from Cu_3 to Cu_6 are around 5.0 eV. At a glance, they are close
to the work function of the bulk (4.65eV). It is, however, an
accidental one; if relaxation (reorganization) energy and electronic correlation effects are properly taken account, the
calculated ionization potential (IP) for small clusters becomes
larger than that of the Koopmans' theorem.

    Except for the band width, the quantities for the d band width
in Table 1 show wrong trend with a cluster size if they are
compared with the reported bulk values; the lowest and highest
orbital energies are too low and the top of the d band relative
to the Fermi level oscillates around 8 eV from Cu_3 to Cu_6. The
d band begins 2 eV below the Fermi level for metal Cu. The calculated density of states (DOS) with the Koopmans' theorem for Cu_n
is, thus, very different from those of bulk. Although we have
not shown the results, the d band given by the Koopmans' theorem
is separated from that of the s band for Zn clusters as well as
Ni clusters.

5. Band Structure from Symmetry-Broken ΔSCF

In the previous section, we have shown that even if we increase
the number of constituent atoms, the top of the d band is far
from the Fermi level. In Koopmans' theorem it is assumed that
when a 3d electron is ionized, resulting d hole is distributed
over the whole cluster which consists of symmetrically equivalent atoms. We, however, noted that the d electron is very
localized at the atom. For example the Cu-Cu distance in metallic Cu is 4.83 a.u., while the mean distance of r, $\langle 3d/r/3d \rangle$ is
only 0.98 a.u. If we apply the Koopmans' theorem to atomic
ionizations of Cu, $3d^{-1}$ state lies 6.8 eV below the $4s^{-1}$ (Fermi
level), while ΔSCF calculation gives that $3d^{-1}$ is 1.6-1.9 eV
below the Fermi level which is near to experiment. We also
found that relaxation energy is very large for the d electron
ionization. It amounts to 4.9 eV. We have performed symmetry-
broken SCF calculations for Cu_2 and Cu_3 where the localization
of the d electron at one of the constituent atoms is allowed,
if the state is energetically lower. The resulting d hole is
localized as expected, and the top of the d band for Cu_2 and Cu_3

Table 2 Band structure for M_n by symmetry-broken SCF (eV)

Cu clusters

	d band			s band			top of d band (corrected)
	LO	HO	width	LO	HO	width	
Cu_1	-8.1	-7.8	0.3	-6.2	-6.2	0.0	-2.9
Cu_2	-7.4	-6.3	1.1	-5.7	-5.7	0.0	-1.9
Cu_3	-7.3	-5.9	1.4	-7.5	-4.3	3.2	-3.1
Cu_4	-8.0	-6.7	1.3	-7.8	-5.3	2.5	-1.4
Cu_5	-8.7	-7.0	1.7	-9.2	-4.8	4.4	-2.2
Cu_6	-9.0	-7.0	1.9	-10.1	-5.4	4.7	-1.7
Cu_∞	-10.0	-6.8	3.2		-4.7		-2.1

Zn clusters

	d band			s band			top of d band (corrected)
	LO	HO	width	LO	HO	width	
Zn_1	-15.4	-15.4	0.0	-7.6	-7.6	0.0	-7.8
Zn_2	-15.1	-14.7	0.4	-9.6	-6.0	3.5	-8.7
Zn_3	-15.0	-14.4	0.6	-11.3	-6.5	4.8	-8.1
Zn_4	-14.7	-14.1	0.6	-11.7	-4.6	7.2	-9.5
Zn_5	-15.5	-14.5	1.0	-13.5	-4.7	8.9	-9.9
Zn_6	-15.5	-14.8	0.8	-14.8	-4.9	9.9	-9.8
Cu_∞	-15.0	-13.9	1.0		-4.3	>9.6	-9.6

Ni clusters

	d band			s band			top of d band (corrected)
	LO	HO	width	LO	HO	width	
Ni_1	-9.3	-5.9	3.4	-6.3	-6.3	0.0	0.4
Ni_2	-8.3	-4.6	3.7	-5.7	-5.7	0.0	1.1
Ni_3	-8.3	-4.2	4.1	-6.2	-4.2	2.0	0.0
Ni_4	-8.2	-4.6	3.6	-6.2	-5.1	1.1	0.5
Ni_5	-8.8	-4.8	4.0	-7.3	-4.8	2.5	0.1
Ni_6	-9.1	-4.5	4.6	-8.7	-4.4	4.4	0.0
Ni_∞	-9.8	-5.2	4.6		-5.2		0.0

a) Details for Cu_n, Zn_n, and Ni_n are given in references [12], [16], and [17], respectively.

lies 0.6-1.8 eV below the Fermi level. Following the discussions given in reference [12], we have modified the results by adding the atomic correlation of 1.3 eV to the above values; the top of the d band thus becomes 1.6 and 3.1 eV below the Fermi level for Cu_2 and Cu_3, respectively. The relaxation energies from Cu_1 to Cu_3 show strong convergence. We can get reliable relaxation energy for larger clusters by extrapolation. We, therefore, modify the orbital energies of larger clusters by adding the relaxation energy and atomic correlation correction to the orbital energies given in Table 1. The obtained band structure is given in the first part of Table 2.

The top of the d band of Cu_{5-6} relative to the Fermi level is around -2.0 eV which is close to the value of the metal. Similarly, we have obtained the band structure of Zn_n and Ni_n and they are given in the second and third parts of Table 2. The symmetry-broken SCF calculations provide DOS of Zn and Ni clusters parallel to that of corresponding metals. As a whole, so far as the position of the d band and band width are concerned, the electronic structure of the small transition metal clusters resemble that of metal.

6. Comparison of Calculated DOS with Experiment

Figure 2 shows experimental DOS's of metal Cu and Cu clusters (size unknown) by SCHMEISSER and coworkers [13] as well as that of Cu_3 by the present work. From the figure, we realize that calculated DOS on Cu_3 are similar to those of experiment and the problems on the position of the d band relative to the Fermi level are solved both experimentally and theoretically. Finally we comment on the first IP's for small transition metal clusters. Recently, using supersonic technique with laser or oven vaporizations, support-less metallic clusters were generated [14],[15]. Calculated IP's by SCF and experiment are summarized in Table 2. It is shown that SCF gives smaller IPs than experiment; it is,

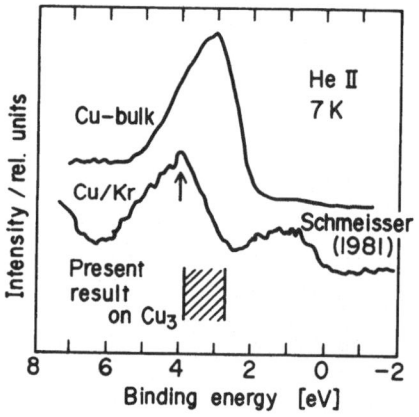

Figure 2.

Experimental DOS for metallic Cu and Cu clusters and calculated DOS for Cu_3

Table 3. IPs for Cu_{1-6} and Ni_{1-6}

	Cu_1	Cu_2	Cu_3	Cu_4	Cu_5	Cu_6
ΔSCF	6.2	5.7	4.1	5.1	4.4	4.3
Exptl.	7.7	7.9	5.0-6.4	6.4-7.9	5.0-6.4	6.4-7.9

	Ni_1	Ni_2	Ni_3	Ni_4	Ni_5	Ni_6
ΔSCF	5.9	4.6	4.2	4.6	4.8	4.4
Exptl.	8.7	6.4	5.0-6.4	5.0-6.4	5.0-6.4	>6.4

The experimental values for Cu_n and Ni_n are from ref. [14] and [15].

however, confirmed that inclusion of the correlation effect provides almost the same value as experiment for the case of Ni_1 and Ni_2 [17], revealing that the inclusion of the correlation effects is indispensable to get the proper IPs. Details on the present calculations are given elsewhere [12],[16],[17].

7. Literature

1. G. A. Ozin, Catal. Rev. 16, 191(1977); Faraday Symp. Chem. Soc. 14, 1(1980)
2. R. P. Messmer, S. K. Knudson, K. H. Johnson, J. B. Diamond, and C. Y. Yang, Phys. Rev. B13, 1396(1976)
3. H. Basch, M. D. Newton, and J. W. Moskowitz, J. Chem. Phys. 73, 4492(1980)
4. C. Bachmann, J. Demuynck, and A. Veillard, Gazz. Chim. Ital. 108, 389(1978); Faraday Symp. Chem.Soc. 14, 170(1980)
5. C. C. J. Roothaan and P. S. Bagus. Method in Computational Physics, Vol.2, 215(Academic, New York,1963)
6. H. Kashiwagi, T. Takada, E. Miyoshi, S. Obara and K. Ohno, Contrib. Group on Atoms Mol. 12, 5(1976)
7. K. Tanaka, T. Nomura, T. Noro, and H. Tatewaki, Contrib. Group on Atoms Mol. 16, 8(1980)
8. A. D. McLean and B. Liu, J. Chem. Phys. 58, 1066(1973)
9. H. Tatewaki, K. Tanaka, F. Sasaki, S. Obara, and K. Ohno, Int. J. Quantum Chem. 15, 533(1979)
10. H. Tatewaki and S. Huzinaga, J. Chem. Phys. 71, 4339(1979)
11. Y. Baer, P. F. Heden, J. Hedman, M. Klasson, C. Nordling, and K. Siegbahn, Phys. Scr. 1, 55(1970)
12. H. Tatewaki, E. Miyoshi, and T. Nakamura, J. Chem. Phys. 76, 5073(1982); ibid., 78, 815(1983)
13. D. S. Schmeisser, K. Jacobi, and D. M. Kolb, J. Chem. Phys. 75, 5300(1981)
14. E. A. Ròhlfing, D. M. Cox, and A. Kaldor, J. Phys. Chem. 88, 4497(1984)
15. D. E. Powers, S. G. Hansen, M. E. Geusic, D. L. Michalopoulos and R. E. Smalley, J. Chem. Phys. 78 , 2866(1983)
16. M. Tomonari, H. Tatewaki, and T. Nakamura, J. Chem. Phys. 80, 344(1984)
17. M. Tomonari, H. Tatewaki, and T. Nakamura, J. Chem. Phys. 85, 2875(1986)

Theoretical Study of the Electronic Structure of CO Adsorbed on Small Cu Clusters

K. Tanaka[1], Y. Mochizuki[1], T. Kawaguchi[1], K. Ohno[1], and H. Tatewaki[2]

[1]Department of Chemistry, Faculty of Science, Hokkaido University, Sapporo 060, Japan
[2]Research Institute for Catalysis, Hokkaido University, Sapporo 060, Japan

1. Introduction

The study of molecular adosorbates on solid surfaces mostly has dealt with their structure and bonding to the substrate in their electronic ground state. In recent years, it has been realized that electronically excited states of adsorbates also play an important role in a variety of surface processes.

Carbon monoxide(CO) had long been a prototype molecule in chemisorption studies. The ground state of the molecule itself has the electronic configuration

$$\ldots (4\sigma)^2(1\pi)^4(5\sigma)^2(2\pi\star)^0 \qquad (1)$$

The assignment of the excitation spectra of the lower excited states of this molecule is well established. The lower valence-type excited states are composed of $5\sigma \rightarrow 2\pi\star$, $1\pi \rightarrow 2\pi\star$, and $4\sigma \rightarrow 2\pi\star$ transitions. The Te values of gas-phase CO[1] are compared with excitation bands obtained by electron energy loss spectroscopy (EELS) of solid CO. The triplet and singlet $5\sigma \rightarrow 2\pi\star$ transitions are located around 6 eV and 8 eV, respectively: The $1\pi \rightarrow 2\pi\star$ transitions overlap with the singlet $5\sigma \rightarrow 2\pi\star$ transition and range from 7 eV to 10 eV.

Electronic excitation spectra of the metal-CO system have been investigated mostly by EELS using several metal surfaces[2a-d]. Two major loss bands are observed around 5-7 eV and 13-14 eV in most systems. Other than these bands, two bands are found at ∿8.5 eV and ∿12 eV in certain cases. All four bands are observed in the CO-Cu metal systems. The bands observed by SPITZER and LÜTH[2-e] are shown in Fig.1. Several speculations on the assignment of the spectra have been put forward[2a-c]. They are summarized in table 1. The assignment of CT, $5\sigma \rightarrow 2\pi\star$, and $4\sigma \rightarrow 2\pi\star$ transitions based on the calculation on NiCO by FREUND et al.[2-d] is also included. There are significant disagreements regarding the assignment of the spectra. In this work, we carry out ab initio computations on model systems of CO plus a small Cu cluster; hoping a small cluster could be a good substitute of metal[3]. Our attention is concentrated on the intra CO valence-type excited states and possible CT excitations are not considered in this report.

2. Method of Calculation

We use Gaussian split valence basis set; (3,3,3,2,1/3,3/4,1) for Cu given by HUZINAGA et al.[4] and MIDI-4 by TATEWAKI and HUZINAGA[5] for C and O. A diffuse p function is added to each atom. Orbital exponents are 0.034, 0.059, and 0.09 for C, O, and Cu, respectively. A self-consistent field (SCF) calculation is carried out for the ground state of the model system of Cu_nCO, (n=1,..,4) whose geometry is shown in Fig. 2. It is well known

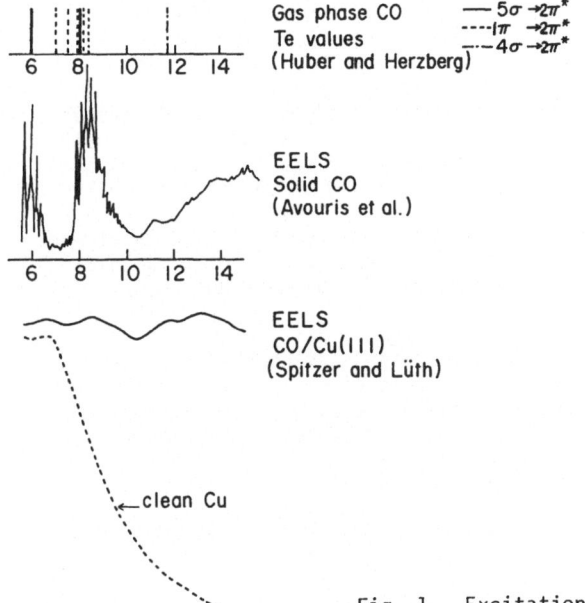

Fig. 1. Excitation spectra of CO and CO/Cu

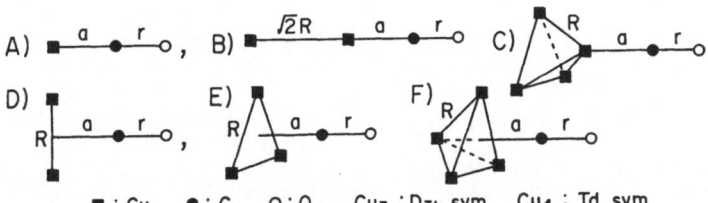

■ : Cu, ● : C, o : O, Cu₃ : D₃ₕ sym., Cu₄ : Td sym.

Fig. 2. Geometry of the model system

Table 1	Previous		Assignment		
	5-7 eV		~8.5 eV	~12 eV	13-14 eV
ref.2a	5σ → 2π*, CT		5σ → 2π*	——	4σ → 2π*, 5σ → ∞
ref.2b	5σ → 2π*, 1π → 2π*, CT				
ref.2c	5σ → 2π*, CT		5σ → 2π*	4σ → 2π*, 1π → 2π*	
ref.2d	CT		5σ → 2π*	4σ → 2π*	

that CO binds with the C end, and the CO axis is perpendicular to the surface in the ground state of Cu-metal systems[6]. The internuclear distance between Cu atoms (R) taken to be the nearest neighbour value of metal Cu (2.556A). The CO distance (r) is fixed to 1.15Å which is observed in CO-Cu metal system[7] and the cluster-C distance (a) is fixed to 2.05Å which is obtained by BAGUS et al.[8] in the SCF calculation for Cu_5CO. Symmetry of the ground state of the model systems are $^2\Sigma^+$ for A, $^1\Sigma^+$ for B, $^1A_1$ for D, ^2E for E, and $^3A_2$ for C and F.

Configuration interaction(CI) calculations are carried out to obtain excited state wave functions. We use only configurations arising from single excitations from the ground state configuration, i.e. Tamm Dankov CI. The K shell electrons of C and O and K,L,3s and 3p shells of Cu are kept frozen. If we use all molecular orbitals(MO's) obtained by the SCF calculation, a great number of configurations due to intra Cu excitations would be generated. As we are interested in only intra CO excitations, virtual orbitals of the cluster side are rejected by the use of the corresponding orbital method[9]. Number of virtual orbitals for CO-Cu cluster is equal to that of CO and the overlap between the space spanned by the CO-Cu cluster virtual orbitals and virtual orbitals of CO is maximized. In CI calculations, only those truncated orbitals are used. The biggest number of configurations of Cu_4CO is 770.

Influence of the truncation of the virtual orbitals is checked for the two Cu_2CO systems(B and D). Truncation error is within 0.2 eV for $4\sigma \rightarrow 2\pi^*$ and $1\pi \rightarrow 2\pi^*$, and 0.4 eV for $5\sigma \rightarrow 2\pi^*$. This error is not serious to the present problem. To test reliability of the Tamm Dankov CI, approximate Te values of CO are calculated and compared with experimental Te values. The maximum error is 0.7 eV. This accuracy is acceptable in discussing the present problem.

3. Results and Discussion

Let us define interaction energy of the ground state, E_{int}, by using total energies, E, of Cu_n, CO and Cu_nCO,

$$E_{int}(Cu_nCO) = E(Cu_n) + E(CO) - E(Cu_nCO). \qquad (2)$$

E_{int}'s calculated at SCF level and net charge by Mulliken's population analysis are given in Table 2.

Table 2 E_{int} (eV) and net charge of atoms
Cuf : the first layer, Cus : the second layer

		CO	A	B	C	D	E	F
E_{int}		——	-0.34	0.06	0.60	-1.54	-0.94	0.07
net charge	C	0.29	0.21	0.21	0.18	0.15	-0.19	-0.03
	O	-0.29	-0.23	-0.22	-0.23	-0.21	-0.23	-0.23
	Cuf	——	0.02	0.18	0.15	0.06	0.14	0.05
	Cus	——	——	-0.17	-0.03	——	——	0.10

Fairly large binding energy is obtained for the model C(on-top Cu_4CO). The binding energy happens to be in good agreement with the experimental value of 0.52 eV for the CO/Cu(111) system[10]. This is in accord with the result obtained by BAGUS et al.[8] who obtained a good agreement of the binding energy for the on-top Cu_5CO model with that of the CO/Cu metal system. The model of on-top with a few second layer atoms seems to be the smallest realistic model for the CO-Cu metal adsorption.

The on-top Cu_2CO(model B) is stabilized but only a little. A comparison of net charge of B and C with those of A indicates that second layer Cu in the models C and B takes electron a little from on top Cu and it seems to play an important role in binding CO with on-top Cu. Small stabilization is also obtained for the model F. We are using the cluster-C distance

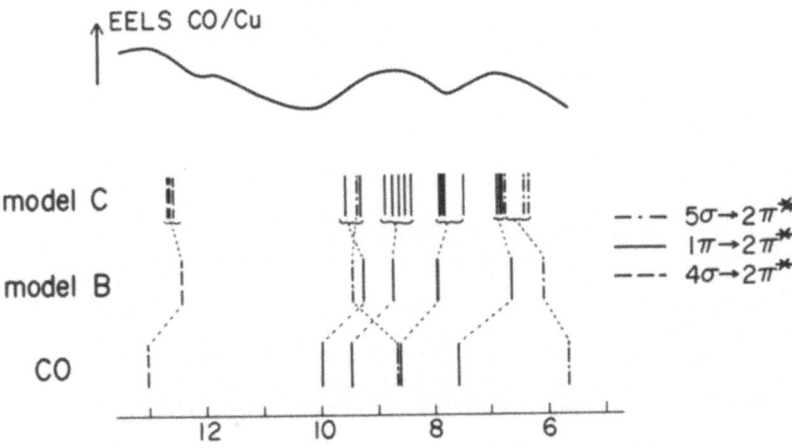

Fig. 3. Vertical excitation energies (eV) and EELS feature

determined in on-top Cu_5CO and the distance might be inappropriate for the hollow Cu_4CO.

Let us discuss excitation energies. Vertical excitation energies of the model C are compared with the spectra obtained by EELS in Fig. 3. Vertical excitation energies of the model B and CO are also included in the figure. In the following we mostly use the result for C to discuss excitation spectra. The excitation energies of $5\sigma \to 2\pi*$ transitions are located around 6.5 eV for the triplet transition and at 9.3 eV for the singlet transition. The excitation energies of $1\pi \to 2\pi*$ transitions spread from 6.8 eV to 9.5 eV. By comparing these excitation energies with the observed bands, it is reasonable to consider that the $5\sigma \to 2\pi*$ and $1\pi \to 2\pi*$ transitions contribute to the lowest two bands. (5-7 eV and ~8.5 eV). The present results support the previous speculation about location of the $5\sigma \to 2\pi*$ transitions, but do not support the assignment by NETZER et al[2-c] who have claimed that the $1\pi \to 2\pi*$ states are blue shifted to contribute to 12-14 eV region. We concur with ISHI and OHNO[2-b] who have suggested that the $1\pi \to 2\pi*$ transitions would contribute to the lowest band.

The triplet $4\sigma \to 2\pi*$ transitions are located around 12.5 eV. These transitions would contribute to one of the higher two bands(12 eV and 13-14 eV). As these two bands are so close to each other, it is difficult to give a clear assignment.

It is noted that the $1\pi \to 2\pi*$ excitation energies are smaller than those of CO, whereas those of $5\sigma \to 2\pi*$ are greater. These two types of excited states energetically overlap. This behaviour could be explained by the following simple picture. The 5σ orbital is essentially a carbon lone pair orbital and its orbital energy in the ground state is stabilized by about 3 eV relative to that of CO. This stabilization is mainly caused by positively charged on-top Cu atom. On the other hand, the 1π orbital is spread considerably to the oxygen atom. The 1π orbital energy is stabilized only by 1 eV. They are the primary origin of the change in the excitation energy of the $1\pi \to 2\pi*$ and $5\sigma \to 2\pi*$ transitions due to the on-top Cu.

The excitation energies obtained by the model B are close to those of the model C. This information might be useful for calculating excitation energies of adsorbed systems which include heavier atoms.

4. Summary

By the use of systems of CO plus a small Cu cluster(Cu_nCO, n=1,2,3,4), the electronic structure of the adsorbate is studied by ab initio SCF and CI calculations. Among model systems used in this work, the on-top Cu_4CO system is the smallest realistic model for CO/Cu metal adsorption. The second layer atoms seem to play an important role in binding CO to Cu.

Vertical excitation energies due to intra CO transitions are calculated by limited CI calculations. But CT transitions are not considered in this work. The results indicate that the $5\sigma \rightarrow 2\pi*$ and $1\pi \rightarrow 2\pi*$ transitions contribute to the lowest two bands(5-7 eV and $\sim$8.5 eV) observed by EELS. The $4\sigma \rightarrow 2\pi*$ transition contributes to one of two higher bands($\sim$12 eV and 13-14 eV).

We would like to appreciate Mr. Shin-ichi Ishi for his kindly bringing our attention to this interesting problem and useful discussions. Thanks are due to memebers of quantum chemistry group of Hokkaido University for their discussions and comments.

5. Literature

1. K. P. Huber and B. Herzberg: Molecular Spectra and Molecular Structure. IV. Constants of Diatomic Molecules (Van Nostrand, New York, 1979)
2. For recent reviews; a) Ph. Avouris and J. E. Demuth: Ann. Rev. Phys. Chem. 34, 49 (1984), Ph. Avouris, N. J. DiNardo and J. E. Demuth: J. Chem. Phys. 80, 491 (1984), Ph. Avouris, P. S. Bagus and A. R. Rossi: J. Vac. Sci. Technol. B, 3, 1484 (1985); b) S. Ishi and Y. Ohno: Surface Sci. 139, L219 (1984), c) F. P. Netzer, J. U. Mack, E. Bertel and J. A. D. Matthew: Surface Sci. 160, L509 (1985), d) H. -J. Freund and R. P. Messmer, W. Spiess, H. Behner, G. Wedl and C. M. Kao: Phys. Rev. B 33, 5228 (1986), e) for CO/Cu spectra, A. Spitzer and H. Lüth; Surface Sci. 102, 29 (1981)
3. H. Tatewaki, E. Miyoshi, and T. Nakamura: J. Chem. Phys.76, 5073 (1982)
4. Physical Sciences Data 16, - Gaussian Basis for Molecular Calculations - , Ed. S. Huzinaga, Elesevier, (1984)
5. H. Tatewaki and S. Huzinaga: J. Compt. Chem. 1, 205 (1980)
6. For example, The Nature of the Surface Chemical Bond, ed. by T. N. Rhodin and G. Ertl, North-Holland, New York, 1974
7. S. Andersson and J. B. Pendry: Phys. Rev. Lett. 43, 363 (1979)
8. P. S. Bagus, K. Herman, and M. Seel: J. Vac. Sci. Technol., 18, 435 (1981)
9. A. T. Amos and G. G. Hall: Proc. R. Soc. A263, 483 (1961)
10. P. Hollins and J. Pritchard: Surface Sci. 89, 486 (1979)

Part III

New Experiments

Optical and Dynamical Properties of Metal Clusters

M. Broyer[1], *G. Delacrétaz*[2], *P. Fayet*[2], *P. Labastie*[1], *W.A. Saunders*[2],
J.P. Wolf[2], *and L. Wöste*[2]

[1]Laboratoire de Spectrométrie Ionique et Moléculaire,
(associé au CNRS No. 171), Université Lyon I, Bât. 205,
43, Bd. du 11 Novembre 1918, F-68622 Villeurbanne Cedex, France
[2]Institut de Physique Expérimentale, Ecole Polytechnique Fédérale
de Lausanne, PHB-Ecublens, CH-1015 Lausanne, Switzerland

The spectroscopy of Na_3 has been systematically investigated with different techniques: Two-Photon Ionization (TPI) and Depletion Spectroscopy (DS). Four excited electronic states have been found in the range from 850 nm to 330 nm. The lifetime measurement of these states suggests that the highly excited ones are partially or totally predissociated. For this reason depletion spectroscopy has been used, and reveals the complete structure of the predissociated C-state. The Na_3 ground state has been investigated by hot band analysis. This allows a precise comparison with the calculations.

Photofragmentation patterns of Al_n^+ are measured in a size-selected sputtered cluster ion beam. The results exhibit Al^+ as the most probable fragment for the cluster sizes n = 3, 5 and 7. This shows that the monomer has a lower ionization potential than the cluster, as explained by recent calculations. Ion-molecule reactions of nickel clusters and carbon monoxide allow the formation of carbonyl compounds. The stoichiometry of these products correlates extremely well with simple electron counting rules.

1. INTRODUCTION

The interest for metal clusters lies at the interface of various disciplines, and aspects of molecular spectroscopy, analytic and synthetic chemistry, surface science, solid state physics, nuclear physics, etc., are incorporated in the explanation of the diverse properties of aggregated matter the size of a few atomic diameters. To the benefit of the field, the richness of chemical and physical properties displayed by metal clusters and the great promise for useful applications experiments must be performed, that provides size-specific informations. Since there is a correlation between spectroscopic data and chemical reactivities of metal clusters [1], the investigation of optical and dynamical properties of these particles is useful.

In the first section of the paper we discuss spectroscopic results obtained from alkaline clusters in a molecular beam. The work, focused mainly on Na_3 provides a rigorous test of electronic structure calculations. In the second section, we discuss results from recent experiments on size-selected cluster ions. Photofragmentation studies of Ag^+- and Al^+-cluster ions reveal information about their electronic structure, and thus relate to both their chemical and physical properties. In comparison, gas-phase reactivity studies of size-selected nickel clusters inform directly about the chemical properties of the clusters, from which their structural and electronic properties can be determined.

2. SPECTROSCOPY OF Na_3

Despite intense interest in the spectroscopy of clusters, the only clusters larger than dimers for which gas phase spectra have been measured so far are Na_3 [2] and Cu_3 [3]. This is mainly due to dissociation processes, which easily occur during Two-Photon-Ionization (TPI). In order to overcome this difficulty and to obtain size specific informations also from bound-free transitions, we developed complementary to the TPI-techniques the method of depletion spectroscopy (DS). Both experimental schemes are indicated in figures 1a and 1b.

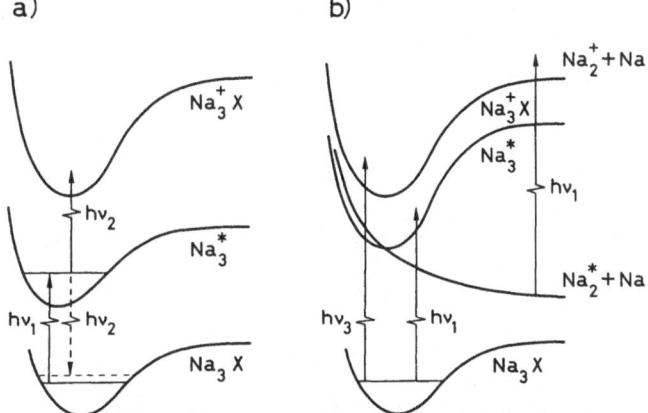

Fig. 1 Scheme of the experimental techniques used for the spectroscopy of Na₃. (a) Two-Photon Ionization (TPI). (b) Depletion Spectroscopy (DS). The dashed line in figure 1a indicates the possible use of Stimulated Emission Pumping for hot band analysis.

The experimental apparatus has been described in detail elsewhere [4]. Briefly, the clusters are produced in supersonic expansion of Ar (nominal pressure 10 bar) seeded with 10 to 100 mbar of sodium through a 50 µm diameter nozzle. The cluster and laser beams intersect at right angles at the entrance to a quadrupole mass spectrometer. By setting the quadrupole to pass the appropriate mass, the relating ion signal can be recorded as a function of the laser wavelength.

As indicated in figures 1a and 1b, the molecules are electronically excited with a tunable laser $h\nu_1$. In the resonant TPI scheme (Figure 1a), the excited molecules are ionized with a second laser, $h\nu_2$ and the photoions are detected directly. In the DS experiment (Figure 1b), which is used to probe the dissociative states, an ultra-violet laser, $h\nu_3$ (308 nm) directly ionizes Na₃ and monitors the remaining population in the molecular beam. The photofragmentation products Na and Na₂ can also be detected by increasing the power of $h\nu_1$ sufficiently to ionize them by multiple step processes.

Results

Four electronic excited states have been observed by resonant TPI spectroscopy in the wavelength range 330 to 850 nm (Figure 2). They are labeled the A-, B-, B'- and C-states, in ascending order. Of these, only the C-state shows any evidence of being predissociative. The A-state shows only a short vibrational progression, thus implying it probably has a geometry similar to that of the

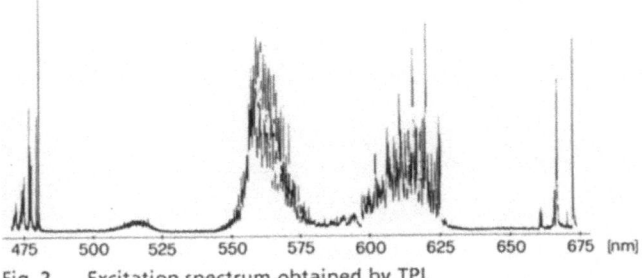

Fig. 2 Excitation spectrum obtained by TPI.

89

ground-state. Lifetime measurements, which were obtained by delaying $h\nu_2$ with respect to $h\nu_1$ indicate it is not predissociated (Table I) [5].

Table I: Na$_3$ Lifetimes

A state	$60 \pm 10\,nS$
B state	$14 \pm 5\,nS$
B' state	$7 \pm 3\,nS$
C state	$7 \pm 3\,nS$

Of the excited electronic states, the B-state is, so far, the best understood and the features are well described in reference 6. Briefly the band can be classified as :(1) a long progression of nearly equally spaced bands ($\omega_0 = 128$ cm^{-1}) which are split into doublets and (2) accompanying each member of the main progression there is a series of closely-spaced bands fanning out from the doublet. It can be shown that this spectrum corresponds to the vibronic pseudorotation spectrum predicted by Longuet-Higgins in the case of large Jahn-Teller distortions. In that case, a simple pattern emerges given by

$$E\,(u,j) = (u+1/2)\,\omega_o + A\,j^2$$

where $u = 0, 1, 2, \ldots$ corresponds to the distortion amplitude and $|j| = 1/2, 3/2, \ldots$ to the internal pseudorotation. The quantum number j is half integral because the electronic wavefunction changes sign in a degenerate state (adiabatic sign-change theorem) upon each complete internal rotation.

The B' state is very complex and not yet well understood. The slow transition between the structure of the B system and the B' system suggests that the B' system corresponds to transitions toward a second Born-Oppenheimer surface of the same electronic excited state.

The C-state is not fully observed TPI because it is partially predissociated (Figure 3). Lifetime measurements of the C-state (Table I) suggest it is predissociated. The complete structure of the state has been observed by DS and an interpretation of it as a weakly distorted state is presently under investigation.

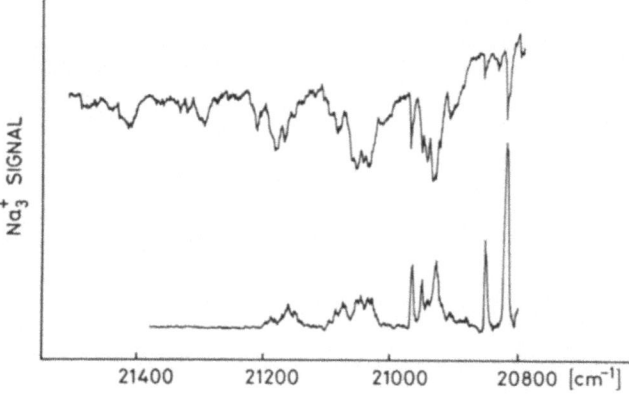

Fig. 3   C-state spectrum obtained by TPI (Lower) and DS (upper).

Another state, the dissociative D-state, has been observed by both DS and as structure on the Na_2^+ signal [7]. No two-photon ionization signal is found for this state, implying the Rydberg levels of Na_3 may be mostly predissociated. Future efforts will concentrate on detecting dissociative transitions of larger aggregates, which cannot be observed by simple TPI-experiments.

Theoretically, the most interesting state of the trimer is its ground-state. By altering the cluster source conditions we have been able to measure the ground-state vibrational frequencies by observing the hot-bands in the spectroscopy of the B-state [8]. We determine fundamental vibrational frequencies $\omega_1 = 139$ cm^{-1}, $\omega_2 = 49$ cm^{-1}, and $\omega_3 = 87$ cm^{-1}. These compare favorably with the values $\omega_1 = 142$ cm^{-1}, $\omega_2 = 58$ cm^{-1}, and $\omega_3 = 94$ cm^{-1} calculated from results of Martins et al. by Thompson [9]. However, hot band measurements are useful only for the lowest lying energy levels of the ground-state. To explore the higher lying levels, we have begun stimulated emission pumping experiments using the C-state as an intermediate state. The results will be published later [10].

3. EXPERIMENTS ON SIZE-SELECTED METAL CLUSTER IONS

Cluster spectroscopy in molecular beams is handicapped in two regards: the range of materials which can be studied with them is limited, and fragmentation phenomena are difficult to measure.

For these reasons, we have developed a triple-quadrupole arrangement coupled with a SIMS cluster ion source to study the properties of metal cluster ions. In this section, we shall discuss photofragmentation and chemical reactivity studies of size-selected metal clusters. The photofragmentation data are shown to be consistent with recent calculations of cluster ionization energies [11]. The ionization energies, in turn, are known to reflect on the chemical properties of the clusters [1]. From the chemical reactivity studies, on the other hand, we are able to determine a correspondence with the physical structures of the clusters.

Apparatus

The apparatus has been described in detail elsewhere [12]. Briefly, metal clusters sputtered from an ambient temperature target (typically by 17 keV Ar$^+$ ions with an on-target current of 2 mA) are first energy filtered and then introduced to the first quadrupole, where they are mass filtered. The size-selected clusters then interact with reactant gas or a laser beam in the second quadrupole, which is operated in a non-discriminative ion-trapping mode. The third quadrupole is used to analyze fragmentation processes or reaction products.

Photofragmentation of Ag$^+$- and Al$^+$-clusters

As shown above for Na_3, photofragmentation is a reliable spectroscopic tool for studying cluster properties. The relative distribution of photofragments also gives useful information about electronic structure. For instance, the results of Bloomfield et al. [13] show that Si_6^+ is an especially abundant fragmentation product, in agreement with recent calculations of Si cluster stabilities [14]. We have recently measured the photofragmentation patterns for size-selected Al clusters. The results for Al_N^+, with $N = 3, 5, 7$ and 9 are shown in Figure 5. For these measurements, the laser beam was chopped and the ion signal was recorded phase-sensitively. Thus, the parent ion signals are negative, corresponding to their depletion by the laser beam, while fragment ion signals are positive. It is evident from the data that Al$^+$ is a preferred fragmentation product for $N = 3, 5$ and 7.

For comparison, we display the fragmentation curves obtained for Ag_N^+, $N = 3, 5, 7$ and 9 in Figure 6 [12]. Ag$^+$-clusters preferentially fragment into a charged product containing an even

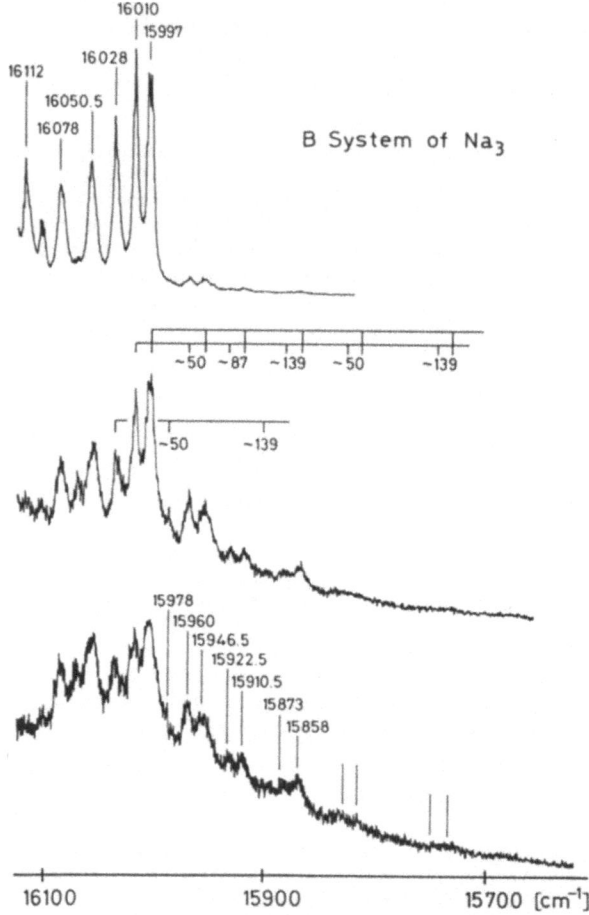

Fig. 4   Hot band progressions recorded at the band head of the B-system. The experiment was carried out at a carrier gas pressure of 9 bar (top), 1.9 bar (middle) and 1 bar (bottom), which corresponds to vibrational NA_3-temperatures of 20 K, 60 K and 100 K.

number of electrons, and except for Ag_3^+, the most abundant product is the next lowest odd cluster. These observations are in accord with the known odd-even alternations in the binding energies of noble metal clusters [15].

The striking difference between the results for the Ag and Al clusters is explained by recent calculations [11] which show that the Al atom has a lower ionization energy than any cluster up to Al_6. Thus, it is energetically favorable for an Al atom to carry away the charge from a fragmenting cluster, in contrast to the cases of Ag and Fe, where the ionization energies of the atoms are higher than those of the clusters [16,17].

Ni cluster carbonyls

The chemical reaction of size-selected clusters also tell about their structural properties. In collaboration with M.J. McGlinchey [18], we have recently measured the reaction products of

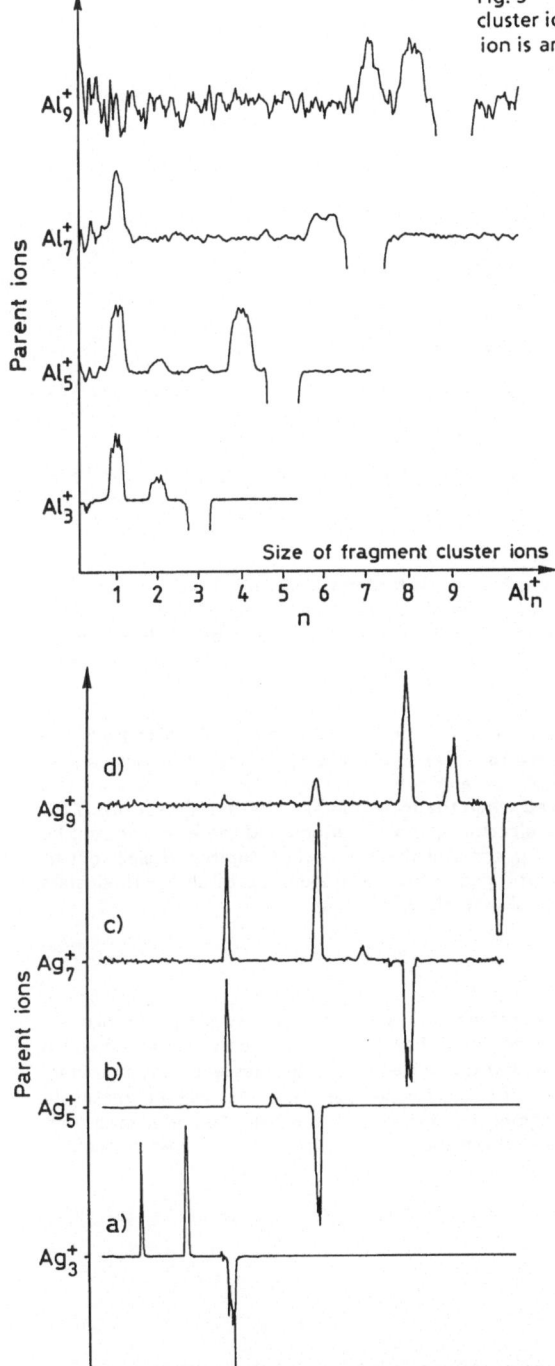

Fig. 5   Fragmentation signals of aluminum cluster ions, Al_n^+, n = 3, 5, 7 and 9. The atomic ion is an abundant product for the smaller clusters.

Fig. 6   Fragmentation signals of silver cluster ions, Ag_n^+, n = 3, 5, 7 and 9. Fragments with even numbers of electrons are favored.

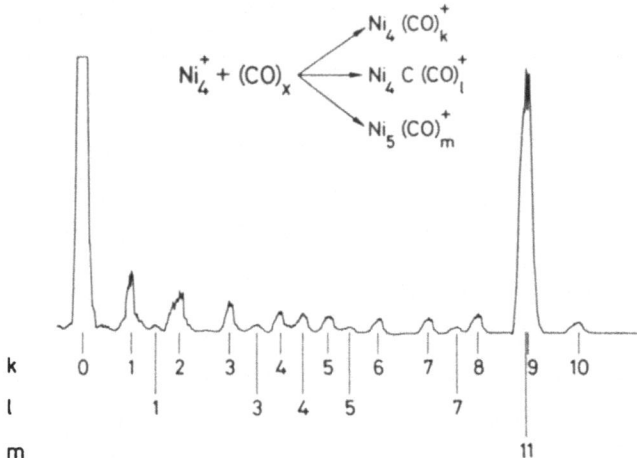

Fig. 7   Resulting mass spectrum when $Ni_4{}^+$, selected by the first quadrupole, is reacted with CO gas in the ion trap.

size-selected Ni cluster ions with CO. Shown in Figure 7 is the mass spectrum which results when $Ni_4{}^+$ is reacted with CO at a pressure of 1.3×10^{-3} mbar in the ion trap. Both carbonyls, $Ni_4(CO)_k{}^+$ and carbides, $Ni_4C(CO)_1{}^+$ are observed. The mass spectrum shows all the carbonyls ($k = 1$ to 10) with an especially enhanced abundance at $k = 9$. No carbonyl is observed above $k = 10$, indicating that 10 CO ligands saturate the cluster.

Similar mass spectra have been recorded for the $Ni_x{}^+$, with $x = 2$ through 13. With $x = 6$, the maximum $k = 13$, for $x = 10$, the maximum $k = 18$. The stoichiometry of the carbonyls can be understood using methods based on empirical electron-counting rules.   For example, a tetrahedral cluster will be maximally stabilized when the total number of valence electrons is 60. This total is made up from the metal valence electrons (10 per Ni atom) and those contributed by the ligand (2 per CO molecule). Thus, $Ni_4(CO)_{10}$ contains the appropriate number of electrons to be stabilized in a tetrahedral configuration. The geometrical arrangements of all the Ni clusters as determined by similar methods are to be published elsewhere [18].

4. CONCLUSIONS

In this paper we have discussed some recent experimental results on various systems. The richness of cluster properties demands a battery of experimental and theoretical tools, not all of which can be applied to any single cluster system with a guarantee of success. However, the boundary between the physical and chemical properties of clusters is not rigid. Thus, knowledge about the electronic states of Na_3 or from the photofragmentation of cluster ions helps to understand their chemical properties. Conversely, what we learn from the chemical properties of clusters helps to understand their physical properties.

ACKNOWLEDGEMENTS

The authors wish to thank the Swiss National Science Foundation for supporting this project. One of us (WAS) acknowledges support by NSF grant INT-8603428.

REFERENCES

[1]   R.L. Whetten, D.M. Cox, D.J. Trevor and A. Kaldor, Phys. Rev. Lett. $\underline{54}$ (1985) 1494.
[2]   A. Herrmann, S. Leutwyler, E. Schumacher and L. Wöste, Chem. Phys. Lett. $\underline{62}$ (1979) 216.
      G. Delacrétaz and L. Wöste, Surf. Sci. $\underline{156}$ (1985) 770.
[3]   M.D. Morse, J.B. Hopkins, P.R.R. Langridge-Smith and R.E. Smalley, J.Chem. Phys. $\underline{79}$ (1983) 5316.
[4]   G. Delacrétaz, Thesis $\underline{603}$, EPFL (1985).
[5]   M. Broyer, Ni Guoquan, J.P. Wolf and L. Wöste, to be published.
[6]   G. Delacrétaz, E.R. Grant, R.L. Whetten, L. Wöste and J. Zwanziger, Phys. Rev. Lett. $\underline{56}$ (1986) 2598.
[7]   M. Broyer, G. Delacrétaz, P. Labastie, J.P. Wolf and L. Wöste, Phys. Rev. Lett. $\underline{57}$ (1986) 1851.
[8]   M. Broyer, G. Delacrétaz, P. Labastie, J.P. Wolf and L. Wöste, J. Phys. Chem., in press.
[9]   J.L. Martins, R. Car and J. Buttet, J. Chem. Phys. $\underline{78}$ (1983) 5646.
      T.C. Thompson, G. Izmirlian, S.J. Lemon, D.G. Truhlar and C.A. Mead, J. Chem. Phys. $\underline{82}$ (1985) 5597.
[10] M. Broyer, G. Delacrétaz, Ni Guoquan, R.L. Whetten, J.P. Wolf and L. Wöste, to be published.
[11] T.H. Upton, Phys. Rev. Lett. $\underline{56}$ (1986) 2168.
[12] P. Fayet and L. Wöste, Surf. Sci. $\underline{156}$ (1985) 134.
[13] L. Bloomfield, R. Freeman and W. Brown, Phys. Rev. Lett. $\underline{54}$ (1985) 2246.
[14] D. Tomanek and M.A. Schlüter, Phys. Rev. Lett. $\underline{56}$ (1986) 1055.
[15] G. Blaise and G. Slodzian, C.R. Acad. Sci. Paris, Ser. B $\underline{266}$ (1968) 1525, I. Katsuke and T. Ichihara, Int. J. of Mass. Spect. and Ion Proc. $\underline{67}$ (1985) 229, and W. Begemann, K.H. Meiwes-Broer and H.O. Lutz, Phys. Rev. Lett. $\underline{56}$ (1986) 2248.
[16] J.L. Martins and W. Andreoni, Phys. Rev. A $\underline{28}$ (1983) 3627.
[17] E.A. Rohlfing, D.M. Cox, A. Kaldor and K.H. Johnson, J. Chem. Phys. $\underline{81}$ (1984) 3846.
[18] P. Fayet, M.J. McGlinchey and L. Wöste, J. Amer. Chem. Soc., to be published.

Molecular Surfaces: Chemistry and Physics of Gas Phase Clusters

A.P. Kaldor, D.M. Cox, and M.R. Zakin

Corporate Research Laboratory, Exxon Research and Engineering Co., Annandale, NJ 08801, USA

1. Introduction

Studies of gas phase metal cluster chemistry start with a fully coordinatively unsaturated, "naked" cluster free of support interactions or ligands. This offers a unique opportunity to explore a number of key scientific issues in chemistry and physics as the number of constituent atoms is increased, and bulk properties as well as stable surface structures evolve. These include questions such as: What are the structures of the various clusters - Are they unique? How do the electronic, optical and chemical properties of the clusters change as a function of the number and type of constituent atoms? How large does the cluster have to be before solid state theoretical description applies? What are the magnetic properties of "naked," and "dressed" clusters? Is there catalytic chemistry possible on such clusters, etc?

This area lies between traditional inorganic cluster chemistry and surface chemistry. In the former the clusters are "clothed", i.e. the ligated or partially unsaturated clusters are stable. In surface chemistry long-range order, subsurface atoms, as well as support interactions play a significant role. This approach may be able to connect two areas where the common thread is the identification of the localized surface bond. While clusters mimic the surface where localized bonding is prevalent, small clusters in addition mimic the surface with the greatest disorder and least amount of long-range order. While the structure of these clusters is not known, it is reasonable to assume that many structural isomers coexist, with a range of energy barriers among them. One could characterize them as fluxional; a surface analogue would be surface reconstruction and faceting. Such restructuring during exothermic chemical reactions is very probable. Beyond the obvious implications for bringing together two important areas of chemistry, and the theoretical foundations underlying them, equally important is that developments in this field may force solid state theory and molecular electronic structure theory to resolve conflicts, whether in language or in fact.

Recent advances in experimental techniques have made it possible to study in the gas phase the physical and chemical properties of clusters of virtually any material. [1] Although clusters consisting of hundreds of atoms have been synthesized, the majority of studies to date have concentrated on clusters of fewer than 30 atoms. [2] In this paper we briefly review selected recent results from our laboratory on a few systems:

the nature of large carbon clusters, the remarkably selective reactions of aluminum with a number of reagents and the reactivity of transition metal clusters with H_2 and CO.

2. Experimental approach

The advent of the laser vaporization technique to generate intense beams of clusters of even the most refractory materials has solved the limitations of conventional oven-beam sources for many experiments. [1] In our laboratory the clusters are synthesized from supersaturated vapor produced by a doubled Nd-YAG pulsed laser focused on a continuously rotating and translating target rod inside the throat of a high pressure, pulsed nozzle. A 5kHz repetition rate copper vapor laser has also been used successfully for quasicontinuous metal vapor production. [3] The hot metal vapor is entrained in the high pressure helium pulse; it is cooled and nucleation and cluster growth occurs in a narrow orifice extending from the vaporization zone. The cooled clusters immersed in the helium carrier, next enter a larger diameter reactor tube, and in a turbulent encounter meet a pulse of helium either seeded, or not, with reactant gas. The partial pressure of the reactant ranges from 0.0001% to 50%, though some experiments with neat reactants have also been performed. The clusters, either bare or with chemisorbed reactant, leave the reactor and undergo supersonic expansion and adiabatic cooling. The helium-shrouded packet traverses a vacuum zone, where it is skimmed to form a molecular beam and five hundred microseconds later enters a chamber equipped with a photoionization time-of-flight mass spectrometer.

The clusters are ionized by either an excimer laser operating with a variety of gas mixtures to yield 7.87, 6.42, or 5.0 eV photons, or by a tunable UV laser source which is able to produce radiation at photon energies lower than 6.5 ev. We take care in the experiments to minimize the laser fluence used to avoid fragmentation of the clusters. Often this means that we have to avoid multi-or multiple photon processes and use gentle, highly efficient single-photon ionization. The experimental apparatus has been described in a number of publications, where details should be sought. [2c,d, 4]

3. Carbon Clusters

Rohlfing et al. reported the first mass spectrum of carbon clusters which exhibited a unique bimodal distribution. [5] Only even clusters were observed for those containing more than forty atoms. Certain mass peaks were observed to be significantly more intense then others, and Kaldor et.al. [6] wondered about the "magic" nature of certain mass peaks in the spectrum, specifically C_{50}^+, C_{60}^+, and C_{70}^+. More recently at Rice University Smalley, Curl, Kroto and their collaborators performed a series of experiments which indeed brought the magic forth. [7-10] They reported that by providing more efficient cooling during cluster production they were able to shift conditions to maximize C_{60}^+ yield. [7] They achieved a contrast of about 100 compared to other ion peaks in the distribution.

They reported that more than 50% of the total carbon cluster yield was C_{60}^+. They also observed odd carbon species in this large cluster range. In reactivity studies of the large carbon clusters they showed that C_{60}, as well as all even C_n clusters, are inert, while the odd clusters react away. [8] From these observations they concluded that large C_n clusters are especially stable, and specifically C_{60} is an ultrastable cluster. [7-10] They proposed that a "soccerball"-like structure could explain the ultrastability. In further experiments they argued that the interior space of such a cluster would be large enough to accommodate only one metal atom, such as lanthanum, or barium, and argued that the observation of only $C_n La^+$, for n>40, and specifically of only the monoadduct, $C_{60}La^+$, was further demonstration of the validity of the proposed "soccerball" structure. [9]

In a series of experiments we have re-examined the carbon cluster system, C_n, and have observed that C_{60}^+ is only a minority constituent of the total carbon cluster ion signal; it is less than 0.1%. [11-12] It is produced by more than one process, one of which may involve the fragmentation of larger clusters. In contrast to what we reported earlier [5] C_n^+, n>40 exhibit a quadratic laser fluence dependence for 6.42 eV photoionization when the more efficient cooling approach, demonstrated by Smalley et al [7] is used. The ionization potential for

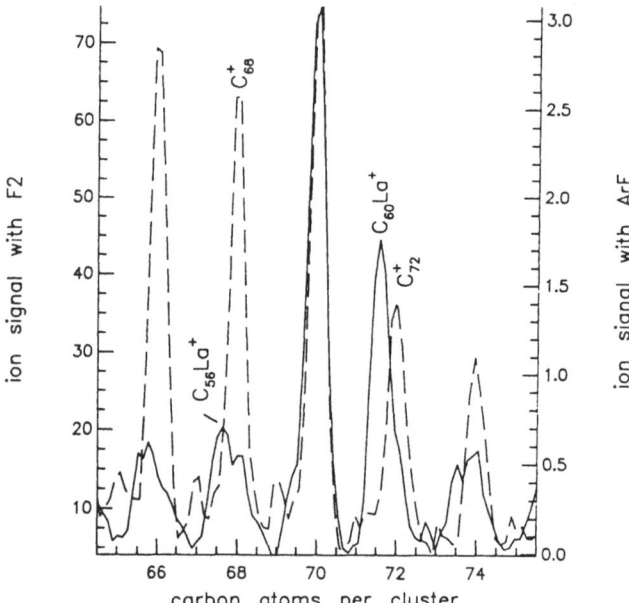

Figure 1   Photoionization time-of-flight mass spectra obtained from pulsed-laser vaporization of a lanthanum- impregnated graphite rod. The solid line and right-hand ordinate refer to the spectrum taken with ArF ionization and the dashed line and left-hand ordinate refer to the spectrum taken with F_2 ionization.

these clusters lies above 6.42 eV. This last observation has an important impact on the metal doping experiments as well. As shown in Figure 1, the sensitivity for the detection of the naked carbon clusters is quite different from that for the lanthanum-doped ones. The ionization potential for the doped clusters more closely resembles that of the naked metal; lying below 6.42 eV they are single-photon detected. With 7.87 eV photons we can single-photon detect all the clusters, and a more realistic measure of their absolute concentration is achieved. The lower concentration of C_{60}^+ obtained with 7.87 eV ionization alluded to above yields a low concentration of $C_{60}La^+$, and an undetectably low concentration of $C_{60}La_2^+$.

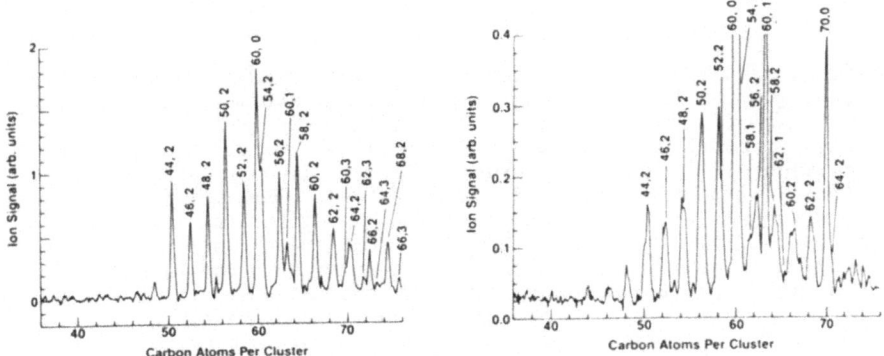

Figure 2   Time-of-flight mass spectrum obtained when a potassium-treated graphite rod is vaporized. Ionization is with 6.42 eV (ArF) photon energy at 0.2 mJ/pulse. (left panel) and 1.3 mJ/pulse (right panel). The peak labels n,m indicate CnKm.

In experiments where the graphite target was loaded with KOH to introduce potassium onto the clusters, we were able to detect carbon clusters containing multiple potassium atoms [11,12], including $C_{60}K_3^+$, shown in Figure 2.   This suggests that the metal loading is not on the interior of the "soccerball-like" structure, rather it is on the exterior.   The more probable suggestion is that an edge site of an open struc-ture is being decorated, something often observed for potassium loaded carbon surfaces. [13] It is interesting to note in Figure 2 that the observed loading shows strong fluence depen-dence. At higher fluence level there are fewer metal atoms observed on the carbon clusters, strongly suggesting that significant laser "cooking" is taking place, something that the Rice group [9] has also noted with respect to the lanthanum experiment, but attributed to the generation of stable species. The metal-loading experiments strongly suggest that the struc-ture of the carbon clusters are not spherical closed shells.

Time-of-arrival studies of different carbon clusters point to two different sources for the production of the clusters. [12] One cluster source is the vaporization zone and the other source produces clusters from deposits formed during the experiment on the walls of the nozzle system. This latter source was discovered by accident when LaC_n^+ signal continued to be observed after a lanthanum-doped graphite target was replaced with one that contained no lanthanum. By adjusting the experimental conditions in the source the relative ion signal of different clusters can be dramatically altered. Similarly. the relative ion signal depends on the arrival time chosen for detection. Such measurements support the fragmentation arguments above.

From our observations we conclude that while the proposal for an ultrastable structure of carbon is very appealing and has a groundswell of theoretical support [14], experimental evidence is still needed. The existence of a spheroidal structure for C_{60}, or other carbon clusters, is yet to be proven.

4. Metal Cluster Reactivity

Reactivity of metal clusters is a probe of their electronic properties, particularly of those electronic states involved in the formation of chemical bonds. Reactivity is probed in these experiments through measurement of the bare metal cluster ion intensity when either pure helium or a helium reactant mixture is injected into the reactor tube. [4a] The experimental measure of reactivity is the survival fraction of the metal cluster ion signal detected in the TOF mass spectrometer. Reactivity is measured as a function of cluster size and reactant. The essential assumptions that this procedure leads to an accurate measure of reactivity are: a) the collisional stabilization rate in the reactor is faster than the unimolecular decomposition rate of the chemisorbed molecule-cluster complex, b) the photoionization process is nondestructive, or at least does not produce bare metal cluster ions, and c) the reaction rates are fast enough compared to the residence time of the cluster in the reactor. It is important to recognize the limits of these assumptions when one considers statements regarding size-selective reactions.

If the assumptions above are valid, it has been shown that the rate constant k_n for the initial chemisorption step can be calculated from equation 1

Equation(1) $k_n = -\ln\{I_f/I_o\}/[R]t$

where I_f and I_o are the final and initial ion signals, respectively, [R] is the concentration of the reactant, and t is the residence time of the cluster in the reactor. In the pulsed reactor used here the absolute concentration of the reactant and the interaction time can only be estimated. This seriously limits the absolute accuracy of the measured rate constants, but since all the clusters experience the same conditions, relative rate constants are more accurate. An important underlying assumption is that all clusters have about the same temperature. In the high-pressure helium carrier gas this is probably reasonably accurate. It would also be most surprising if the internal energy of the clusters would vary non-monotonically.

One of the most interesting systems we have explored is the reactivity of aluminum clusters with a number of gaseous species. [15] The chemisorption of deuterium on aluminum clusters is the most cluster size-specific of any system we have studied. The six atom cluster is the most reactive and no reaction is observed for clusters containing fewer atoms. Reactivity falls off rapidly for Al_7 and Al_8. Under the most severe conditions reaction is observed up to Al_{20}, but not beyond. Similar results are observed for H_2, but the relative rate constants are about $1.7^+_- 0.4$ higher than for D_2. The primary products of the reaction are mono molecular adducts. There is no evidence for further reaction. [14] Unlike other hydrogen chemisorption systems we have studied, there is no further hydrogen addition and saturation coverage is achieved with only one molecule of hydrogen.

This is not a detection problem. If one adds the reactant to the carrier gas which sees the laser-produced plasma to vaporize the metal, deuterium is incorporated into the cluster in a variety of different mechanisms, including reaction. Under these conditions and using 7.87 eV photons to ionize the products for mass analysis, stable products are produced and detected of the generic formula $Al_n D_y$, with n=6-7, y=1,2; n=8-12, y=1-5; n=13-21, y=1-3. These observations suggest that the detection of the chemistry products is a reliable indicator of neutral stabilities. [15]

The reactivity pattern for chemisorption of different molecules by aluminum clusters exhibits in each case a different and unique size dependence. It is interesting that Al_6 is the most reactive cluster with CO, but in contrast to the hydrogen case both smaller and larger clusters will react with CO at high enough pressure. [15] For O_2 the most reactive cluster is the dimer. The reactivity drops dramatically for Al_3-Al_6 followed by a slow, monotonic rise towards larger clusters. Above 20 atoms, the rate is comparable to that of the dimer.

For water, Al_{10} is found to be the most reactive cluster, while smaller clusters are relatively inert and larger ones are reactive. Methanol exhibits a surprising reactivity pattern which suggests that there may be "magic numbers" for the minimum in reactivity at Al_7, Al_{14}, and Al_{20-21}. The primary reaction of H_2S is dissociative sulfidation, producing a species Al_nS as the product. This species is quite facile for the further adsorption of H_2S, but the system is poisoned towards further dissociative adsorption, as evidenced by lack of dehydrogenation. Methane is found to be inert with Al_n under all reaction conditions. The relative reactivity of these molecules is ordered as follows:

$$O_2 > CH_3OH > CO > D_2O > D_2 > CH_4$$

with O_2 about a thousand times more reactive than D_2.

There is significant theoretical activity to understand the electronic structure and reactivity of aluminum clusters. [16] At this symposium results of calculations of the ionization potential and the stability of aluminum clusters using the Jellium model have been presented. [17] The results suggest that a shell model may be useful in explaining some of the observations. Upton has performed extensive calculations on Al_n and Al_nH_2 to obtain relative stabilities for several geometrical structures. [18] While he did not do an exhaustive search of all possible structures, he did find that the heat of adsorption of H_2 increases with cluster size. There is only one structure among those considered for Al_6 for which chemisorption is exothermic. His findings are consistent with the observations reported above. Why adsorption of D_2 terminates abruptly beyond Al_6 remains an open question.

Recent studies of transition metal clusters also revealed that the chemisorption of hydrogen is remarkably sensitive to both the number of metal atoms in the cluster as well as the metal type. [19] For example Nb_8, Nb_{10}, and Nb_{16} are significantly less reactive toward molecular hydrogen than other niobium clusters. [4c, 19c] For cobalt, Co_6 and Co_7 are the least reactive. [19c] For iron the reactivity is found to be oscillatory, depending on the number of iron atoms in the cluster. [19a-c] Fe_{6-8} are the least reactive small clusters, while Fe_{15-17} are the least reactive of the mid-size clusters. About three orders of magnitude variation is observed in reaction rate for Fe_8 to Fe_{25}, but there is little change for larger clusters. We believe that this is a measure of the onset of band structure, or bulk-like behavior. Vanadium achieves the same kind of behavior at V_{17}-V_{18}, while niobium reaches this state for clusters with more than 28 atoms.

Oscillations in reactivity correlate well with the metal cluster ionization thresholds for clusters containing eight or more atoms for iron, [19a] niobium, [4c, 19d] and vanadium. [19d, 20] Thus the electronic character of the cluster is one important parameter in controlling hydrogen chemisorption. Very encouraging results on theoretical calculations of the ionization potential variation with cluster size for iron were presented in a paper by K. H. Bennemann and coworkers at this conference, which suggest that the nonmonotonic variation is associated with size-dependent changes in the width and position of the local electron density of states caused by structural changes in the cluster.

On iron surfaces hydrogen chemisorbs dissociatively. [21] The cluster behavior is most likely similar, as evidenced by the measured desorption energy of about 1.3 eV [22], observation of a kinetic isotope effect indicating an activated process [19b], and the recent measurement of significant increase in the ionization potential of the Fe_xH_y species relative to the bare clusters. [23] The exothermicity of the reaction is sufficient to break a single metal-metal bond in the cluster and allow as a result local restructuring of the cluster to optimize the overall energetics of the cluster-adsorbate complex.

One model to explain the correlation of ionization potential variation and reactivity is based on electron accepting and donating interactions. [4c, 19a, 20] As the hydrogen molecule approaches the cluster, the long-range interaction will be repulsive, creating a barrier to the reaction. As the hydrogen molecule gets closer the cluster acts as an electron donor, interacting with the sigma star antibonding orbital of H_2, and as an acceptor with the sigma bonding molecular orbital of the adsorbate. Both of these interactions weaken the adsorbate bond and strengthen the bonding to the cluster. If the attractive interaction overcomes the repulsive one, dissociative chemisorption will occur with a small activation barrier [19a] in the entrance channel. The magnitude of the activation energy depends on the quantitative resolution of the competition between these two forces. The lower IP clusters are better donors than acceptors, and thus have smaller activation energies. This simple model predicts that for a sufficiently low IP cluster, bonds can be broken without a barrier and that metals on the left end of the periodic table would be most facile, in agreement with observations. [19]

For the smallest clusters the IP has become sufficiently large that donation from H_2 sigma to M_n will dominate. The stabilization due to this interaction will increase with increased IP and to a first approximation the chemisorption rate should also increase. Symmetry of the orbitals involved in forming the bonds should play an increasingly important role since the density of states near the Fermi energy is reduced as the cluster size decreases. Thus the reactivity pattern as a function of cluster size for small clusters, n<8, is quite perverse. For example, the pentamer of vanadium is very reactive but Fe_5 is not. Fe_4 is reactive, but V_4 is not. Niobium clusters with n=4-8 are nearly as reactive as the larger clusters. In each instance the IPs are sufficiently large that little reactivity would be predicted from the metal to hydrogen charge transfer model used with the larger clusters. Further theoretical work on these problems is needed.

The hydrogen chemisorption rate as a function of cluster size reaches a maximum and levels off for the transition metal clusters studied to date. The specific size varies from metal to metal. At this point we are dealing with a non-activated reaction. Harris and Anderson examined this case from a theoretical perspective. [24] Theirs is one of a large number of theoretical studies of the activation of hydrogen by metal clusters and surfaces.

Upton has studied the electronic factors important in the dissociative chemisorption of hydrogen on aluminum clusters, specifically on Al_6 in some detail. [16,18] He explored the role of the charge state of the cluster, be it anion, neutral, or cation. He found that the charge had only a small effect on the activation barrier. Similar size-selective behavior might be expected from charged and neutral clusters, particularly if the orbital ionized does not participate directly in the reaction. While the calculations were for aluminum, conceptually they explain why the size-selective reactivity pattern [25] for Nb_7^+, Nb_8^+, and Nb_9^+ with H_2 is similar to their neutral counterparts.

The chemisorption of CO has been measured for first row transition metals [26] V, Fe, Co, [19c, 25] Ni, and Cu; second row metals [26] Nb, [19c, 26] Mo, Ru, and Pd; and third row metals [26] W, Ir, and Pt. Our results show that CO readily chemisorbs on most transition metal clusters containing three or more atoms, and that the reactivity of clusters larger than four atoms is within a factor of 5 as a function of cluster size and

metal. This is in direct contrast with the hydrogen examples
discussed above. CO chemisorption probably has no, or only a
slight barrier and is not an activated reaction. For one set of
experimental conditions we find that reactivity is first observed for atoms of Pt and Co; dimers of Pd, trimers of V, Ni, Nb,
Ru, and Ir; tetramers of W, pentamers of Cu and Mo; and hexamers
for Fe and Al. Through a series of stability calculations,
assuming unimolecular decomposition is inhibiting us from seeing
reaction for smaller clusters, metal-CO relative bond strength
can be ordered:

Co, Pt, Pd > V, Ni, Nb, Ru, Ir, W > Mo, Fe, Cu, Al

This ordering is consistent with heats of desorption of CO from
surfaces of the various metals. [26]

5. Summary

In this paper we have provided examples of stimulating new
results that are emerging in the experimental studies of clusters. Ionization potential studies offer a powerful tool to
probe the properties of such systems. Ionization potentials of
carbon clusters were used to resolve experimental issues relating to cluster stability. The results, when combined with other
findings, raise serious doubts about the experimental "support"
for ultrastable spherical,"soccerball-like" structures of
carbon. Ionization potentials are also useful in explaining observed reaction rate measurements for a number of transition
metal systems. They are probes of electronic structure, of both
reactants and products. There is an emerging picture that
charge transfer models are useful as a first approximation to
explain cluster reactivity, and with extensions may provide a
general model for cluster reactivity. Size-selective chemistry
appears to be associated with chemisorption that requires bond
breaking, i.e. with an activated process. Geometrical structure
of bare clusters is an important piece of information missing
from our general picture and chemical probes are most likely
first tools to gain insight and information about structure from
surface adsorption data.

The field of gas phase cluster research is expanding rapidly. We have only seen the tip of the iceberg. Strong interaction between experiment and theory is essential for rapid progress in the field, especially in light of the fact that only
limited kinds of experimental data are currently available. It
is important to focus the effort where we can interact effectively.

References

1a. T. G. Dietz, M. A. Duncan, D. E. Powers, and R. E. Smalley: J. Chem
Phys. 74, 6511 (1981)
b. S. J. Riley, E. K. Parks, C. R. Mao, L. G. Pobo, and S. Wexler: J.
Phys. Chem. 86, 3911 (1982)
c. E. A. Rohlfing, D. M. Cox, R. Petkovic-Luton, and A. Kaldor: J. Phys.
Chem. 88, 6229 (1984)
d. E. A. Rohlfing, D. M. Cox, and A. Kaldor: J. Phys. Chem. 88, 4497
(1984)

2. For recent reviews, see: R. L. Whetten, D. M. Cox, D. J. Trevor, and A. Kaldor: Surf. Sci. 156, 8 (1985); A. Kaldor, D. M. Cox, D. J. Trevor, and M. R. Zakin: Z. Phys. D, in press

3. R. L. Whetten, D. M. Cox, D. J. Trevor, M. Zakin and A. Kaldor, to be published

4a. R. L. Whetten, D. M. Cox, D. J. Trevor, and A. Kaldor: J. Phys. Chem. 89, 566 (1985)

b. D. M. Cox, D. J. Trevor, R. L. Whetten, E. A. Rohlfing, and A. Kaldor: Phys. Rev. B32, 7290 (1985)

c. R. L. Whetten, M. R. Zakin, D. M. Cox, D. J. Trevor, and A. Kaldor: J. Chem. Phys. 85, 1697 (1986)

d. M. R. Zakin, R. O. Brickman, D. M. Cox, K. C. Reichmann, D. J. Trevor, and A. Kaldor: J. Chem. Phys. 85, 1198 (1986)

5. E. A. Rohlfing, D. M. Cox, and A. Kaldor: J. Chem. Phys. 81, 3322 (1984)

6. A. Kaldor, D. M. Cox, D. J. Trevor, and R. L. Whetten: Catalyst Characterization Science, ACS Symposium Series No.288, ed. M. L. Deviney and J. L. Gland, p.111, (1985)

7. H. W. Kroto, J. R. Heath, S. C. O'Brien, R. F. Curl, and R. E. Smalley: Nature 318, 162 (1985)

8. Q. L. Zhang, S. C. O'Brien, J. R. Heath, Y. Liu, R. F. Curl, H. W. Kroto, and R. E. Smalley: J. Phys. Chem. 90, 525 (1986)

9. J. R. Heath, S. C. O'Brien, Q. Zheng, Y. Liu, R. F. Curl, H. W. Kroto, F. K. Tittel, and R. E. Smalley: J. Am. Chem. Soc. 107, 7779 (1985)

10. Y. Liu, S. C. O'Brien, Q. Zhang, J. R. Heath, F. K. Tittel, R. F. Curl, H. W. Kroto, and R. E. Smalley: Chem. Phys. Lett. 126, 215 (1986)

11. D. M. Cox, D. J. Trevor, K. C. Reichmann, and A. Kaldor: J. Am. Chem. Soc. 108, 2457 (1986)

12. D. M. Cox, K. C. Reichmann, and A. Kaldor: J. Chem. Phys., to be submitted.

13. T. R. Baker, private communication.

14. See for example:

a. A. D. J. Haymet: J. Am. Chem. Soc. 108, 319 (1986); Chem. Phys. Lett. 122, 421 (1985)

b. R. C. Haddon, L. E. Gries, and D. Raghavachari: Chem. Phys. Lett. 125, 459 (1986)

c. M. D. Newton and R. E. Stanton: J. Am. Chem. Soc. 108, 2469 (1986)

d. M. Osaki and A. Takahasi: Chem. Phys. Lett. 127, 242 (1986)

e. D. J. Klein, T. G. Schmalg, G. E. Hite, and W. A. Sertz: J. Am. Chem. Soc. 108, 1301 (1986)

f. R. L. Disch and J. M. Schulman: Chem. Phys. Lett. 125, 465 (1986)

15. D. M. Cox, D. J. Trevor, R. L. Whetten, and A. Kaldor, in preparation.

16. See, for example, T. H. Upton: Phys. Rev. Lett. 56, 2168 (1986)

17. M. L. Cohen, this proceedings

18. T. H. Upton: J. Chem. Phys., submitted, and references therein

19a. R. L. Whetten, D. M. Cox, D. J. Trevor, and A. Kaldor: Phys. Rev. Lett. 54, 1494 (1985)

b. S. C. Richtsmeier, E. K. Parks, K. Liu, L. G. Pobo, and S. J. Riley: J. Chem. Phys. 82, 3659 (1985)

c. M. D. Morse, M. E. Geusic, J. R. Heath, and R. E. Smalley: J. Chem. Phys., 83, 2293 (1985)

d. D. M. Cox, R. L. Whetten, M. R. Zakin, D. J. Trevor, K. C. Reichmann, and A. Kaldor: AIP Conference Proceedings No.146, Adv. in Laser Science I, Nov. 1985, eds. W. C. Stwalley and M. Lapp, AIP, New York (1986)

e. M. E. Gausic, M. D. Morse, and R. E. Smalley: J. Chem. Phys. 82, 590 (1985)

20. M. R. Zakin, D. M. Cox, R. L. Whetten, D. J. Trevor, and A. Kaldor, to be published.

21. E. Shwtorovich, R. C. Raetzold and E. L. Muetterties: J. Phys. Chem. 87, 1100 (1983)
22. K. Liu, E. K. Parks, S. C. Richtsmeier, L. G. Pobo, and S. J. Riley: J. Chem. Phys. 83, 2882 (1985)
23. M. R. Zakin, D. M. Cox, R. L. Whetten, D. J. Trevor, and A. Kaldor: Chem. Phys. Lett., submitted.
24. J. Harris and S. Anderson: Phys. Rev. Lett 55, 1583 (1985)
25a. P. J. Brucat, C. L. Pettiette, S. Yang, L-S. Zheng, M. J. Craycraft, and R. E. Smalley: J. Chem. Phys. 85, 4747 (1986)
  b. J. M. Alford, F. D. Weiss, R. T. Laaksonen, and R. E. Smalley: J. Phys. Chem. 90, 4480 (1986)
26. D. M. Cox, K. C. Reichmann, D. J. Trevor, and A. Kaldor, to be submitted.

Multiphoton Ionization of Sodium Halide Clusters

K. Sattler

Department of Physics, University of California, Berkeley, CA 94720, USA

Multiphoton ionization (MPI) studies are reported which show sets of magic numbers for sodium halide clusters being different if the conditions for condensation are changed. "Warm" clusters show the magic numbers of neutral cuboids whereas "cold" clusters show those of charged cuboids. This is explained by electronic excitation of sodium halide admolecules desorbing from cubic cores, either before or after ionization, respectively.

1. Introduction

This paper deals with the question of whether small clusters of atoms are portions of the infinite solid or whether they have their own individual properties. In particular we ask if the crystal structure of the bulk is found for small systems as well.

For most materials this question can rarely be answered at present because the techniques for cluster analysis require that the particles are ionized, and the structure in mass spectra may be caused by the stability distribution of the charged clusters alone. Therefore, even if the geometrical structure of clusters can be determined from mass spectra, the distribution of the neutral ones could be either structureless or show another set of irregularities.

With clusters from ionic materials however, the situation is different. A cluster consisting of positive and negative ions M^+ and X^- is held together predominantly by Coulomb forces. If the numbers of the positive and the negative ions are equal, the cluster is neutral, and it is positively or negatively charged if one X^- or one M^+ is missing, respectively. That is, in principle there is no difference in the bond conditions for either neutral or charged ionic clusters. Therefore, a tendency for a certain geometrical structure should be the same for both charge states.

In this work we report on various studies on metal halide clusters which show that this is indeed the case. Despite the conclusions from the results having been taken from mass spectra of ionized clusters, we get information about the geometrical structure of both the neutral and the charged clusters.

We have found that the structure in mass spectra of metal halide materials depends on the generation and ionization conditions /1/. Most important seems to be the original temperature of the clusters when they leave the generation source, as well as electronic excitations in the laser field. We find that a set of parameters can be chosen where neutral cluster distributions are observed. With another set of parameters we observe ion cluster distributions. Both distributions are generated after multiphoton desorption processes.

2. Experimental

The clusters are generated by quenching the metal halide vapour (typically 20 Pa at 1000 K) in helium atmosphere of typically 100 Pa at 70 K ("cold") or 300 K ("warm"). For time-of-flight analysis the particles are positively ionized alternately by electrons or laser photons (1000 electron pulses and 50 laser pulses per second, respectively). For each type of ionization simultaneously one spectrum is accumulated in the memory of a multichannel analyser. The laser pulses (duration 10 nanoseconds) are generated by an excimer-pumped dye laser producing some 4 mJ per pulse, or 0.1 mJ after frequency doubling (UV). For a focal diameter of 0.1 mm this corresponds to a peak power of some 5 GWcm^{-2} (visible) or 150 MWcm^{-2} (UV).

3. Results

First we consider ionization by electron impact. Fig.1 shows a mass spectrum of NaCl-clusters condensed in "warm" He gas and ionized by 40 eV electrons. Nonstiochiometric cluster ions $Na_nCl_{n-1}^+$ are observed because one chlorine ion splits off after having been neutralized by ionization. Local maxima are observed at n=14, 23, 38, and 63, corresponding to cuboids containing 3×3×3, 3×3×5, 3×5×5, and 5×5×5 single ions. Less distinct maxima at n=32 and 53 correspond to less compact cuboids with 3×3×7 and 3×5×7 single ions. Small intermediate lines are caused by doubly charged clusters, which are discussed in detail elsewhere /2/.

The same sequence of irregularities has been found in SIMS experiments for various ionic materials /3,4/ where clusters being generated as ions have been detected. Furthermore it has been shown that the original cluster distribution is structureless /5/. The highly excited clusters cool down by evaporation within less than 100 microseconds. Only after this evaporation are the intensity anomalies observed.

The observed local maxima in Fig.1 are generated from particles with odd numbers of single ions Na^+ and Cl^- and therefore are due to the charged clusters. In this case all three cuboid sides have to be given by odd numbers. In contrast, for neutral NaCl-clusters at least one side of the cuboid must contain an even number of atoms. This type of magic numbers is found with MPI mass spectra from "warm" clusters.

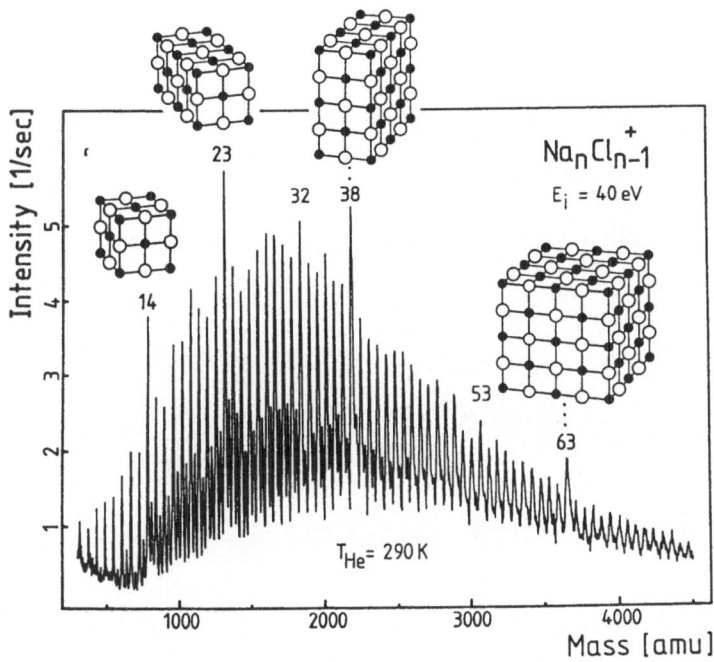

Fig.1: Time-of-flight mass spectrum of NaCl-cluster ions condensed in H_2O-cooled He gas and ionized by 40 eV electrons

Figures 2 and 3 show MPI mass spectra (a-c) and electron impact reference spectra (d) of NaCl clusters condensed in water-cooled ("warm", Fig.2) or LN_2-cooled helium gas ("cold", Fig.3). <u>Warm</u> clusters show distinct magic numbers n=18, 20, 24, 30, 40, 48, 50, 56, 60, and 72 corresponding to neutral $(NaCl)_n$ cuboids of 3×3×4, 2×4×5, 3×4×4, 3×4×5, 4×4×4, 4×4×5, 4×4×6, 4×5×5, 4×4×7, and 4×6×6 single ions, respectively. By variation of the photon energy between 2.3 eV and 4.4 eV this set of magic numbers does not change. If instead <u>cold</u> clusters are ionized by MPI the spectra are completely different (Fig.3).

The observations can be explained by the assumption that the neutral NaCl-clusters originally consist of cuboid cores with Na^+Cl^--admolecules being attached to their surface. These cores are observed in mass spectra after their admolecules have been evaporated. The first excited electronic state of these admolecules is a Van-der-Waals state with the electron being transferred from the Cl^- to the Na^+ /6,7/. Thus the ion pairs are neutralized on the cluster surface after photon absorption.

If absorption channels exist in a cluster for excitation energies far below the ionization limit, these will be populated with much higher probability than the ionization

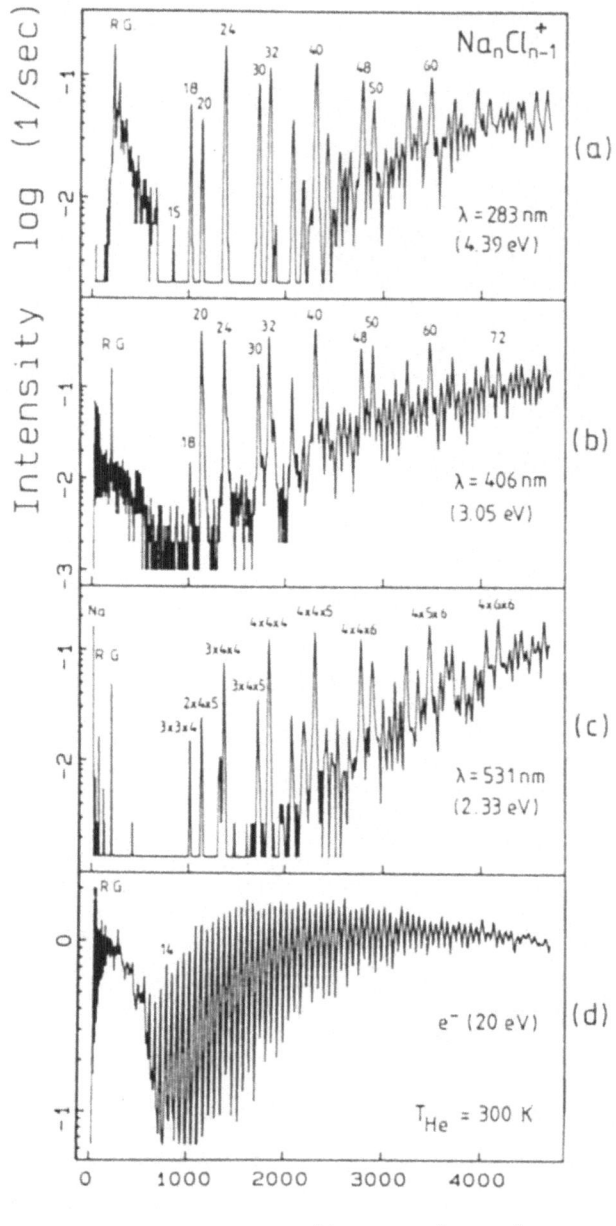

Fig.2: MPI mass spectra (a-c) and electron impact reference spectrum of NaCl clusters condensed in water-cooled helium gas ("warm" clusters)

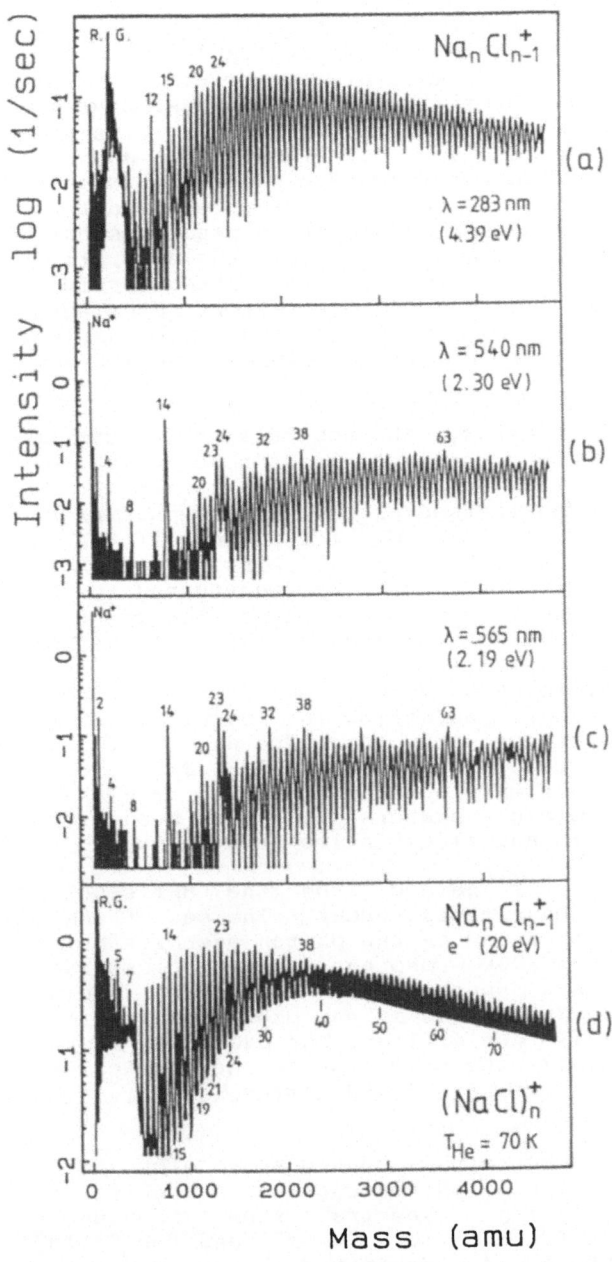

Fig.3: MPI mass spectra (a-c) and electron impact reference spectrum (d) of NaCl clusters condensed in LN$_2$-cooled helium gas ("cold" clusters)

channels because the number of photons being needed is smaller. In this case most of the clusters absorb the radiation from the laser pulse without being ionized. For NaCl-clusters this leads to a neutral cluster distribution which is different from the original one. Therefore we have to deal with three kinds of distributions: the original distribution D_0, the intermediate distribution D_i, and the final distribution D_f^+. The model described here takes into account that D_0 has been observed to be structureless and that structure is generated only after desorption processes have occurred /5/. For weakly bound surface atoms the desorption probabilities depend on the cluster temperature and therefore are different for <u>warm</u> and for <u>cold</u> clusters. From warm clusters the Na and Cl atoms desorb because their sublimation energies are low (typically 40 meV for chlorine and 400 meV for Na).

For <u>cold</u> clusters, in contrast, the neutralized adatoms do not desorb at once. Therefore the particles can be ionized at the Na-site instead of the Cl^--site. For ionization of the Na adatom about 5 eV are needed instead of 10-12 eV (depending on the position in the cluster) for Cl^- ionization. If the Na adatom is transformed into a Na^+ ion, its binding energy to the cluster increases by 5 eV, which is transferred to the cluster as vibrational energy. As a result the cluster heats up and the remaining neutral admolecules as well as the unpaired Cl atom evaporate. Additional heat can be transferred by NaCl admolecules recombining to Na^+Cl^-. The cluster becomes excited enough to overcome redistribution barriers and to transform into a new geometrical structure. The number of ions is odd and therefore the cuboid structures 3×3×3, 3×3×5, etc. are formed. This explaines why in the MPI spectra <u>originally cold</u> clusters show those magic numbers which have been found for <u>warm</u> clusters with SIMS and electron impact studies.

Structural information is gained from the MPI studies because no resonance enhancements occur. The set of magic numbers is found not ot depend on the photon energies being applied (one exception: UV photons, cold clusters). Furthermore, the MPI mass spectra from different sodium halides, NaCl, NaBr and NaI, taken at <u>one</u> photon energy (for instance 2.81 eV, Fig.4) are similar. The anomalies in the mass spectra are essentially the same despite these materials having different electronic structure and different ionization potentials.

In summary, MPI mass spectra have been measured for sodium halide clusters at different photon energies and different cluster temperatures. The spectra show pronounced irregularities, being different for "warm" and for "cold" clusters. The observations are described by a model as follows: The original neutral cluster beam consists of cuboid cores with admolecules attached to the surface. These admolecules are electronically excited and thereby neutralized in the radiation field of the laser beam. Then, because of their low sublimation energies, their evaporation rates depend on the cluster temperature. They desorb from warm clusters

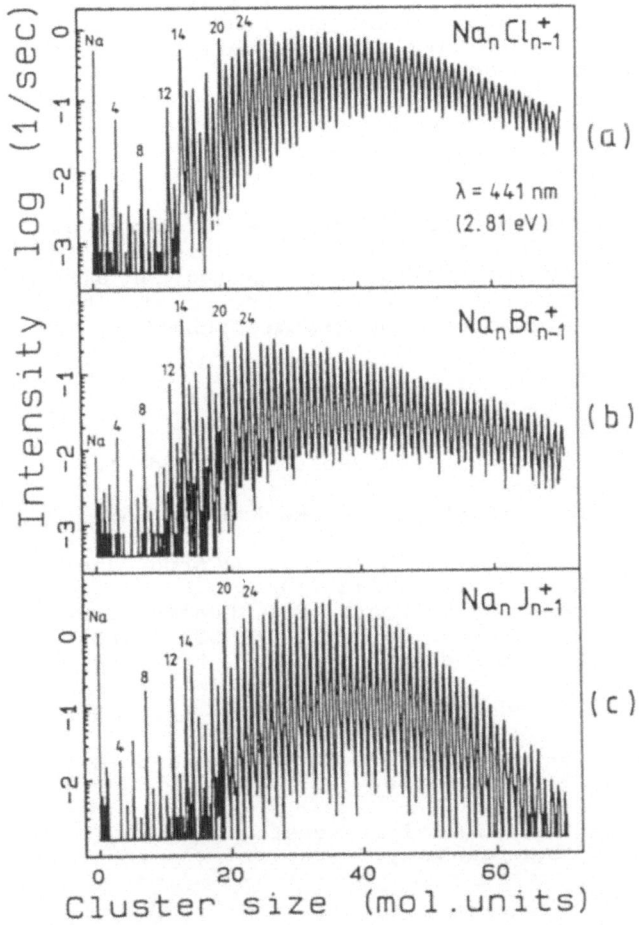

Fig.4: MPI mass spectra of "cold" NaCl, NaBr, and NaI cluster taken at the same photon energy, 2.81 eV

leaving "naked" $(Na^+Cl^-)_n$ cores. The clusters are also cooled by the adlayer desorption . They cannot overcome redistribution barriers after having been ionized. Therefore the <u>neutral</u> cluster distribution is observed in the mass spectra. It is an intermediate distribution which is generated in the laser field.

<u>Cold</u> clusters are ionized at a neutral Na adatom because the excited NaCl pairs do not desorb at once. By relaxation of the Na^+ ion and by recombination of the adlayers ($Na^+Cl^0 \rightarrow Na^+Cl^-$) the clusters are heated up and overcome a redistribution barrier <u>after</u> having been ionized. Therefore the irregularities characteristic for <u>charged</u> clusters are observed in the MPI mass spectra of originally cold clusters.

113

We conclude that MPI studies on sodium halide clusters yield strutural information about both, the neutral and the charged clusters. The tendency to have the rock salt structure of the bulk is found for both charge states.

Acknowledgements

The author wishes to express his gratitude to R. Pflaum from the University of Konstanz, FRG, for making available results of his PhD thesis prior to forthcoming publication. He also acknowledges continuous cooporation with E. Recknagel. This work is supported by the Deutsche Forschungsgemeinschaft.

References

1. R. Pflaum, K. Sattler and E. Recknagel, submitted for publication
2. R. Pflaum, P. Pfau, K. Sattler and E. Recknagel, Surf. Sci. 156, 165 (1985)
3. T.M. Barlak, J.R. Wyatt, R.J. Colton, J.J. DeCorpo and J.E. Campana, J. Am. Chem. Soc. 104, 1212 (1982)
4. T.M. Barlak, J.E. Campana, J.R. Wyatt, B.I. Dunlap and R.J. Colton, Int. J. Mass. Spectr. Ion Phys. 46, 523 (1983)
5. W. Ens, R. Beauvis and K.G. Standing, Phys. Rev. Lett. 50, 27 (1983)
6. P. Davidovits and D.C. Broadhead, J. Chem. Phys. 46, 2986 (1966)
7. P.K. Swaminathan, A. Laaksonen, G. Corongiu and E. Clementi, J. Chem. Phys. 84, 867 (1986)

The Author is on leave from the Universität Konstanz, Fakultät für Physik, D-7750 Konstanz, West-Germany

Dynamical Aspects of Molecular Clusters Studied by Multiphoton Ionization Technique

Y. Achiba[1], *K. Sato*[2], *and K. Kimura*[2]

[1]Department of Chemistry, Tokyo Metropolitan University,
Setagaya, Tokyo 158, Japan
[2]Institute for Molecular Science, Myodaiji, Okazaki 444, Japan

1. INTRODUCTION

The technique of multiphoton ionization combined with a pulsed laser has been demonstrated as a versatile tool in the fields of molecular spectroscopy in the last ten years /1/. Especially the ability of high sensitivity and selectivity of this method has been useful for a photoionization source combined with mass spectrometer to detect very small amounts of sample in the gas phase. Actually so far, most of the recent metal cluster experiments using laser vapourization technique have been performed by multiphoton ionization with a time-of-flight mass spectrometer /2/.

However, it is also true that the multiphoton ionization method has brought some serious problems, such as photodissociation and/or photofragmentation. This is partly caused by ionization process itself, because the multiphoton ionization process always proceeds via certain excited state (resonant intermediate state), and thus in some cases other rapid deactivation processes take place in competition with the ionization transition. This makes the situation much more complex compared with other ionization sources. Furthermore, this technique necessarily requires high-density photon flux, which may cause photofragmentation of parent ions.

In order to elucidate these problems, measuring kinetic energy distributions of photoelectrons generated in the multiphoton ionization is one of the most appropriate methods. Information deduced from photoelectron spectroscopy always simply reflects molecular status immediately after the ionization transition. This fact leads to eliminate the influences of photofragmentation after the ionization.

In the present work, we will demonstrate versatility of measuring photoelectron spectra to understand photoionization processes of van der Waals molecules in multiphoton excitation. We will also present the dynamic behavior of these molecules in the excited states, paying much attention to understanding the internal state distributions of molecules which are produced in the process of photodissociation at electronically excited states.

2. EXPERIMENTAL

In the present work we have used a laser excited-state photoelectron spectroscopic technique previously developed in this laboratory /3/. The apparatus with a pulsed beam source is essentially the same as that previously reported /4/. A pulsed valve with an orifice of 0.7 mm (diameter) was used, its duration being 1 ms and its repetition 10 Hz. A conical skimmer with a 1 mm aperture was used.

An ArF excimer laser (Quanta-Ray EXC-1) was used for 193 nm radiation. Six other far-UV lasers at 199.8, 204.2, 208.8, 217.8, 223.1 and 228.7 nm are the n-th anti-Stokes lines produced by a home-made hydrogen Raman shifter, using the second, third, and fourth harmonics of a Nd-YAG laser (Quanta-Ray DCR-1A). The laser radiation was focused by a quartz lens of $f = 500$ mm.

In order to carry out two-color laser experiments, we used two dye lasers (Quanta-Ray PDL-1) which are pumped simultaneously by the Nd-YAG laser.

Photoelectron measurements were carried out with a pulsed NO jet at a stagnation pressure of 1 atm. Photoelectron kinetic energy spectra were measured with a time-of-flight (TOF) electron energy analyzer described previously/4/. The resolution of the TOF electron analyzer is 0.15 eV in the energy region studied. Mass spectrometric measurements were also carried out with a quadrupole mass filter (ULVAC MSQ-400).

3. RESULTS AND DISCUSSION
3-1. Dissociative Reaction of Dimer Ion

One of the most remarkable aspects appearing in the multiphoton ionization process of van der Waals molecules is that the photoionization often results in the formation of fragment ions rather than the formation of parent ions. In the case of two-photon ionization (TPI) of $(NH_3)_2$, for example, we can only observe NH_4^+ ion signal, even when we carefully excite only $(NH_3)_2$. Thus the question has arisen why the NH_4^+ ion is dominant instead of the $(NH_3)_2^+$ parent ion. In order to explain these experimental findings there have so far been postulated two possible mechanisms; 1)the proton transfer reaction occurring immediately after photoionization of the dimer molecule /5/, and 2)the photodissociation forming neutral NH_4 radical before ionization, and then the photoionization of the NH_4 radical taking place /6/. Since photoelectron spectra always reflect the status produced immediately after the ionization events, measurements of photoelectron energy distributions for the TPI of $(NH_3)_2$ must be quite useful to answer the above question.

Prior to measuring photoelectron spectra associated with the TPI of $(NH_3)_2$, we intended to clarify the TPI mechanism and whether the ionization proceeds resonantly or not. In the present work we used six different laser lines which are produced by an anti-Stokes Raman shifter for the second, third, and fourth harmonics of a Nd-YAG laser as well as ArF laser line. At all wavelengths used, the two photon energies are large enough to ionize $(NH_3)_2$. Therefore if the ionization yields (total ion currents) varies with the excitation wavelength, it is reasonable to consider that the TPI is a resonantly enhanced ionization.

In Fig. 1(b), the NH_4^+ ion yield is plotted as a function of laser wavelength. As we can see in Fig. 1(b), between 193 nm and 208.8 nm the NH_4^+ ion yield curve is almost constant, and at around 217.8 nm the curve suddenly starts to drop. From the TPI ion yield measurement, we can roughly conclude that the first one-photon state is resonant and its electronic state is like a quasi-continuum which is well characterized by a very diffuse excitation spectrum. One of the most plausible interpretations for the characteristic feature is rapid relaxation processes such as photodissociation or intramolecular energy redistributions.

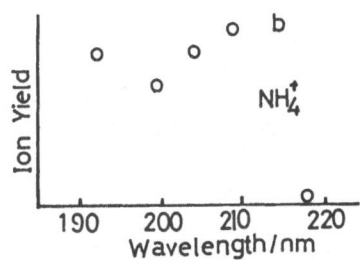 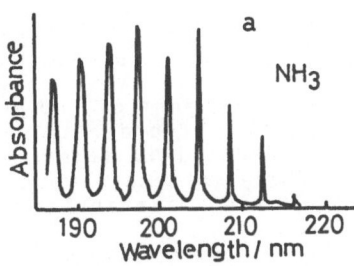

Fig. 1(a) Absorption spectrum of the NH_3 monomer in gas phase. (b) Relative NH_4^+ ion yield obtained by two-photon ionization.

Now let us consider the photoionization dynamics based on the photoelectron spectroscopic measurements. The typical TPI photoelectron spectra obtained for $(NH_3)_2$ are shown in Figs. 2(a), (b), and (c). In Fig. 2(a), the spectrum consists of the three peaks A, B, and C which are attributed to the ionization of monomer, dimer, and much higher order polymers, respectively. It should be noted here that in Figs. 2(b) and (c) there is no peak corresponding to the monomer. Such a spectral difference arises from the fact that the excitation at 208.8 nm is accidentally coincident with the excitation energy for the vibronic level of the excited A state of the monomer.

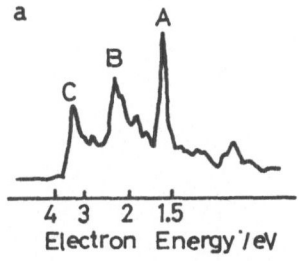 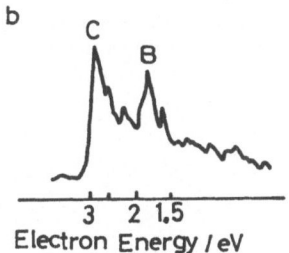 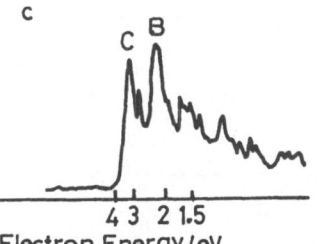

Fig. 2. Photoelectron spectra obtained by two-photon ionization of NH_3 gas in a supersonic jet at different laser wavelengths, (a) 208.8 nm, (b) 217.8 nm, and (c) 199.8 nm.

The results obtained in the present photoelectron measurements strongly suggest that the precursor of the NH_4^+ ion formation is the parent ion $(NH_3)_2^+$ and then very rapid proton transfer takes place forming NH_4^+ ^{3} + NH_2. There has been no evidence found which suggests the ionization of photodissociated neutral species such as NH_4. However, as has been shown already in the ion yield curve, it is also true that the lifetime of the excited dimer or higher plymer is quite short, reflecting very diffuse excitation spectrum. Thus we can finally conclude that under the present laser fluence condition, the up-pumping ionization rate from the one-photon resonant excited state is much faster than the photodissociation process associated with the process of $(NH_3)_2 \rightarrow NH_4 + NH_2$.

3-2. Predissociation of van der Waals Molecule

Two-photon ionization mass spectrum of supersonic-expanded NO gas through a pulsed nozzle shows only NO^+ formation but no $(NO)_2^+$ at all, while the existence of the NO dimer in the ground state has been confirmed /7/. In order to understand the character of the electronically excited NO dimer, we have studied resonant TPI processes of $(NO)_2$ in detail, especially by measuring photoelectron energy distributions with pulsed UV laser sources around 200 nm.

First we have measured a total ion current spectrum of the NO jet as a function of laser wavelength. Here again we have obtained very diffuse excitation spectrum which is quite similar to that of the ammonia dimer. From mass spectrometric measurements, it has been well confirmed that the jet mainly consists of monomer and dimer. It should be noted that among the laser lines used, only 204.2 nm line is accidentally coincident with the excitation energy of $v = 2$ level of NO A $^2\Sigma$. Therefore, the production of NO^+ ion obtained by other laser lines is due to the signals of the TPI of NO dimer through a resonant excited state. Such a situation is very similar to the system of $(NH_3)_2$ mentioned above.

Second we have measured photoelectron spectra at several different laser lines /8/. The present photoelectron measurements have strongly suggested that the UV laser excitation causes a rapid photodissociation of the NO dimer in the excited state, forming the Rydberg A $^2\Sigma$ state of bare NO and subsequent additional one-photon excitation of the A state then leads to the formation of NO^+. Actually the photoelectron peak energies obtained at different laser wavelengths shown in Fig. 3 are fairly coincident with the energies calculated by the assumption that the NO^+ ion is formed from the A-state NO molecule by a one-photon ionization.

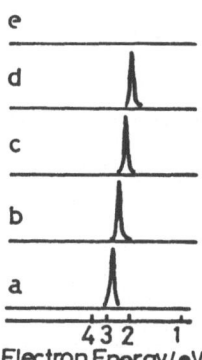

e

d

c

b

a

432   1
Electron Energy/eV

Fig. 3. Photoelectron spectra obtained by two-photon ionization of NO in a supersonic jet at different laser wavelengths, (a) 193 nm, (b) 199.8 nm, (c) 208.8 nm, (d) 217.8 nm, and (e) 223.1 nm.

From the present experimental data indicating very diffuse absorption spectral features of the excited NO dimer and no $(NO)_2^+$ formation suggest the occurrence of rapid predissociation after excitation. This evidence is different in the ionization mechanism from that in the case of the NH_3 dimer, where the up-pumping rate at the excited state is much faster than the one of the predissociations.

3-3. Rotational State Distributions of Photodissociated Fragment Molecule

Using two-color laser multiphoton ionization technique we have investigated rotational state distributions of photofragments /9/. The electronically excited photofragment NO^* is produced by direct UV photodissociation of the rare gas-NO van der Waals molecules. As has been revealed by fluorescence excitation spectra, the Franck-Condon allowed region of the excited states of these van der Waals molecules have a repulsive potential, indicating that after photoexcitation the van der Waals molecule immediately dissociates into electronically excited photofragment molecule and ground-state rare gas atom. The rotational state distributions of the fragment NO were then determined by a two-photon ionization technique using a visible laser at the same time.

The rotational intensity distribution obtained for the Ar-NO complex is shown in Fig. 4. The ion current spectrum shown in Fig. 4 reflects the following electronic transition; $NO^*(A)$, v = 0 → $NO^*(F)$, v = 0, where A and F mean the electronically excited Rydberg states, A $^2\Sigma$ and F $^2\Delta$, respectively. By considering the j dependence of the transition probability, from Fig. 4 we can deduce the relative population of the rotational distribution as shown in Fig. 5.

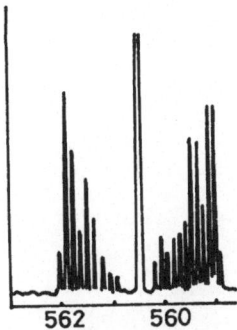

562   560
Wavelength /nm

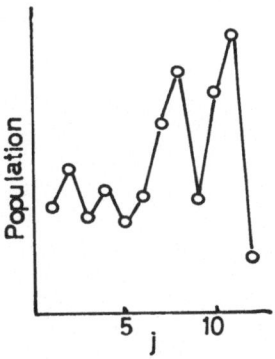

5   j   10

Fig. 4. An ion-current spectrum of the NO^* (A) state in the wavelength region 562-559 nm, obtained with the Ar-NO complex.

Fig. 5. Relative rotational population of the NO^* (A) state.

Experimental results suggest the following interesting features: 1) Two maxima appear in the rotational state distributions. 2) By changing the excess energy E (corresponding to the energy above the dissociation limit of the van der Waals molecule at the excited state) as well as the rare gas partner, the j number at the two maxima (j_{max}) vary in such a way that $j_{max} \propto \sqrt{E}$ and $j_{max} \propto \sqrt{\mu}$, where μ is the reduced mass of the rare gas-NO complex.

This phenomenon is analogous to the rotational rainbow structure already found in inelastic scattering experiments. Therefore, if the present observation is indeed the case where the rotational state distribution appearing in the photodissociation process is much affected by the scattering factor, the present method would be quite appropriate to understand anisotropic properties of a potential surface of a molecular excited state.

REFERENCES

1. For example, K. Kimura: Advan. Chem. Phys. 60 161 (1985).
2. For example, E. A. Rohlfing, D. M. Cox, A. Kaldor: J. Chem. Phys. 81, 3322 (1984).
3. Y. Achiba, K. Sato, K. Shobatake, K. Kimura: J. Chem. Phys. 77, 2709 (1982).
4. Y. Achiba, K. Sato, K. Kimura: J. Chem. Phys. 82, 3959 (1985).
5. H. Shinohara, N. Nishi: Chem. Phys. Lett. 87, 561 (1982).
6. H. Z. Cao, E. M. Evleth, E. Kassab: J. Chem. Phys. 81, 1512 (1984).
7. O. Kajimoto, K. Homma, T. Kobayashi: J. Phys. Chem. 89, 2725 (1985).
8. K. Sato, Y. Achiba, K. Kimura: Chem. Phys. Lett. 126, 306 (1986).
9. K. Sato, Y. Achiba, H. Nakamura, K. Kimura: J. Chem. Phys. 85, 1418 (1986).

Production of Metal Clusters by Nozzle Beam Expansion and Analysis by TOF Mass Spectrometry

Y. Saito

Department of Applied Physics, Faculty of Engineering,
Nagoya University, Nagoya 464, Japan

1. Introduction

Interest in the cluster science lies in various fields such as the study of bond nature in a small system of atoms[1], the analogy between atom clusters and solid surfaces[2], and the development of a frontier field in chemical physics dealing with the experimentally least-known species[3,4]. We have been exploiting an experimental technique to produce cluster beam by non-seeded supersonic nozzle expansion[5,6]. The present work using element metals, mainly lead, is in a first step toward seeking for promising methods to generate an intense beam of metal clusters and also to control the size of clusters contained in the beam. Establishment of this technique is of great importance in order to explore the physical properties of this exotic matter. The purpose of production and analysis of compound(NaCl, KCl, PbS, PbSe) clusters is to obtain information about the stability of the clusters and to clarify the relation between the composition in an extremely small system and that in an infinite bulk system.

2. Experimental

Figure 1 shows schematically the apparatus, which consists of the cluster beam generator and the time-of-flight(TOF) mass spectrometer. The atom clusters are formed by expanding the vapor in a crucible into a vacuum through a nozzle. Figure 2 shows shapes and dimensions of the nozzles used. Each nozzle was machined into a cap of the crucible. The caps and the crucible are made of boron nitride. The crucible is heated by resistive heating of tungsten wire wound around it.

After passing through a beam-defining collimator, the cluster beam is ionized continuously by electron impact. The energy of ionizing electron, E_i, can be varied from 0-150 eV. The ionized clusters are extracted and accelerated to 500 eV from an ionization region. The ionized beam is then chopped electrically by an ion deflector. The chopped ion beam is focused to the ion detector at the end of a drift space of 0.95 m long.

The amount of metal samples charged in the crucible was about 8 g for Pb. For the compound samples, commercially available NaCl, KCl, PbS and PbSe were charged into the crucible without special preparation. The compound vapor was expanded from the single crucible.

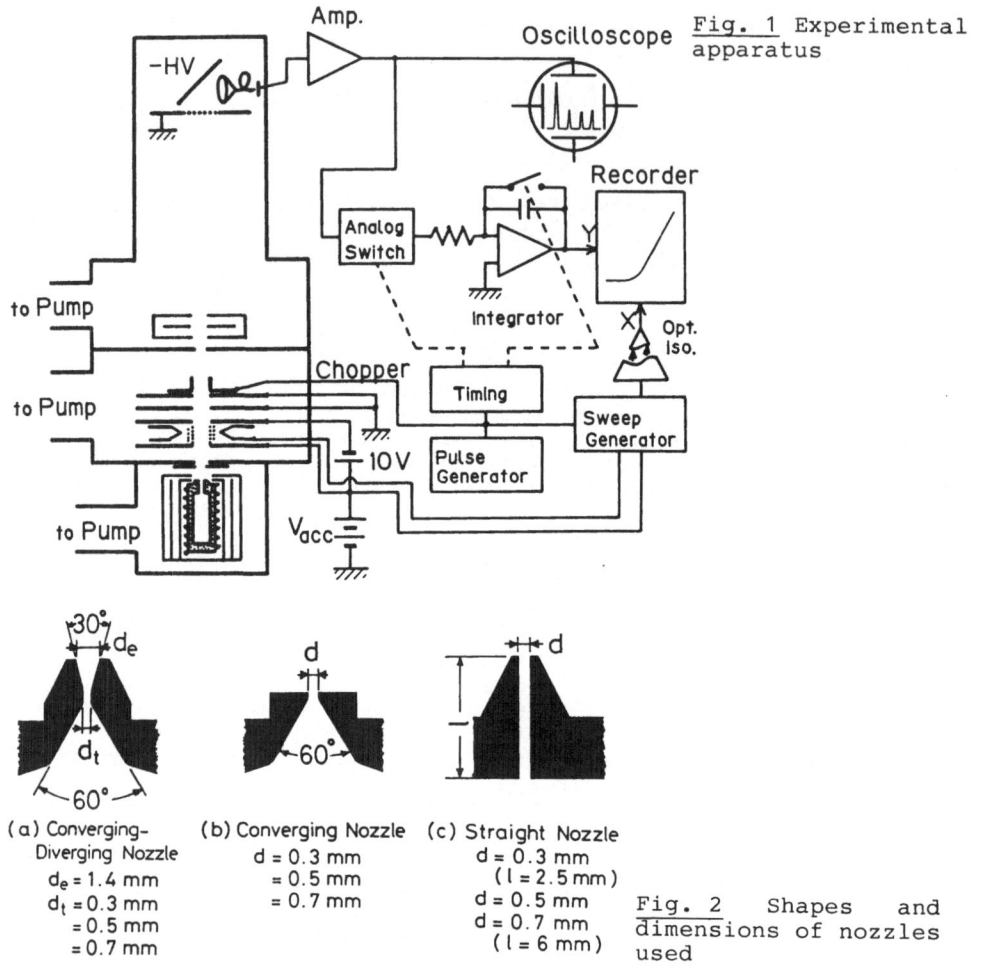

Fig. 1 Experimental apparatus

Amp.

Oscilloscope

Recorder

Analog Switch

Integrator

Opt. Iso.

Chopper

Timing

Sweep Generator

10 V

Pulse Generator

V_{acc}

to Pump

to Pump

to Pump

(a) Converging-Diverging Nozzle
$d_e = 1.4$ mm
$d_t = 0.3$ mm
$= 0.5$ mm
$= 0.7$ mm

(b) Converging Nozzle
$d = 0.3$ mm
$= 0.5$ mm
$= 0.7$ mm

(c) Straight Nozzle
$d = 0.3$ mm
$(l = 2.5$ mm$)$
$d = 0.5$ mm
$d = 0.7$ mm
$(l = 6$ mm$)$

Fig. 2 Shapes and dimensions of nozzles used

3. Element Metal Clusters(Pb, In)

Production of element metal clusters was tried for Pb, In, Ag and Sn. Pb and In formed clusters easily because of their high vapor pressure. Studies shown below were carried out using Pb metal.

3.1. Features in TOF Mass Spectra

Figure 3 shows a TOF mass spectrum of Pb clusters produced from a converging-diverging nozzle with throat diameter d = 0.5 mm. Crucible temperature T_O was 1350 °C. Clusters up to Pb_{11} are observed. For In, those up to In_{10} were detected. The ion intensity of Pb_n^+ does not show a monotonic decrease with the increase of the size but shows a maximum at Pb_7^+ and a step at Pb_{10}^+. For In, a step at In_7^+ is also found.

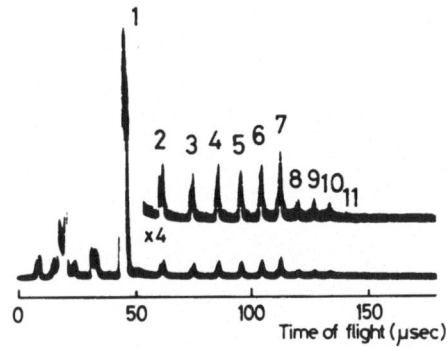

Fig. 3 TOF mass spectrum of lead clusters. E_i = 35 eV

Intensity distribution in the mass spectra is a product of various factors: (1) size distribution of neutral clusters in a beam, (2) cross section for ionization, (3) fragmentation, and (4) detection efficiency of ions(secondary electron yield). If we assume that the cross section for ionization is proportional to the geometrical cross section of the spherical clusters and that the detection efficiency is proportional to a square root of mass of ion[7], the factors (2) and (4) give only a monotonic change in the intensity distribution in mass spectra. We consider that the deviation from the monotonic change in mass spectra is mainly due to the factor (1) and (3). Therefore, the maximum intensity at Pb_7^+ shows high abundance of Pb_7 in neutral beam and/or high stability of Pb_7^+ ion. This may reflect the high stability of the pentagonal bipyramid structure of seven-atom cluster. The great stability of this structure is noted by many authors[8]. The pentagonal arrangement of seven atoms corresponds to the smallest multiply-twinned particle(MTP). The particles have been commonly found in fcc metal particles except for Al[9]. In small particles(diameter less than a few hundred Å), the MTP structure is more stable than a single crystal.

3.2. Effect of Nozzle Shape and Crucible Temperature on the Cluster Formation

Effectiveness of cluster formation for different nozzle shapes and crucible temperatures is investigated in terms of mean cluster size, $\bar{n}$, defined by

$$\bar{n} = \frac{\sum_n nI_n}{\sum_n I_n} , \qquad (1)$$

where n represents the number of atoms contained in the cluster and I_n, the relative abundance of the n-atom clusters. Mass spectra used to calculate $\bar{n}$ were restricted to those taken with E_i = 35 eV in order to eliminate effects of E_i-dependence of the ionization cross section and fragmentation. The E_i-dependence of the relative ion intensity and the fragmentation are described elsewhere[6].

Figure 4 shows measured $\bar{n}$ values for various nozzles shown in Fig. 2 at T_0 = 1250 and 1350 °C. Three factors affecting

123

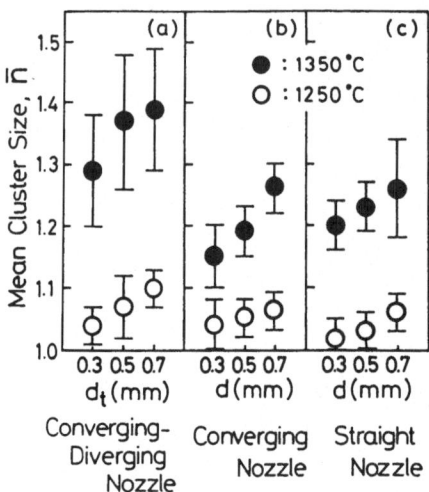

Fig. 4 Mean cluster size, $\bar{n}$, obtained from (a) converging-diverging, (b) converging, and (c) straight nozzles at T_O = 1250 and 1350°C

the cluster formation are found: (i) shape of nozzle, (ii) nozzle diameter, and (iii) crucible tempetature. The effects of these factors are as follows:

(i) The converging-diverging nozzles produce clusters most effectively. This may result from the confinement of the vapor expansion by a diverging section of the nozzle.

(ii) The larger the nozzle diameter is, the more and bigger clusters are produced. Since the flow properties are scaled with respect to downstream distance in terms of nozzle diameter, d, the number of atom collisions as well as the transition time to pass a temperature interval is proportional to d [10]. Therefore, cluster formation is promoted by increasing d. The boundary layer which reduces effective flow diameter needs also to be taken into account for smaller nozzle: the thickness of the boundary layer is estimated to be 0.1 mm at the throat of nozzle for T_O = 1350°C.

(iii) Higher crucible temperature is more effective. In the present experiment, the stagnation pressure is the saturated vapor pressure of the metal equilibrated at the crucible temperature. Therefore, cluster formation depends more strongly on vapor density than on stagnation temperature.

Ion counting rates are estimated from measured ion currents; 10^9 counts/sec for Pb^+, and 10^7-10^8 counts/sec for Pb_n^+ ($2 \leq n \leq 10$) in the order of magnitude. This high intensity will allow us to apply the present technique to measure physical properties of clusters in a free space by electron spectroscopy such as UPS and ELS in coincidence with mass analysis, provided that the cluster generator can supply clusters stably for a long period, say 1 hour. However, the present technique is limited to the metal with high vapor pressure and to small clusters composed of less than about ten atoms. For refractory metals, novel technique such as laser vaporization[1,2] must be used. Methods utilizing gas-condensation will be promising for production of bigger clusters[3], if counting rates can be raised much higher.

3.3. Ionization Potentials

The ionization efficiency curves were recorded to determine the ionization potentials(IP's) by linear extrapolation method. Pb monomers, the IP being 7.415 eV, are used as a standard specimen for calibration of electron energy. Measured IP's of lead clusters up to Pb_7 are shown by open circles in Fig. 5. The large reduction of the IP in the dimer relative to the monomer is first found. Secondly, IP's of the larger clusters have a trend approaching slowly toward the work function of a bulk Pb metal. A solid curve shows the work function W_R of a metal sphere given by the continuum model:

$$W_R = W_\infty + (3/8)(e^2/R) \tag{2}$$

where W_∞ is the work function of the bulk and R, a radius of the metal sphere. The general trend of the change in the measured IP's agrees with that for the spherical metal drop.

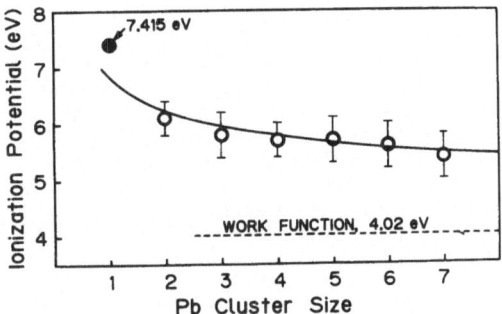

Fig. 5 Ionization potentials of Pb clusters. Open circles: measured ionization potentials, solid curve: work function of spherical metal drop

4. Compound Clusters(NaCl, KCl, PbS, PbSe)

NaCl, KCl, PbS and PbSe are stable compound with stoichiometric composition of 1:1 in solid state. NaCl and KCl are almost purely ionic crystals, whereas PbS and PbSe are largely covalent. Compound clusters from MX to $(MX)_4$ have been observed for all the above compounds, where M represents a cation and X, an anion. A mass spectrum for PbS clusters is shown in Fig. 6, as an example.

Several characteristics are found from mass spectra. The first is concerned with compositions of the clusters. For alkali halides, the clusters formed keep their stoichiometry. On the other hand, for lead chalcogenides, clusters having non-stoichiometric compositions as well as those having stoichiometric composition are formed, though the former clusters are less abundant than the latter. These results show that the composition in bulk solid is most favorably kept even in an extremely small system. This trend is strong for alkali

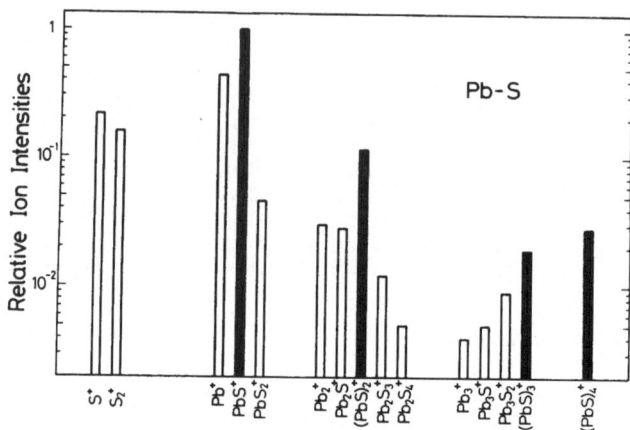

Fig. 6   Relative intensity of $Pb_nS_m^+$ ions.   Nozzle used is a
straight type with diameter 0.7 mm. T_O = 1300°C, E_i = 140 eV

halides.   A second feature is concerned with the most abundant
clusters: MX and $(MX)_2$, having comparable abundance, are the
most abundant for alkali halides,   whereas MX is for lead chalcogenides.   Thirdly,   concerning the variation in intensity of
cluster ions, alkali halides show a monotonic decrease with the
increase of size,   whereas lead chalcogenides do not show monotonic decrease but show $(MX)_4^+$ with higher intensity than   that
of $(MX)_3^+$, probably due to the stability of $(MX)_4$ cluster.

References

1. R. L. Whetten, D. M. Cox, D. J. Trevor and A. Kaldor: Surf.
    Sci. 156, 8(1985).
2. S. C. O'Brien, Y. Lie, Q. Zhang, J. R. Heath, F. K. Tittle,
    R. F. Curl and R.E. Smalley: J. Chem. Phys. 87, 4074(1986).
3. K. Sattler: Surf. Sci. 156, 292(1985).
4. W. D. Knight, W.A. de Heer, K. Clemenger and W.A. Saunders:
    Solid State Commun. 53, 445(1985).
5. Y. Saito, K. Yamauchi, K. Mihama and T. Noda: Jpn. J. Appl.
    Phys. 21, L396(1982).
6. Y. Saito, M. Suzuki, T. Noda and K. Mihama: Jpn. J. Appl.
    Phys. 25, L627(1986).
7. J. Roboz: Introduction to Mass Spectrometry (Robert E.
    Krieger, New York 1979) p. 163.
8. M. R. Hoare and J. A. McInnes: Adv. Phys. 32, 791(1983).
9. T. Hayashi, T. Ohno, S. Yatsuya and R. Uyeda: Jpn. J. Appl.
    Phys. 16, 705(1977).
10. O. F. Hagena: Surf. Sci. 106, 101(1981).

Size Dependence of Photoluminescence in CdSe Clusters

*T. Arai and K. Asai**

Institute of Applied Physics, University of Tsukuba

The size dependence of the energy positions and intensities of the photoluminescence lines of CdSe microcrystals formed by evaporating onto cooled substrates in Ar or He gas at low pressure has been measured.

The two lines originating from the exciton and impurity states were observed
at respectively 1.81 eV and 1.73 eV at liquid N_2 temperature in almost all the
samples. The intensity of the exciton line decreases with decreasing cluster
size, although the intensity of the impurity line decreases, to a decreasing
extent, with decreasing cluster size. The exciton line can hardly be observed
in samples which are less than 6 nm in diameter. This phenomenon is interpreted to be due to the absence of the many-body effect.

1. Introduction

Many researchers have dealt with the size dependence of physical properties
of metallic, ionic and semiconducting materials. For example, the even-odd effect in the ionization energy has been observed in monatomic metals [1] and
splitting and shifting of electronic states have been observed in optical absorption spectra by a number of investigators [2].

It is the objective of this work to determine the size dependences of the
energy positions and intensities of photoluminescence lines in CdSe microcrystals. The results will be interpreted in terms of the absence of the many-
body effect.

2. Experimental

2.1 Sample Preparation and Characterization

CdSe microcrystals were prepared by the gas-evaporation method [3]. After eva-
cuating the vacuum chamber down to 2×10^{-6} Torr, 200 Torr of Ar or He gas was
introduced to clean the chamber. This process was repeated a few times. A
zinc-blende-type CdSe crystal (5N) was evaporated onto a cold SiO_2 glass substrate in Ar or He gas at low pressure using an Al_2O_3 crucible. To avoid condensation of water and vacuum pump oil, the temperature of the substrate was
kept slightly higher than that of the substrate holder, with the heater mounted
on the cold finger fixed to the bottom of liquid N_2 jar. The temperature of the
substrate was ~100 K and that of other parts was ~80 K. To get samples of
various sizes, He and Ar pressures were varied between 1 and 20 Torr and between 0.1 and 0.6 Torr, respectively. The temperature of the source was also
varied between 700° and 800°C. When the evaporation was carried out at lower
pressures and at lower source temperatures, a smaller crystal size was obtained. The sample obtained in He gas was smaller than that obtained in Ar gas.
It can be concluded that a smaller sample can be obtained when the energy loss
in one of the collisions is small and the number of collisions is small. In
order to obtain a sample with a narrow size distribution, the microcrystals

---

*Present address: Electronics Materials Lab., Central Research Lab.,
Sumitomo Metal Ind. Ltd.

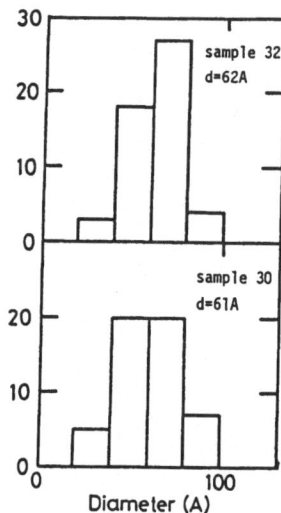

Fig. 1. A typical size distribution

sample 32
d=62A

sample 30
d=61A

Diameter (A)

were collected from a small area in the central part of the substrate. A typical size distribution measured with an electron microscope (J.E.M.-2000 CX, Japan Electron Company) is shown in Fig. 1. The typical evaporation conditions and diameters of samples obtained are listed in Table 1. The crystal structure of the microcrystal obtained was determined by electron diffraction (E = 200 KV, camera length 205 cm). A typical diffraction pattern and electron micrograph are shown in Fig. 2(a) and (b). As is seen in Fig. 2(a), the diffraction pattern for a relatively large sample shows lines originating from both wurtzite and zinc-blende structures. The intensity of the line originating from the wurtzite structure decreases with decreasing size of the microcrystal. There is no line originating from Cd, Se and CdO. From the above phenomena, we can conclude that the surface energy for the zinc-blende-type structure is smaller that that for the wurtzite type, when the volume of the microcrystal is the same. This may originate from the wurtzite structure, which is more close-packed than zinc-blende structure, but the zinc-blende symmetry is more spherical than that of wurtzite. The composition ratio of Cd and Se atoms in the microcrystal was measured with an X-ray microanalyzer (TN-2000 system, Northern Co. Ltd.). The result is shown in Fig. 3. The composition ratio does not differ significantly from the stoichiometric composition of CdSe.

Table 1. Typical evaporating conditions and diameters of sample obtained

Sample No.	Inert gas pressure	Crucible temp. [°C]	Substrate temp. [°C]	diameter [nm]
38	Ar (0.6)	772 ~ 773	-88 ~ -105	20.0
25	Ar (0.6)	783 ~ 811	-100	16.8
28	He (1.0)	720 ~ 730	-96 ~ -100	7.2
32	He (1.0)	697 ~ 703	-97 ~ -100	6.2
30	He (1.0)	696 ~ 700	-70 ~ -95	6.1

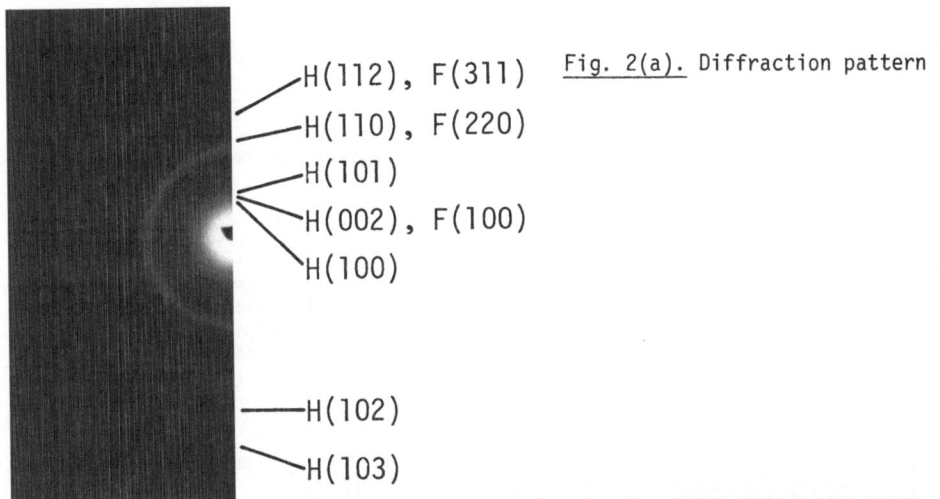

H(112), F(311)

H(110), F(220)

H(101)

H(002), F(100)

H(100)

H(102)

H(103)

Fig. 2(a). Diffraction pattern

H: hexagonal
F: cubic

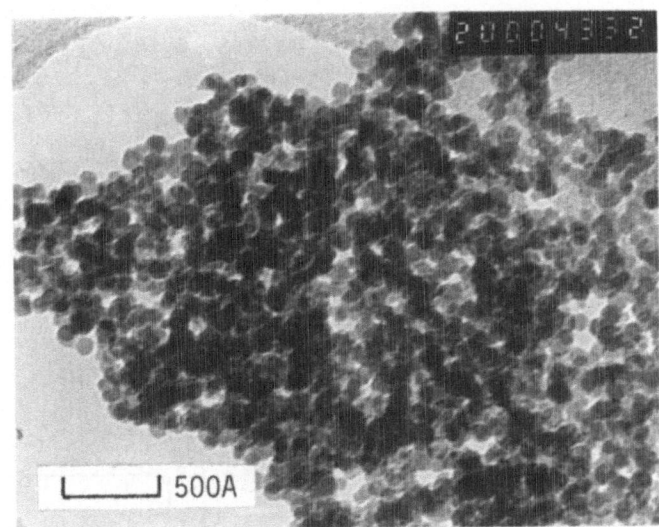

500A

Fig. 2(b).
Electron micrograph

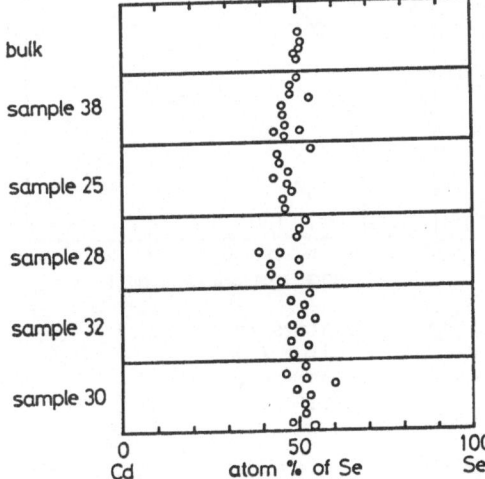

bulk

sample 38

sample 25

sample 28

sample 32

sample 30

0   50   100
Cd   atom % of Se   Se

Fig. 3. Composition ratio of Se and Cd

2.2 Photoluminescence

A photoluminescence measurement was performed with an Ar$^+$ laser (488 nm), a doublepass monochromator (Jobin Yvon U-1000) and a photon-counting system. To avoid damage and coalescence of the sample, the power of the Ar$^+$ laser was kept at less than 10 mW.

The photoluminescence measurements obtained for a bulk crystal and for the microcrystals at liquid N$_2$ temperature are shown in Fig. 4 by solid lines. In the bulk and in the larger microcrystals, two photoluminescence bands located near 1.81 eV and near 1.73 eV were observed clearly. In the bulk crystal, the lines located at 1.81 eV, A$_1$, and at 1.80 eV, K$_1$, originate from exciton recombination [4]. The lines located at 1.73 eV and at energies lower than 1.73 eV originate from impurity recombination and recombination accompanied by LO phonon emission, respectively [4]. In some samples, the line associated

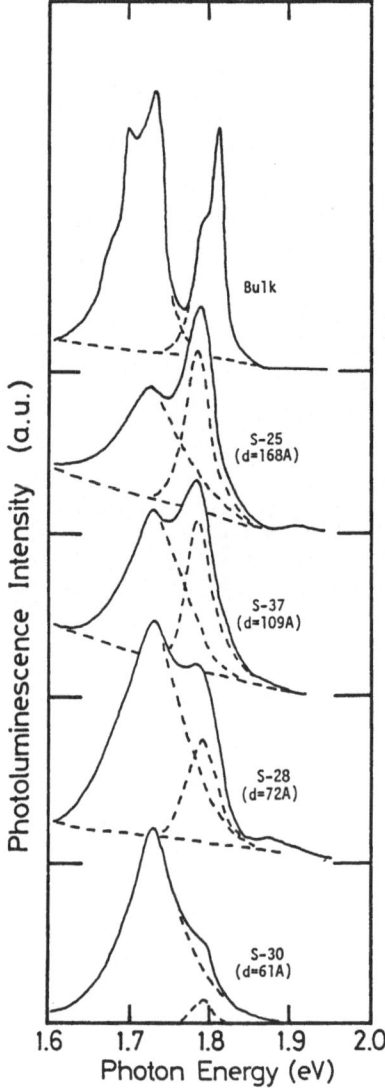

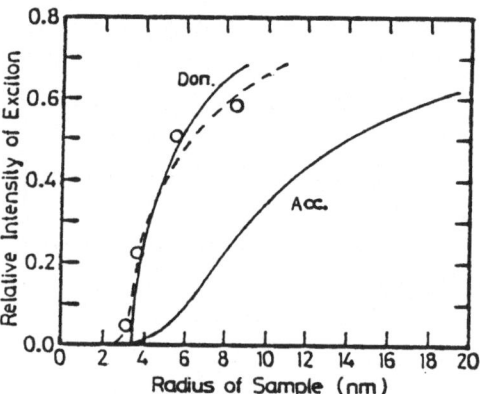

Fig. 4. Photoluminescence spectra at liquid N$_2$ temperature.
_____ : experimental;
------ : decomposed into two bands

Fig. 5. Relative intensity of exciton line.
o, --- : experiment;
_____ : calculated with (3.3)

with LO phonon emission was observed. The LO phonon energy obtained from these lines was 0.0275 eV. This value is equal to the value determined from the lines in the bulk crystal [4]. The position of each line for microcrystals is almost the same as that for the bulk, but the half-width of each line seems to be wider than that in the bulk crystal. The emission band for each sample is de-composed into two bands (due to impurities and the exciton) as shown by the dotted line in Fig. 4. The ratio of the integrated intensity of the exciton band to that of the impurity band decreases with decreasing size of the sample, as shown by the dotted line in Fig. 5. The exciton line was hardly detectable in the sample 30, which is 6 nm in diameter.

3. Discussion

Radiative recombination occurs mainly from the lowest states of impurities and the exciton. We will consider only these lowest states. Impurity states in a medium having a large dielectric constant ε can be treated with a hydrogen-like model. The effective Bohr radius of the impurity state, a^*, is given by

$$a^* = \frac{m}{m^*} \varepsilon\, a_H \quad , \tag{3.1}$$

where a_H is the Bohr radius and m^* is the effective mass. In a first approxi-mation, the optical frequency dielectric constant, ε_∞, can be used as the di-electric constant ε in (3.1). Therefore, $\varepsilon = 7.02$. The effective masses in the conduction band and the valence band are $0.13m$ and $0.45m$, respectively. Thus, the effective Bohr radius of the donor and acceptor states are 2.8 nm and 0.8 nm.

The absorption edge of CdSe is given by direct allowed transitions between conduction and valence bands. Therefore, the radius of the lowest energy state in the exciton of CdSe is also given by (3.1). In this case, the reduced mass should be used instead of the effective mass. The smallest radius of the ex-citon is ~3.6 nm. In a simple model, the impurity state can be considered as if an electron surrounds an atomic core of impurity. An excitonic state can also be described in the same way. When the diameter of the microcrystal is of the same order as the diameter of the state, the electron may be scattered as it orbits. Therefore, the energy states become unstable and the emission lines arising from those states may disappear.

The absorption edge in this size usually does not change as much [5], therefore, the decrease in the intensity of the exciton can be interpreted as indicating the absence of the many-body effect. On the basis of the above considerations, we assume that the exciton cannot exist near the surface of a cluster, which is denoted by the shading in Fig. 6. Therefore, the decrease in the intensity of luminescence originating from the exciton in terms of size reduction is described by

$$I/I_0 \propto \left(1 - \frac{a}{R}\right)^3 \quad , \tag{3.2}$$

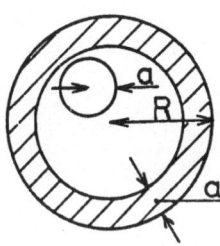

Fig. 6 Model for calculation of luminescence intensity

where a stands for the radius of the exciton and R the radius of the sample. The size dependence of intensity for the impurity has the same formula. Thus, the size dependence of the intensity ratio of the exciton band to impurity level should be written as

$$I_{exc}/I_{imp} = (1 - \frac{a_{exc}}{R})^3/(1 - \frac{a_{imp}}{R})^3 \quad . \tag{3.3}$$

The size dependence of the inensity ratio calculated from (3.3) for donors and acceptors is shown by the solid line in Fig. 5. The coincidence between the experimental and theoretical donor curves is much better than that between the experimental and theoretical acceptor curves. From this result, the impurity luminescence seems to arise from a recombination of an electron in a donor state.

The position of the impurity luminescence line is 1.73 eV. The band gap of CdSe at 77 K is 1.84 eV. Therefore, the impurity state should lie 0.11 eV from the band edge. According to the calculation of the hydrogen-like model, the energy of the acceptor level is ~0.13 eV from the top of the valence band and that of the donor level is ~0.037 eV from the bottom of the conduction band. From this consideration of the energy positions, the luminescence should be related to an acceptor state.

By considering both the intensity ratio and the energy position, one can conclude that the impurity luminescence should arise from a recombination of a donor-acceptor pair. The small disagreement in energy may arise from the difference in energy gap originating from the difference of the crystal field in wurtzite and zinc-blende structures.

To analyze the experimental results in detail, the following two points should be taken into consideration: a) The wave function of the exciton in a small sample should shrink due to the existence of a potential well and the binding energy should become large. This is in fact observed in a two-dimensional quantum well. The peak position of the excitons determined from the dotted curves in Fig. 4 does not vary with decreasing sample size. The disagreement may be due to the existence of the surface layer which includes many lattice defects and lattice distortion. b) The hydrogen model is not correct for the lowest state of the exciton. The radii of the exciton and impurity should be smaller than that calculated from the hydrogen model.

Acknowledgements

We are indebted to Prof. M. Inoue of Tsukuba University for helpful discussions. The work was partially supported by a Grant in Aid for Scientific Research from the Ministry of Education, Japan.

References

1 I. Katakuse, T. Ichihara, Y. Fujita, T. Matsuo, T. Sakurai, H. Matsuda: Int. J. Mass Spectrom. Ion Processes **67**, 229 (1985)
2 W. Shulze, H.U. Becker, H. Abe: Chem. Phys. **35**, 177 (1978)
3 K. Kimoto, K. Kamiya, Y. Nonoyama, R. Uyeda: Jpn. J. Appl. Phys. **2**, 702 (1963)
4 D.C. Reynolds et al.: Phys. Rev. **156**, 881 (1967)
  D.L. Rosen et al.: Phys. Rev. B **31**, 2396 (1985)
5 A.G. Stasenko: Sov. Phys.-Solid State **10**, 186 (1968)
  G.C. Papavosilou: J. Solid State Chem. **40**, 330 (1981)

Part IV

Dynamical Processes

Equilibrium Structures and Dynamical Processes in Microclusters

R. Car[1], *M. Parrinello*[1], *and W. Andreoni*[2]

[1]International School for Advanced Studies, I-34100 Trieste, Italy
[2]IBM Research Division, Zurich Research Laboratory,
CH-8803 Rüschlikon, Switzerland

We discuss the relevance of the Unified Molecular Dynamics — Density Functional approach to the study of cluster physics. We present specific calculations for small Si aggregates, and show that ground-state structures can be efficiently obtained with a dynamical simulated annealing. We find that, while the $T = 0$ equilibrium structures are symmetrical, the atomic configurations which correspond to the local minima close in energy to the ground state are disordered. These appear to be the relevant structures at high temperatures.

1. INTRODUCTION

The evidence for 'magic numbers' in the mass spectra of atomic clusters [1,2], i.e., specific atomic densities N at which the aggregates are especially stable, has by now stimulated several quantum-mechanical calculations of equilibrium structures and cohesive energies for $N \lesssim 10$ [3,4]. Given a cluster of N atoms at the positions $\{R_I\}$, these calculations are meant to explore the potential energy surface $\Phi(\{R_I\})$ which characterizes the interatomic interaction and to find the global minimum, i.e., the ground state of the cluster or in other words the $T = 0$ equilibrium configuration. In practice, however, by following the conventional quantum-mechanical procedures, one is restricted to investigating only very few points, which represent highly symmetric structures and may (or may not) correspond to local minima. As the number of atoms N increases, the number of local minima is expected to increase exponentially, so that the search for the global minimum soon becomes impractical. An additional disadvantage of the conventional methods is that they are restricted to $T = 0$ properties, which are certainly not sufficient to explain experimental data. In particular, the basic assumption one makes in using these results to interpret 'magic numbers' is that they reflect the relative stability of the $T = 0$ equilibrium structures. Although in some cases such a correlation seems to exist, this is not fully understood in view of the $T \neq 0$ processes involved during either the formation or the treatment of these aggregates prior to mass spectroscopy (e.g., photofragmentation).

Some of the above difficulties can be overcome by using either Monte Carlo (MC) or Molecular Dynamics (MD) techniques. These techniques allow simulations at finite temperature through an efficient sampling of the potential energy surface, in which amongst all the atomic configurations contributing to $\Phi(\{R_I\})$, one selects only those which are statistically relevant, as measured by an appropriate Boltzmann factor. In addition, MD procedures also allow studying dynamical processes. Within these schemes, the $T = 0$ equilibrium geometries can be efficiently determined by means of a simulated annealing strategy [5].

Due to intrinsic computational difficulties, MC and MD simulations of clusters [6] have so far been restricted to the use of model potential energy surfaces $\Phi(\{R_I\})$ resulting from empirical or semi-empirical additive few-body potentials. Apart from rare-gas systems, the reliability of these model potentials for real materials is not yet well established. This is especially true for covalent systems, where the local bonding configurations strongly depend on the details of the local electronic structure.

Recently, a new approach has been proposed [7] which allows to perform ab-initio MD simulations where the interatomic interactions are treated at the Density Functional (DF) level, in the local density approximation. This method has already been successfully applied to the study of liquid and amorphous silicon [8].

In this paper, we discuss the new capabilities of this method and their relevance to the study of cluster physics. In particular we show the feasibility, in first-principles DF calculations, of a dynamical search of the equilibrium ground-state geometry through a simulated annealing strategy [7], in which the atoms find their way to the T=0 equilibrium configuration, without any simplifying assumption on the symmetry or on the starting geometry. We give explicit results for small silicon clusters. We further discuss the possibility of extending these calculations to study thermodynamical properties and reaction processes, within the limits of validity of the adiabatic Born-Oppenheimer (BO) separation between nuclear and electronic dynamics.

2. THEORETICAL METHOD

As explained in [7], it is convenient to introduce a fictitious classical dynamical system whose degrees of freedom are given by the atomic coordinates $\{R_I\}$ as well as by the occupied single-particle orbitals used to represent the electronic density in the Kohn-Sham formulation of DF theory. The classical dynamical system is defined by the Lagrangian

$$L = \sum_i \frac{1}{2}\mu \int d^3r |\dot{\psi}_i|^2 + \sum_I \frac{1}{2} M_I \dot{R}_I^2 - E(\{\psi_i\}, \{R_I\}) + \sum_{ij} \Lambda_{ij} \left(\int d^3r\, \psi_i^*\psi_j - \delta_{ij}\right). \tag{1}$$

Here the dot indicates time derivative, the M_I's are the atomic masses, μ is a fictitious 'mass' associated with the electronic degrees of freedom, and Λ is an Hermitian matrix of Lagrangian multipliers introduced to impose the orthonormality constraints for the orbitals ψ_i. The classical potential energy E is defined in [7], and contains both quantum-electronic kinetic energy and electrostatic and exchange-correlation interactions within DF theory. Here we adopt a pseudopotential formulation, so that the ψ_i's refer to the valence states only.

The Newton equations of motion derived from the Lagrangian in (1) are also reported in [7]: They can be solved and the time evolution of the dynamical system can be studied as in standard MD simulations. In particular, a temperature can be associated with the average value of the classical kinetic energy corresponding to the degrees of freedom in (1). By varying this temperature, it is possible to make the system defined in (1) undergo various thermal treatments, such as annealing and quenching. During such processes, all relevant degrees of freedom, both electronic and atomic, are relaxed simultaneously. In particular, we can set up a dynamical process that brings the system to its equilibrium state at $T=0$. This constitutes a practical implementation of a simulated annealing strategy to minimize the function E having the meaning of a classical potential energy in (1).

So far, we have not discussed the relevance of the classical dynamical system to the physical problem we have in mind, i.e., the efficient sampling of the potential energy surface $\Phi(\{R_I\})$ of a cluster. The connection is easily established, however, by noting that Φ and E are related by

$$\Phi(\{R_I\}) = \underset{\{\psi_i\}}{\text{Min}}\ E(\{\psi_i\}, \{R_I\})\ . \tag{2}$$

Therefore, the absolute minimum of E and that of Φ coincide and, by optimizing the function E, we obtain the ground state of a cluster. In addition, by performing numerical simulations with system (1), it is possible to study finite T and dynamical processes for system (2). This is the case since, although the atomic trajectories of system (1) and those of system (2) may be quite different in general, they can be led to coincide under special conditions. For sufficiently small values of the 'mass' μ, the electronic dynamics proceeds much faster than the atomic dynamics and the temporal evolution of (1) is adiabatic, i.e., the energy transfer between atomic and electronic subsystems is negligible over a typical MD observation time. In such conditions, if the classical dynamical system is initially prepared in a state in which E is minimized with respect to the ψ_i's and the $\dot{\psi}_i$'s vanish, its subsequent trajectory will mimic a BO trajectory of the cluster with potential-energy surface $\Phi(\{R_I\})$ very closely. Note that a small μ requires a small time step in the numerical integration of the coupled differential equations of motion for system (1): The simulation of adiabatic motion may be costly. In our experience, however, this procedure has always been significantly more convenient than the alternative way of completely decoupling electronic and atomic motions and performing a separate electronic minimization at any new atomic step.

When one is only interested in finding a local minimum of a potential function, a Steepest-Descent (SD) procedure may conveniently replace MD. In analogy with second-order Newton's equations, we may give the following first-order differential equations:

$$\sqrt{\mu}\ \dot{\psi}_i = -\ \delta E/\delta \psi_i^* \ + \text{constraints}$$

$$\sqrt{M_I}\ \dot{R}_I = -\ \partial E/\partial R_I\ , \tag{3}$$

which define a mass-weighted SD trajectory on the E surface that moves downhill from the starting point to a nearby local minimum. This trajectory can be considered as resulting from Newton equations in the limiting case in which the instantaneous velocities of the various degrees of freedom are kept constantly equal to zero, as in a $T=0$ MD run. When using (3), optimal relative relaxation rates between electrons and atoms can be achieved by adjusting the 'mass' μ. SD optimization is considerably faster than simulated annealing strategies when searching for a local minimum. This is the case, for instance, when we are interested in solving the standard electronic problem for a fixed atomic configuration, which is defined in (2). In this case, it appears that there is only one minimum, and the SD approach applied to the electronic degrees of freedom is an efficient procedure to minimize E, provided we are able to start from an initial electronic configuration with the same symmetry as the final one. In such a case, a very efficient integration of (3) can usually be achieved by adapting the procedure given in [9] to this purpose.

However, when we are interested in a global rather than a local sampling of a potential surface, a MD procedure must be followed. In a finite T MD run, the velocities of the degrees of freedom are different from zero and the complexity of the system brings in random thermal motion. By varying the temperature, the dynamics of the system can switch continuously from a random walk in parameter space to an SD approach to the closest local minimum. At high T, the system samples a large portion of the potential surface by diffusing through different configurations. By slowly reducing the temperature, jumps between different configurations become less and less likely and, for sufficiently slow cooling rates, the system is eventually caught in the deeper or absolute minimum. At this point, one can switch on the SD procedure and quickly find the optimal relaxation. An important caveat should be added: The above procedure is not always bound to work successfully since, in some cases, the cooling rates required may be so slow as to be practically impossible. Even in this case, however, one expects to obtain a statistically relevant sampling of the potential energy surface from the annealing procedure.

An additional remark should be added, since two strategies are possible for finding the global minimum of $\Phi(\{R_I\})$: In one strategy, one follows a trajectory lying on the Φ surface, whereas in the other one does not care about the actual path followed by the system in the course of the optimization procedure. Since in the latter case one does not require an adiabatic evolution of the system, the parameter μ can be adjusted to fully optimize the relative relaxation rate of electrons and atoms. This may lead to a very efficient scheme. In what follows, however, we have not explored this possibility and all annealing runs have been performed with paths lying on the Φ surface. Therefore, the high-T configurations explored during the annealing runs are representative of physical high-T configurations. The annealing procedure was supplemented by SD relaxation to obtain quick geometry optimization, when the region of attraction of the relevant minimum had already been attained.

3. APPLICATION TO SILICON CLUSTERS

We have adopted a pseudopotential framework and used a first-principles non-local pseudopotential [10] for the electron-ion interaction. Exchange-correlation effects were described within the local-density approximation using the parametrized form from [11]. A plane-wave representation was used for the electronic states in (1), with an energy cutoff of 6 Ry. Note that the convergence of the calculated equilibrium geometries is already very good for a cutoff of 5 Ry. The use of periodic boundary conditions implicit in the plane-wave scheme forces us to adopt a periodically-repeated supercell approach. Although this is probably not the most efficient geometry for representing a cluster in vacuum, the advantages of a plane-wave representation are overwhelming and the method of [7] is well suited for dealing with a very large number of plane waves. Only the Γ point was used to represent the Brillouin zone. We have explicitly checked that the interaction between the periodic images belonging to different cells is negligible, given the sizes of the cell and the clusters used in the present study. For clusters of up to five Si atoms, we have used a simple cubic cell with an edge of 19 a.u. This corresponds to approximatively 2000 plane waves in the expansion of the wave functions.

In the dynamical simulations, we have used a value of μ = 300 a.u. and a time step Δt = 7 a.u. to integrate the Newton equations with the Verlet algorithm and ensure adiabatic motion. The control of temperature was achieved by constant rescaling of velocities. The initial atomic velocities for the MD runs were always taken to be zero. A time step Δt = 100 a.u. and μ = 390000 a.u. were used to integrate the SD equations (3) with the lowest-order finite difference scheme.

A typical annealing run for the Si_4 cluster is illustrated in Fig. 1, which shows the variation of Φ in the course of the run in which an approximatively linear annealing schedule was used. No attempt was made to optimize the number of steps in the annealing procedure. Faster procedures are likely to be possible. The rightmost segment of the curve, corresponding to SD relaxation, does not occur in the time scale of the figure and is shown for illustrative purposes only.

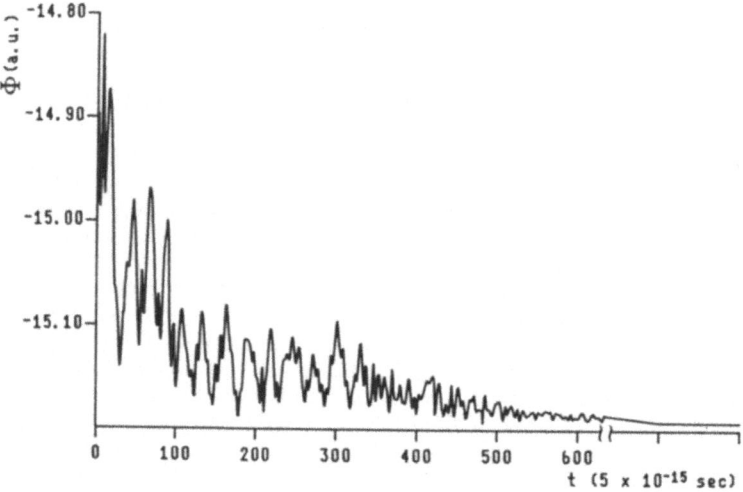

Fig. 1. Variation of the potential energy of Si_4 during cooling from T $\simeq$ 5000 K to T $\simeq$ 600 K

The simulation was started with a geometry corresponding to a local minimum of the potential energy surface. This geometry was obtained by applying a SD relaxation, i.e., an infinitely fast quench, to a high-T structure generated with MD. Figure 2(a) shows the integral n(r) of the radial distribution function. A realistic view of this geometry is displayed in Fig. 3(a). The structure consists of a triatomic subunit, in which each atom is bonded to the other two, plus an appended atom bonded to one atom only. This is a disordered structure where no two bond lengths are equal, as a consequence of the perturbation induced by the somewhat loose additional atom to the triangular subunit. The system was then submitted to a thermal treatment consisting of rapid heating followed by equilibration and slow cooling (Fig 1).

Si_4

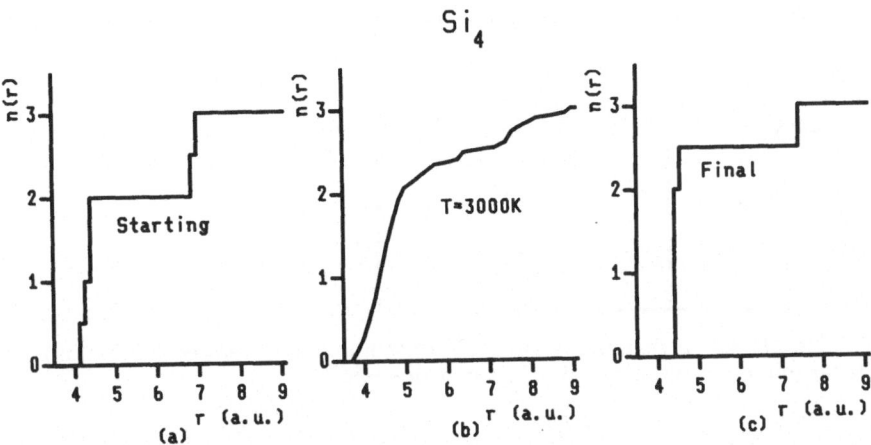

Fig. 2. Si_4: Integral $n(r)$ of the radial distribution function for (a) starting geometry (local minimum), (b) snapshot of a high-T configuration, (c) final geometry (global minimum)

At high T, memory of the starting configuration is lost. The function $n(r)$ averaged over a period of $\sim 3 \times 10^{-13}$ sec at $T \approx 3000$ K, Fig. 2(b), is reminiscent of a 'molten' state. Figure 3(b) displays a snapshot of this run. A new structure becomes stable at lower temperatures in correspondence with the switching of a bond and the formation of an additional weak bond so that all the atoms are now bonded at least twice (Fig. 2). This symmetrical geometry is shown in Fig. 3(c), and corresponds to the absolute minimum of $\Phi(\{R_I\})$. Note the rather small difference in binding energy of only 0.4 eV/atom between the initial disordered structure and the true ground-state.

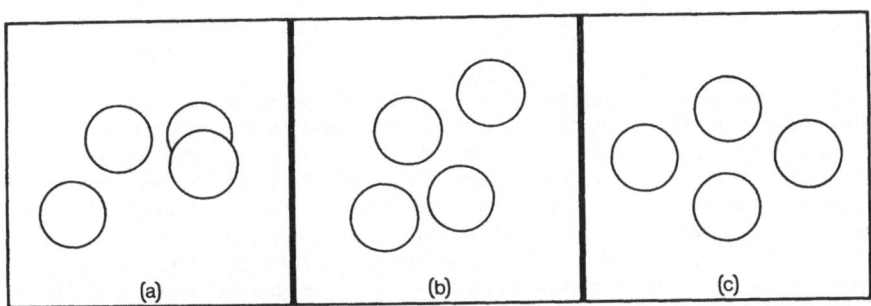

Fig. 3. Si_4: Structures corresponding to (a), (b) and (c) in Fig. 2, respectively

A similar thermal treatment was applied to the five-atom cluster. The resulting $T=0$ equilibrium geometry and the corresponding $n(r)$ are shown in Figs. 4(a) and 5(a), respectively. The number of time steps necessary to form the stable low-T structures was similar in the two clusters studied. The most crucial part of the entire scheme corresponds to the formation process: Too rapid cooling in this region will inevitably lead to a metastable local minimum as the one shown in Fig. 5(b) for Si_5. The difference in binding energy between this structure and the ground state is only 0.1 eV/atom.

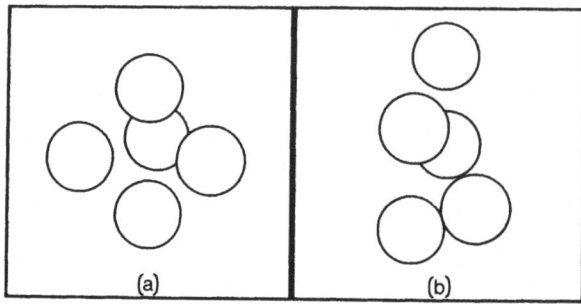

Fig. 4. Si$_5$: Structures corresponding to (a) global minimum (final configuration) and (b) local minimum obtained after a rapid quench

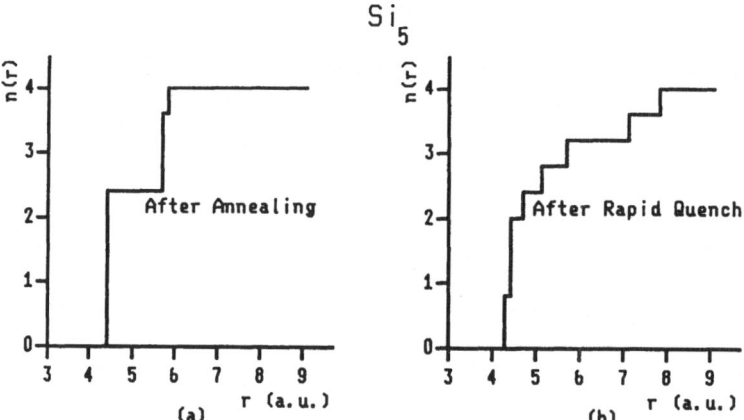

Fig. 5. Si$_5$: Integral n(r) of the radial distribution function corresponding to (a) and (b) in Fig. 4, respectively

A number of interesting considerations can be drawn from the analysis of our cluster data. The stable low-T structures correspond to symmetrical forms, and are essentially identical to those found in previous ab-initio quantum-mechanical calculations. On the other hand, the high-T structures as well as those resulting from too fast quenching treatments may be qualified as disordered defective structures. They differ from the ground state in that one or two bonds are broken, inducing disorder in the structure. Their binding energy is quite large, almost comparable to that of the ground state, and, in general, significantly larger than the binding energy of the stable symmetrical isomers usually considered in the literature as possible competitors of the ground state [4]. It is an interesting observation that already for N as small as four, the potential energy surface of a covalent system like silicon is so complex as to show the presence of metastable local minima corresponding to disordered structures. Further investigations of this point may help clarify the role played by chemical bonding in the ability of different elements to form amorphous structures.

We also remark that for both clusters studied, the high-T geometries consist of interacting smaller subunits, i.e., the (N-1)-cluster plus one atom. This suggests that a likely path for fragmentation at high T of the four- and of the five-atom Si clusters will involve the formation of an (N-1)-cluster plus a monomer, as indicated by experimental data [2].

Note that at least in the cases considered here, the most likely high-T geometries are structurally different from those encountered at low T. Therefore, the interpretation of high-T processes in terms of the T=0 structures does not appear to be valid.

In conclusion, equilibrium geometries of atomic aggregates can be obtained by means of a strategy that combines the careful treatment of the ground-state electronic structure, customary in ab-initio local-density calculations, with the capability of achieving global optimization characteristic of a simulated annealing scheme based on statistical mechanics. We are presently extending the calculations presented here to larger clusters. A study of thermal properties and dynamical processes, such as association and dissociation, is also under way.

REFERENCES

1. See for instance: O. Echt, K. Sattler, E. Recknagel: Phys. Rev. Lett. **47,** 1121 (1981); T.P. Martin: Physica **127B,** 214 (1984); W.D. Knight, K. Clemenger, W.A. de Heer, W.A. Saunders, M.Y. Chou, M.L. Cohen: Phys. Rev. Lett. **52,** 2141 (1984); E.A. Rohlfing, D.M. Cox, A. Kaldor: J. Chem. Phys. **81,** 3322 (1984); S.C. O'Brien, Y. Liu, Q. Zhang, J.R. Heath, F.K. Tittel, R.F. Curl, R.E. Smalley: J. Chem. Phys. **84,** 4074 (1986), and references therein
2. L.A. Bloomfield, R.R. Freeman, W.L. Brown: Phys. Rev. Lett. **54,** 2246 (1985); W. Begemann, K.H. Meines-Broer, H.O. Lutz: Phys. Rev. Lett. **56,** 2248 (1986)
3. J.L. Martins, J. Buttet, R. Car: Phys. Rev. **B31,** 1804 (1985); W. Andreoni, J.L. Martins: Surf. Sci. **156,** 635 (1985); M.H. McAdon, W.A. Goddard III: Phys. Rev. Lett. **55,** 2563 (1985); B.I. Dunlap: J. Chem. Phys. **84,** 5611 (1986); T.H. Upton: Phys. Rev. Lett. **56,** 2168 (1986)
4. K. Ragavachari: J. Chem. Phys. **84,** 5672 (1986); D. Tomanek, M.A. Schlüter: Phys. Rev. Lett. **56,** 1955 (1986)
5. S. Kirkpatrick, C.D. Gelatt, Jr., M.P. Vecchi: Science **220,** 671 (1983)
6. See, e.g., J. Jellinek, T.L. Beck, R.S. Berry: J. Chem. Phys. **84,** 2783 (1986); E. Blaisten-Barojas, D. Levesque: Phys. Rev. **B34,** 3910 (1986); R. Biswas, D.R. Hamann: Mat. Res. Soc. Symp. Proc. **63,** 173 (1986)
7. R. Car, M. Parrinello: Phys. Rev. Lett. **55,** 2471 (1985)
8. R. Car, M. Parrinello: Proc. of the XVIII Int. Conf. on the Physics of Semiconductors, Stockholm, Sweden, 1986 (in press)
9. M.C. Payne, J.D. Joannopoulos, D.C. Allan, M.P. Teter, D.H. Vanderbilt: Phys. Rev. Lett. **56,** 2656 (1986)
10. U. von Barth, R. Car: (unpublished)
11. J.P. Perdew, A. Zunger: Phys. Rev. **B23,** 5048 (1981)

Many Body Effects on Static and Dynamical Properties of Si, C, and K Microclusters

C. Satoko

Institute for Molecular Science, Myoudaiji, Okazaki, Aichi 444, Japan

Many-body interaction effects on the static properties and dynamical properties of metal and semiconductor microclusters are
investigated. The many body effects contribute to the bond distances ,the geometries, and the melting and dissociation process of
the microclusters which are calculated by a proper Hamiltonian.

1. Introduction

Much theoretical work on the interactions between atoms have been
developed. The first-type is a simple pair potential such as
Lennard-Jones and Born-Mayer potential which is suitable for the
neutral rare-gas Ar and Xe microclusters. The second type is an
effective pair potential which has been used in liquid metals.
This includes effectively the many body effects such as the
repulsive force between the second nearest neighbor sites. These
potentials have been used for the study of the dynamical properties because the total energies are calculated very easily by
the sum of these pair potentials. However,the use of these two-
body interactions is questionable for small systems such as metal
and semiconductor microclusters.

The calculated normalized bond distances of rare gas microclusters are shown in Fig.1 as a function of the inverse cube root
of the number of atoms N. These are calculated by using the
Lennard-Jones potential. The bulk bond distance obtained is 97
percent of the diatomic molecule, which agrees with the experimental value of solid Ar and Xe. The less the number of atoms, the
longer the bond distance. This positive slope is characteristic
of the simple pair potential where the forces between the nearest
neighbor atoms are repulsive and the other forces are always
attractive.

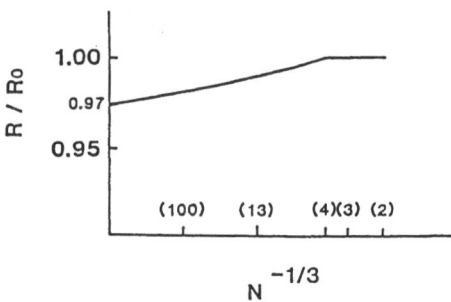

Fig.1. The bond distances
from a Lennard-Jones potential.

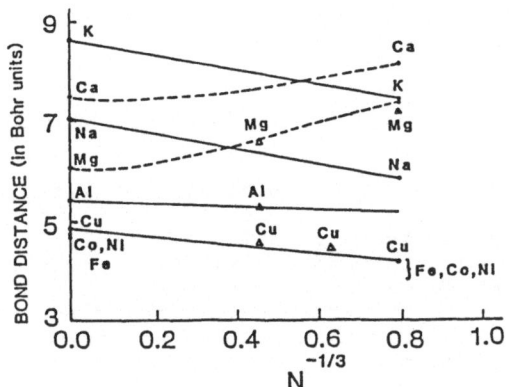

Fig.2. The bond distances of metals

The bond distances of metal clusters are shown in Fig 2 as a function of the inverse cube root of the number of atoms N. Some of them are obtained from the experiments(black circle),and the others are calculated(white triangles)[1,2]. There are two kinds of slopes in the size dependence of the clusters. Alkali-metals (K and Na) and transition metals (Fe, Co, Ni, and Cu) show the negative slopes. These negative slopes are due to the many body interaction. On the other hand, the bond distances of alkali-earth metals (Mg and Ca) show the positive slopes in the range of small size. In the large size range we expect the negative slope due to their metallic bonds. These negative and positive slopes mean the importance of the many-body effect which strongly depends upon the electronic structure.

The many-body interaction is also important for giving the dynamical properties. The self-consistent harmonic approximation was adapted to study the melting of Pb microclusters[3]. The interaction was approximated by use of the Born-Mayer pair interaction where parameters were determined to fit the bulk properties. The melting temperature obtained was too high to fit with experimental values. When we studied the melting temperature in the molecular dynamics with the same pair interaction, it was comparable with that of the self-consistent harmonic approximation. Therefore the disagreement with the experiment is not due to the self-consistent harmonic approximation , but due to the use of the pair potential approximation.

2. Model Hamiltonian for the many-body interaction

To study metallic and semiconductor microclusters, available best methods are to use the ab-initio Hartree-Fock Hamiltonian plus configuration interactions or the local-density functional Hamiltonian. These methods,however,are so laborious for the study of large clusters that they are not always realistic for dynamical studies. Therefore we take a simple model Hamiltonian where the parameters are determined from the comparison with the calculated electronic structure of the small clusters.

We introduce the following Hamiltonian for the valence electrons of alkali-atom and semiconductor microclusters:

$$H = \Sigma \ t_{\alpha 1, \beta m} c^{\dagger}_{\alpha 1 \sigma} c_{\beta m \sigma} + \Sigma (\varepsilon_{\alpha 1} + U Q_{\alpha}) c^{\dagger}_{\alpha 1 \sigma} c_{\alpha 1 \sigma}$$

$$+ \Sigma \ \Phi_{\alpha \beta} + \Sigma Q_{\alpha} Q_{\beta} / \varepsilon R_{\alpha \beta} . \tag{1}$$

The fist term is the transfer energies between orbital 1 at site α and orbital m at site β. The $\varepsilon_{\alpha 1}$'s in the second term are the orbital energies of orbital 1 at site α , which are shifted by the Coulomb repulsive energies UQ_{α} between the electrons. The Q_{α}'s represent the difference between the electron number and the nuclear charge at site α. The third term is the Morse potential between the ion-cores. The last term is the screened-Madelung potential with the screening constant ε .

To study the electronic structure of covalent materials and alkali-metals, we consider the interaction between the s-,px-,py- and pz-orbitals. The transfer energies t's are written by the Koster-Slater parameters $(ss\sigma)$,$(pp\sigma)$,$(sp\sigma)$,$(pp\pi)$ which are assumed to have the exponential damping dependence on the distance R between the atoms:

$$(ss\sigma) = (ss\sigma)_0 \ exp(-\lambda \ R + \lambda R_0) \ and \ so \ on. \tag{2}$$

The values of these parameters are shown in Table 1. These parameters are determined so that the calculation reproduces the electronic structures and geometries of two and three clusters which are calculated by the LCAO-Xα-force method[4,5]. We also introduce parameter Δ representing the energy difference between the s and p orbitals. The transfer energies are the values at each bond distance R_0(shown in the parentheses in table 1) which is the equilibrium bond distance of the diatomic molecules. Damping factors λ in the Koster-Slater parameters are taken to be about 30-60 percent of those in the Morse potentials[6]. The ratio of transfer energies to Δ in the C case is larger than that in the Si case. As pointed out in section 4, such ratio determines whether the clusters are linear or bent structures.

Table 1. Parameters for the Si,C and K clusters.

For Si clusters(at R_0=2.25 A),C clusters(at R_0=1.00 A)

$\Delta = \varepsilon_p - \varepsilon_s$ = 2.53eV,5.27eV

$(ss\sigma)/\Delta$ = 0.18, 0.44 , λ = 0.67/A, 1.31/A
$(pp\sigma)/\Delta$ = 0.61, 0.88 , λ = 0.67/A, 1.70/A
$(sp\sigma)/\Delta$ = 0.49, 0.78 , λ = 0.67/A, 1.31/A
$(pp\pi)/\Delta$ = 0.22, 0.66 , λ = 0.67/A, 2.26/A
U/Δ = 0.67, 1.11

For K cluster(at R_c=3.92 A)

$(ss\sigma)$ = 0.35 eV, λ =0.47/A
U= 1.08 eV

The values in Table 1 do not give the best fit and some are flexible. However, the following qualitative results are not sensitive to these values.

3. Bond distance analysis by atomic stress

I discuss the size dependence of the bond distances by use of the atomic stress tensor which is defined as the following:

$$g_\mu{}^{ij} = \Sigma_\tau \; f_{\mu\tau}{}^i (R_\mu{}^j - R_\tau{}^j)/2 \; , \tag{3}$$

where the pair force , $f_{\mu\tau}{}^i$, is the i-th component of an effective pair force between the μ-th and τ-th atoms. The term in the parentheses is the j-th component of the distance between the μ-th and τ-th atoms. The components of this atomic stress parallel to the surface are related with the surface tension. The components perpendicular to the surface are related with the pressure of the particle.

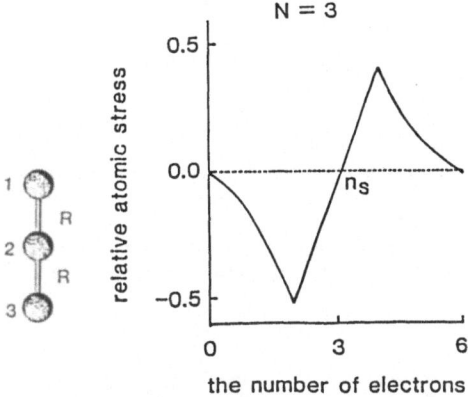

Fig.3. The geometry and the atomic stress of the first atom

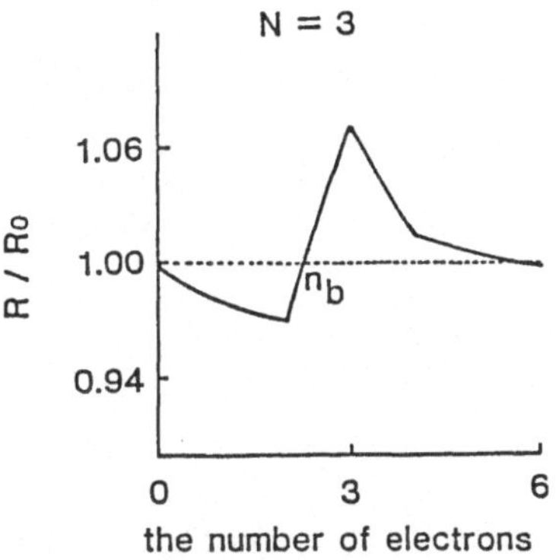

Fig.4. The bond distance R normalized by R_0

We consider a simple example of the linear chains with three atoms as shown in Fig 3. When the force acting on the first atom is zero ($f_{12} + f_{13} = 0$), the atomic stress on the first atom parallel to the axis is given by

$$g_1 = -R\, f_{12}/2 = R\, f_{13}/2. \tag{4}$$

This shows that if the pair force between the nearest neighbor atoms f_{12} is repulsive (attractive) ,then the atomic stress is negative(positive). When the atomic stress is negative, the linear structure is clearly unstable in spite of the zero force acting on the first atom because the pair force between the first and third atoms is attractive.

Next, in Fig.3 we show the dependence of the atomic stress g_1 upon the number of electrons in the cluster. These values are calculated by using the model Hamiltonian of the K parameters in

145

table 1. Since the atomic stress is negative in the number of electrons from zero to $n_s (\sim 3)$, the linear structure is not stable. We may conclude that K_3+ is triangle. The g_1 becomes positive at the electron number $\bar{n}_s$. We may conclude that K_3- is linear. It is very sensitive to the damping factor λ whether the sign of the atomic stress in the half occupation number($= 3$) is positive or negative. With our values of parameters in Table 1, the neutral K_3 cluster is bent because the g_1 is negative. Figure 4 shows the calculated bond distance normalized by the equilibrium bond distance of diatomic molecules. For the number of electrons from zero to $n_b (\sim 2.3)$ and from some value(~ 5.7) to 6 the bond distance of the triatomic molecule is shorter than that of the diatomic molecule, while for the number of electrons from n_b to some value(~ 5.7) the bond distance becomes longer. The former corresponds to the occupation number of the Mg,Ca clusters, while the latter to the Na,K clusters. The bond distance of Fe, Co and Ni clusters behaves like the latter, since the number of d-electrons n is considered to be equivalent to the number of s-electrons n/5.

4. Stable geometries of Si and C microclusters

The static properties are given by solving the model Hamiltonian, but it is very complex to find the minimum points of the the total energy in a large cluster. We obtain the geometries of the Si,C clusters and their cations with the minimum energy by using the dynamical method by Car and Parrinello combined with the model Hamiltonian(Appendix I)[7]. This is very useful to study both the electronic state and the geometries of the large cluster.

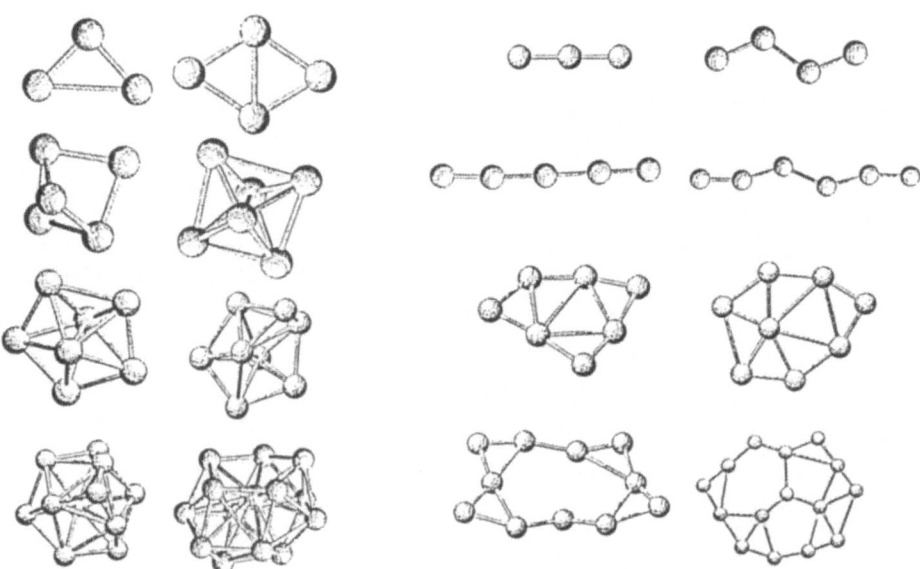

Fig.5. The geometries of Si_n.   Fig.6. The geometries of C_n.

The optimized stable geometries of several silicon clusters are shown in Fig 5. The minimum distance of each cluster increases with the size of the cluster. The distorted octahedron structure of the Si_6 cluster is different from that of the sp^3 bond which has the ring structure[8]. Triangles in the cluster contract by 20% in length as compared to those in the crystal (111) surface. This contraction attributed to the interaction between dangling bonds is important for silicon clusters of any size. The stable cluster Si_7 has the distorted-pentagon structure which may grow to the amorphous silicon in the crystal growth. We calculate the fragmentation energies which are defined as the total energy difference between the clusters of size n and n-1. The large fragmentation energies occur in the clusters, Si_4, Si_6, Si_{10}, and Si_{14}, which are more stable than the others[9,10]. The high stability of the clusters Si_4, Si_6 and Si_{10} corresponds to the experimental results[9,10].

Figure 6 shows the stable geometries of several carbon clusters. The C_3, C_4, C_5, and C_6 clusters which are linear show the bond alternation. The planer structure is more stable than the linear structure for C_n with $n \geq 7$. The carbon atoms in the ground state grow up in the rugged planar shape and tend to form the shape like the graphite structure. The large fragmentation energies in the neutral C clusters occur in the number of the atoms 2,3,5,7,12,15. These number are not always corresponding to the experiments of the ionized clusters[11] because the geometries of the C clusters are very sensitive to the ionicity of the clusters as you will see later.

Figures 7 and 8 show the doubly ionized clusters of Si and C atoms. The geometries of the Si^{2+} clusters are similar to those of the neutral clusters. However the C^{2+} clusters contain triangles being quite different from the neutral clusters.

Fig.7. The geometries of Si_n^{++}.   Fig.8. The geometries of C_n^{++}.

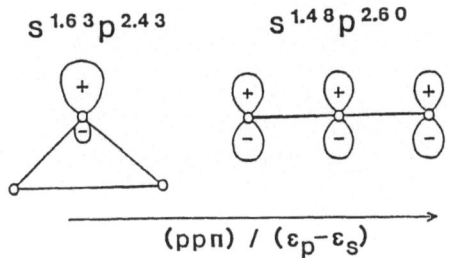

$$s^{1.63}p^{2.43} \qquad s^{1.48}p^{2.60}$$

(ppπ) / ($\varepsilon_p - \varepsilon_s$)

Fig.9. The localized orbitals and the occupation number of the s and p orbitals.

Figure 9 shows the occupied localized orbitals of the Si_3 in the bent and linear geometries. The other localized orbitals are disregarded as they are common in both the geometries. The s and p orbital occupations of the centered atoms are shown at the top of the figure. The occupancy of the p orbital in the linear shape is larger than that in the bent shape. This means the bent shape gains the total energy by the decrease of p- and s- orbital energy difference. On the other hand, the linear structure gains the total energy by the $(pp\pi)$ interaction. The parameter ratio $(pp\pi)/\Delta$ of the Si is in the bent region, while that of the C in the linear region. When the cluster is doubly ionized, the localized orbital is not occupied. So the $p\pi$ orbital perpendicular to the plane of this figure induces the geometry change from the linear to the bent structure to gain the transfer energy. When the length of the C cluster increases, the C cluster becomes bent at a certain length to gain the $pp\pi$ interaction energies perpendicular to the plane of Fig.8 . The beginning of the bent structure is related to the damping range λ .

5. Dynamical process

Electronic structures of the alkali-metal clusters have been calculated by the jellium model which seems to account for the stability of the microclusters[12]. The total energies per one atom show the minima at the observed magic numbers 2, 8, 20, and so on. However,I think the spherical jellium model is very questionable for the small microclusters less than 10 atoms. In Fig.10 the solid line shows the calculated total energies per atom of the K clusters with the assumed fcc structure which has the bond distance 4.5A. When the structure is optimized, the calculated total energies go down as shown in the dotted line. Before the optimization the total energy per atom of the K_4 is higher than that of the K_2. However, after the optimization its calculated total energy is lower than that of the two atoms. This shows the magic number is not always corresponding to the minima in the total energies per atom, but is related to the formation process. The dynamical process should be considered to explain the observed magic numbers.

To see the fragmentation motion of K_8 I have tried to carry out the calculation of the molecular dynamical processes on the adiabatic potential surface obtained by the transformation matrix method(Appendix II). The following characteristic features are obtained from the temperature dependence of the calculated pair distribution function as shown in Fig. 11. To see the many-body

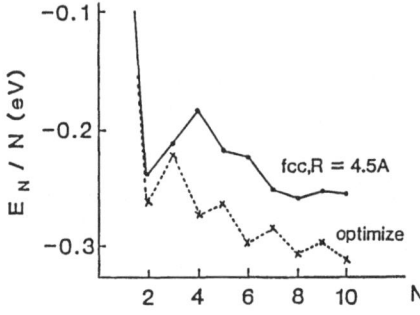

Fig.10. The binding energies per atom of K clusters.

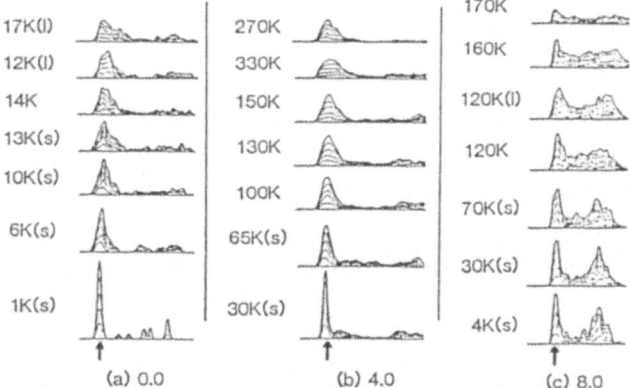

Fig.11. The temperature dependence of the pair distribution function for K_8. The number of electrons (a)0.0 (b)4.0 (c)8.0. The solid and liquid phases are denoted by the (s) and (l), respectively. The peak positions by the nearest neighbor pairs are shown by the arrows.

interaction we change the total number of electrons of the cluster as 0(case a),quarter occupation 4(case b) and half occupation 8(case c). The curves are obtained by summing up the pair distribution in every time step 1psec of the molecular dynamics. At 0 K the number of main strong peaks in the cases (a) and (b) is only one for the nearest neighbor pair because the structures are round. On the other hand the second peak for the next nearest neighbor pair in the case (c) is also large as well as the first peak for the nearest neighbor pair because the structure is linear. When the temperature is raised,most peaks in the case (a),(b) and (c) become broader and shift to the longer distance side(expansion). However, case (c) shows a different behavior. To the left hand side of the first peaks a new peak grows up with increasing temperature,that is to say,the contraction of the bond distance. This bond contraction is due to the collective movements of K_2 atoms which we call subcluster. The bond distance of the subcluster K_2 is smaller than that of the K_8 cluster(see Fig 2). This behavior is considered to be characteristic of the many-body interaction between the metal atoms. With the increasing temperature, the clusters are dissociated into these subclusters. The further details will be published elsewhere[13].

6. Conclusion

In summary,we have discussed the many body effects in microclusters. First,we have pointed out that there are negative and positive slopes in the change of the bond distance with that of the size. The negative(positive) slopes have been ascribed due to the positive(negative) atomic stress. Second,we have discussed geometries of Si and C clusters. The ratio $(pp\pi)/(\varepsilon_p - \sigma_s)$ is important for giving their geometries. Finally,the dynamical properties were treated in the adiabatic approximation. The results show the appearance of the subcluster in the heated clus-

ters. The cluster is dissociated into the subcluster in the high
temperature.

These calculations were performed on the HITAC S180 and M680
Computer Systems at the Institute for Molecular Science.

Appendix I   Dynamical method of Car and Parrinello

Total energy $E(\mathbf{R}, \phi)$ is the functional of the wave functions
which are given by the linear combination of the atomic orbitals:

$$\phi_i = \sum_\alpha c_{i\alpha} \chi_\alpha . \tag{A1}$$

The vector $\mathbf{R}$ represents the atom coordinates which are given as
functions of time t. When the total energy is minimized with
respect to the coefficients of (A1), these functions are the
eigenstates. The dynamical method uses the following dynamical
equation to minimize the total energy with respect to the wavefunctions:

$$\mu \ddot{c}_{i\alpha} = -\sum_\beta h_{\alpha\beta} c_{i\beta} + \sum_j \varepsilon_{ij} c_{j\alpha} , \tag{A2}$$

and at the same time looks for the minimum energies with respect
to the atom positions by using the equation of the molecular
dynamics:

$$m \ddot{\mathbf{R}}_\alpha = - \left(\frac{\partial E}{\partial \mathbf{R}_\alpha} \right)_\phi . \tag{A3}$$

the ε_{ij}'s are undefined lagrangian multipliers to conserve the
norm of the wavefunctions, which are given by the following equation:

$$\varepsilon_{ij} = \sum_{\alpha\beta} c_{i\alpha} h_{\alpha\beta} c_{j\beta} - \mu \sum_\alpha \dot{c}_{i\alpha} \dot{c}_{j\alpha} . \tag{A4}$$

These equations (A2) and (A3) are solved to find the minimum
energy. The friction terms are added to the above dynamical equations to slow down the velocity of the atom and to stop the atom
at the minimum point. As the minimum point is not always the
lowest minimum point, then we repeat the calculation with
different initial conditions. Finally we obtain the true eigen-
vector and the stable geometries in the ground state.

Appendix II   Transformation matrix method

At proper initial conditions, the following dynamical equation
are solved:

$$m\ddot{\mathbf{R}}_\alpha = - \frac{dE}{dR_\alpha} \tag{A5}$$

where the forces acting on the atoms are obtained by the Hellmann-Feynman theorem. In each time step of molecular dynamics, we
need look for the eigenfunctions (A1) of the Hartree-Fock equation. When the difference between the time step is very
small(about 10^{-15} sec), the change of the eigenfunction is small.
We solve the transformation matrices V between the old eigenfunction at time $t-\delta$ and the new eigenfunction at time t:

$$C_{i\alpha}(t) = \Sigma_{j} C_{j\alpha}(t-\delta)V_{ji}(t-\delta,t). \qquad (A6)$$

The transformation matrices are approximately unit matrices because of the smallness of the time difference δ. In this method every trajectory moves on the real adiabatic potential surfaces.

References

1. B.Delley,D.E.Ellis,A.J.Freemann,E.J.Baerends and D.Post: Phys.Rev.**B27**,2132(1983).
2. C.Satoko,B.Delley,and D.E.Ellis:unpublished.
3. E.Matsushita and T.Matsubara:Prog.Theor.Phys.**59**,15(1978).
4. C.Satoko:Phys.Rev.**B30**,1754(1984).
5. C.Satoko:In Dynamical Processes and Ordering on Solid Surface ed. by A.Yoshimori and M.Tsukada p104(Springer-Verlag 1985).
6. F.Ducastelle and F.Cyrot-Lackmann: J.Phys.Chem.Sol.**32**,285(1971).
7. R.Car and M.Parrinello:Phys.Rev.Letters **55**,2471(1985).
8. S.Saito,S.Ohnishi,C.Satoko and S.Sugano:J.Phys.Soc. of Japan **55**,1791(1986).
9. L.A.Bloomfield,R.R.Freeman and W.L.Brown:Phys.Rev.Letters **54**,2246(1985).
10. J.R.Heath,Yuan Liu,S.C.O'Brien,Qing-Ling Zhang,R.F.Curl, F.K.Tittel and R.E.Smalley:J.Chem.Phys.**83**,5520(1985).
11. E.A.Rohlfing,D.M.Cox and A.Kaldor:J.Chem.Phys.**81**,3322(1984).
12. W.D.Knight,K.Clemenger,W.A.Heer,W.A.Saunders,M.Y.Chouand and M.L.Cohen:Phys.Rev.Letters **52**,2141(1984).
13. C.Satoko:to be published.

Total Energy Surfaces: $(S_8)_n$

T.P. Martin, T. Bergmann, and B. Wassermann

Max-Planck-Institut für Festkörperforschung, Heisenbergstr. 1,
D-7000 Stuttgart 80, Fed. Rep. of Germany

1. Introduction

Although clusters are now routinely observed in mass spectro-
meters, their most fundamental property, their structure, is
still essentially unknown /1/. This is a particularly frustrat-
ing state of affairs for the theorists, who, if given the
structure, can use their sophisticated methods to calculate the
electronic and vibrational properties of clusters. Since experimental investigations have not yet provided much structural
information, the theorists have been forced to attack the
problem themselves. However, it may be a task which is just
beyond the present state of the art. This situation can be con-
trasted to that which exists in solid state physics where precise structural data is available. One can imagine the confusion which would exist today if the solid state physicist would
still have to rely on total energy calculations for the determination of complicated crystal structures. It was the development of diffraction techniques that finally allowed rapid
advances in solid state physics. Cluster science awaits a
similar breakthrough. In the meantime the cluster theorist
faces a challenge even more difficult than that of the solid
state physicist. Rather than examining a small set of structures, a vast multidimensional total energy surface must be
mapped out and examined /2-4/.

Any attempt to determine the properties of a total energy
surface encounters two fundamental difficulties. Suppose one
trys to systematically define the energy surface at the points
on a coordinate grid. Consider the seemingly modest goal of
constructing a grid with only ten intersection points along
each of the $3N-6$ axes. This means that for a 10 atom cluster
the total energy must be evaluated 10^{24} times. Even if this
were possible, which of course it is not, such a course mesh
of points would leave many important features hidden. The sheer
expanse of multidimensional space poses the first fundamental
computational problem.

The second difficulty is no less formidable. Each of the
total energy calculations must be carried out with a high degree of precision, because many features on the surface, e.g.
well depth and barrier heights differ by only a small fraction
of the total energy. An ab initio calculation of the total
energy of a ten-atom metal cluster, including the effects of
correlation, is a state-of-the-art task /5-7/ even if it is
performed for only one point on the energy surface. A complete
mapping of the surface using such techniques is unthinkable.

Fortunately, this second difficulty can be avoided for certain materials. If the atoms in a cluster interact with one another isotropically, it is possible to define a pair potential that depends only on the distance between the atoms. This is the case for rare gas clusters /8-12/ and alkali halide clusters /13-15/. Pair potentials have also been defined and used for certain molecular clusters /16-21/.

Careful investigations indicate that sulfur condenses to form at least 15 different solid allotropes and that liquid sulfur contains at least 30 different types of molecules /22,23/. Recent computer simulation studies have prepared the way for a better understanding of the transition from one form to another /24/. Under certain experimental conditions large sulfur clusters are known to exist in the vapor phase and to possess a building block containing 8 atoms /25/. In this paper, techniques for dealing with total energy surfaces will be applied to sulfur clusters composed of 8-atom, ring molecules.

2. Calculational Procedure

Sulfur has six outer electrons. However, four of these electrons form lone pairs which are important in determining the configuration of sulfur molecules but which do not participate in chemical bonding. The remaining two electrons form chemical bonds. This strong tendency to bond with two neighboring atoms explains why sulfur condenses into chains and rings. It also explains why the rings are not bonded with one another chemically but merely with weak van der Waals forces.

In these calculations we have assumed that the sulfur, 8-atom, ring molecule is a rigid unit. Each atom in the molecule interacts with each atom in all other molecules through a van der Waals potential with the form,

$$V_{ij} = A(a^{12}/r_{ij}^{12} - 2a^6/r_{ij}^6) \tag{1}$$

The Parameters A and a have been chosen to be 1.28×10^{-21} J and 3.5×10^{-8} cm, respectively. A similar potential has been fitted to the lattice constant and sublimation energy of orthorhombic sulfur and has then been used successfully to predict the vibrational frequencies, the elastic constants and the Grüneisen parameters /26/.
The most important property of the total energy surface is the location of the minima. This information was extracted in the following way: Each rigid sulfur molecule is assigned six coordinates, three center of mass coordinates and three Euler angles. The initial cluster configuration thus defined will not be a reasonable one because in these calculations the initial 6N-6 coordinates (i.e. the initial point in configuration space) are chosen using a random number generator. From this starting point a minimization routine is initiated. It is of great importance that this routine be not only fast, but also precise. Speed is essential in order to allow minimization from a large number of starting configurations. Only in this way can the multidimensional surface be adequately searched. Precision is essential to avoid the danger of falsely identifying a

saddle point as a true minimum. Therefore, the minimization routine first performed a course minimization using the steepest descents method which was followed by a refinement using a so-called Davidson-Flecher-Powell method /27/. Starting from a variety of initial configurations, the energy of a given minimum was reproducible to six digits and the vibrational frequency to four digits.

3. Energy Minima

If one tries to visualize a likely configuration of two ring molecules, without actually making a calculation, the first and perhaps the only candidate that comes to mind is the "symmetric stack", configuration B shown in Fig. 1. Intuition plays no tricks here. The symmetric stack does turn out to be highly stable. Three other less stable configurations are also shown in Fig. 1. However, the symmetric stack does not represent the global minimum on the $(S_8)_2$ surface. The most stable form is the "staggered stack" shown in Fig. 2. Notice that one sulfur atom in each ring nests in the center of the other ring. Not only does this configuration have the highest binding energy, but it is reached 600 times in 1014 minimizations from random initial configurations, whereas the symmetric stack is seldom the end product of the minimization procedure. It is reached only 15 times in 1014 minimizations. The number of times each of the five stable configurations is reached is shown in Fig. 3. Why is the symmetric stack "found" so few times even though it has such a high stability? In order to find the symmetric configuration, the two rings must slide over one another. However, such paths in configuration space encounter the four minima represented by the staggered stack. The catchment area of the symmetric stack is in this manner contained and limited in extent. The physical significance of relative catchment areas will be discussed below. First, consider the total energy surface of the trimer, $(S_8)_3$.

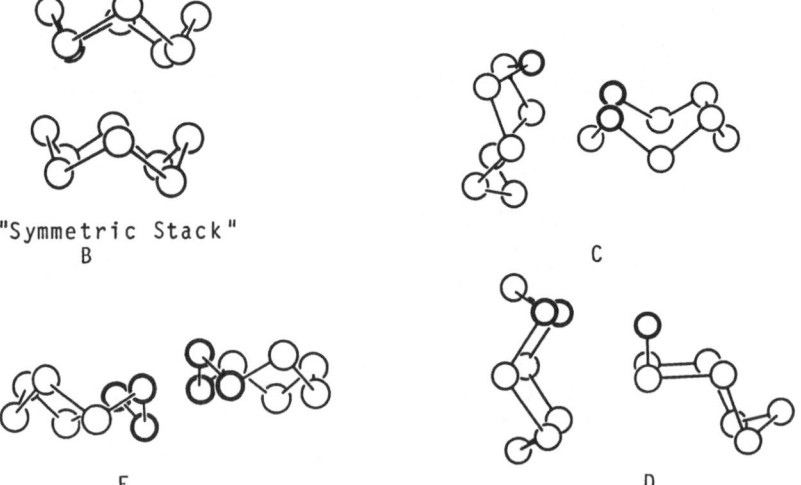

"Symmetric Stack"
B
C

E
D

Fig. 1 Dimer $(S_8)_2$ configurations at four local minima. Letters correspond to those in Fig. 3.

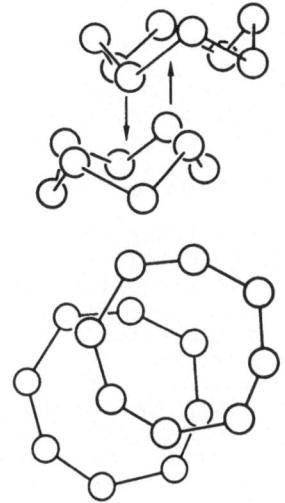

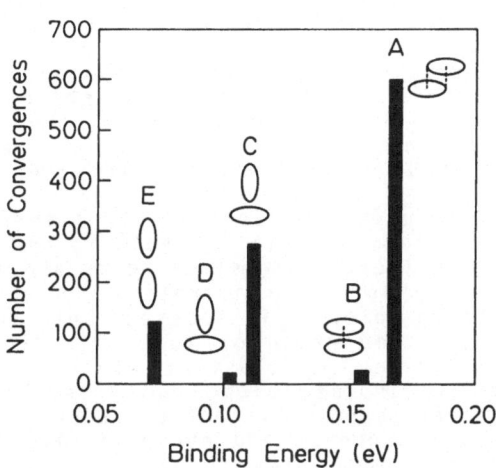

Fig.2 "Staggered Stack" is the most stable form of the sulfur dimer.

Fig.3 The number of times each stable $(S_8)_2$ configuration is reached after minimization from 1014 random starting configurations.

When the total energy of $(S_8)_3$ was minimized from 2200 random starting configurations, 60 distinct local minima were found. The number of times a given configuration was reached is plotted as a function of the binding energy in Fig. 4. The most stable configuration is once again the staggered stack. However, it was obtained as an end configuration, only about fifty times in 2200 tries.

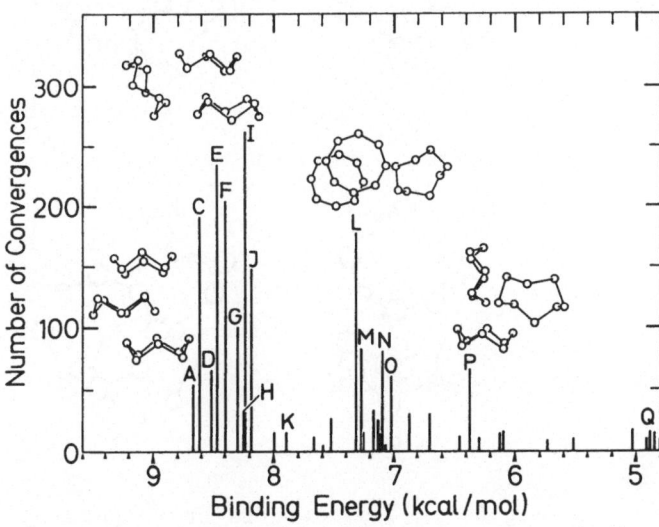

Fig.4 The number of times each stable $(S_8)_3$ configuration was reached after minimization from 2200 random starting configurations.

155

It would not be particularly instructive to present all 60 stable configurations of the trimer. However, certain generalizations can be made. Comparing the binding energies of the configurations, it becomes apparent that they can be divided into three groups. The most stable, 8-9 Kcal/mol, use the staggered-stack dimer as a seed. The third molecule can then be "pasted on" at various points. For example, the stack can be lengthened one more unit. This can be done in two different ways, Fig. 5. It becomes apparent that the staggered-stack tends to grow as a spiral with a periodicity of either four or eight molecules. The "A" form is the most stable configuration of the trimer. These stack forms are actually the exception for the group under consideration. In all other cases the face of the third molecule is "pasted" onto the side of the dimer seed, Fig. 6. The group of local minima below 5 Kcal/mole can be characterized as clusters containing three molecules bonded with one another along the molecular edges. Only small displacements are sufficient to cause these structures to collapse into more compact and energetically favorable forms. That is, their catchment area is small. Therefore, open structures are rarely the end-product of a minimization.

We have seen that the staggered-stack configuration is the most stable form of $(S_8)_2$. In addition, perfect crystals of orthorhombic and monoclinic sulfur are known to contain staggered-stacks of infinite length, Fig. 7. Therefore, it would be easy to think of cluster growth in terms of staggered stacks growing longer and longer. However, this picture is

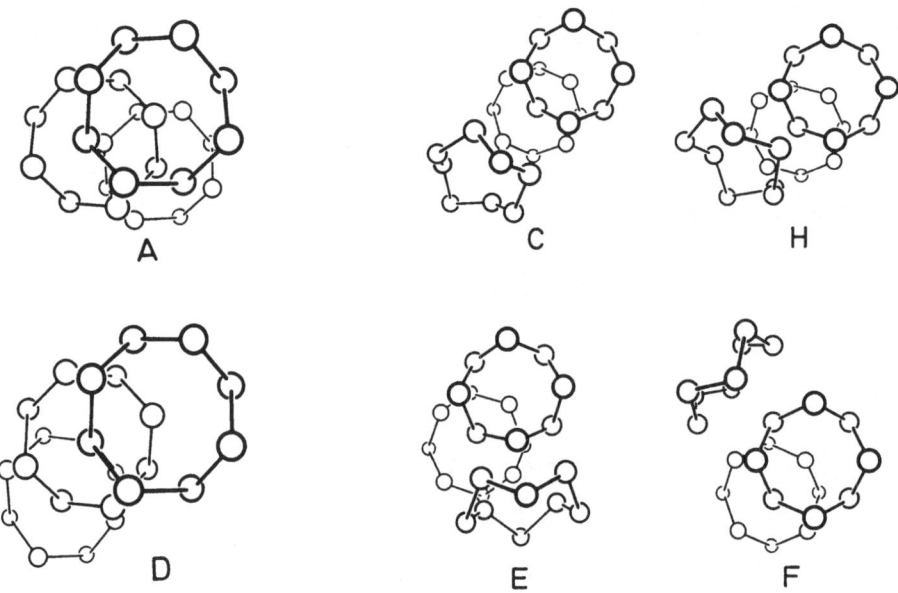

Fig.5 The "staggered stack" can be continued in two ways.

Fig.6 $(S_8)_3$ structures corresponding to local minima.

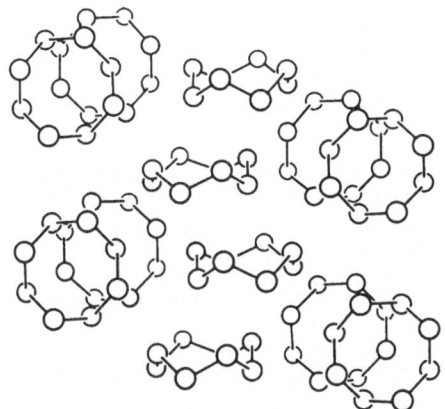
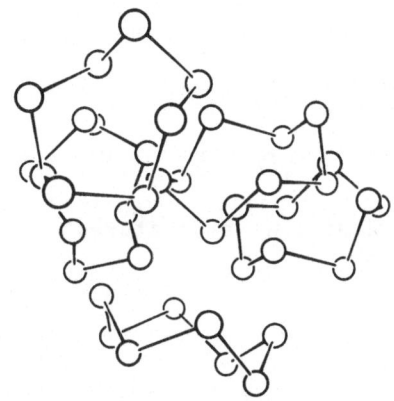

Fig. 7 Orthorhombic
       crystalline sulfur

Fig. 8 Most stable configuration
       of $(S_8)_5$

incorrect. "Linear" growth stops at the trimer. Large clusters
tend to be more close-packed. The most stable form of the pentamer is shown in Fig. 8. It is important to remember that the
structure of clusters does not necessarily resemble that of
crystals.

4. Catchment Area

Another important characteristic of a total energy surface
is often called the "catchment" region /4/. Imagine rain falling onto the surface. Every raindrop, no matter where it hits
the surface, will eventually find its way to a minimum. That
area of the surface which collects rain drops for a specific
minimum is said to define a catchment area. All of configuration space can be uniquely divided up into a set of such areas.
In this way each point in configuration space can be said to
belong to a given minimum (except, of course, those points
defining the boundary line between catchment areas). In the
above discussion the concept of catchment area proved useful in
a qualitative description of a total energy surface. We will
now attempt to use the concept in a more quantitative way.

Catchment area has a particularly simple meaning for the
case of a microcanonical ensemble. In such an ensemble each
point in configuration space is equally probable. Therefore,
the probability of the cluster having a given configuration is
simply proportional to the length of a constant energy contour
in the corresponding catchment area. This interpretation must
be only slightly modified when considering the canonical ensemble. In this case the probability of reaching a given point in
phase space must be weighted by exp(-H/KT). Specifically, the
probability that a cluster occupies catchment region "a" relative to the probability that it occupies region "b" is just,

$$P_a/P_b = \int_a \exp(-H/KT)dq^3dp^3 / \int_b \exp(-H/KT)dq^3dp^3 \qquad (2)$$

For the purposes of this discussion we will assume that the integration over momentum coordinates is independent of the configuration since both configurations have the same mass. However, the configurations will not, in general, have the same moment of inertia.

The integration over configuration coordinates is easily carried out if it is further assumed that each catchment region has a parabolic form. This "harmonic" assumption is certainly valid at low temperatures.

$$V_a = E_a + \sum_i \omega_a^2(i) q_a^2(i)/2m \qquad (3)$$

where $\omega_a(i)$ is the ith vibrational frequency in the parabolic catchment region belonging to minimum a. The corresponding displacement amplitude has been denoted with $q_a(i)$. Using this expression the integral in Eq. 2 is easily evaluated to give

$$P_a/P_b = \exp[(E_b - E_a)/KT] \prod_i \omega_b(i)/\omega_a(i) \qquad (4)$$

Here we have used the classical expression for the energy of a harmonic oscillator in order to retain a simple geometrical description of the probabilities. The quantum mechanical result is obtained if each ω in the product in Eq. 4 is replaced with a corresponding $1-\exp(-\hbar\omega/KT)$. Notice that the relative probability of a cluster being in a given minimum is dependent on the depth of the minimum (through the Boltzmann factor) and on the extent of the catchment region (through the product of inverse frequencies). Just one very low frequency vibration can, at non-zero temperature, stabilize an energetically unfavorable configuration.

Now it is possible to state and understand a rather unusual observation. Cluster configurations corresponding to shallow minima tend to become more favorable at high temperatures. Cluster configurations corresponding to deep minima are less likely to be found at high temperatures. The reason for this behavior is the following. Energetically unfavorable clusters sitting in shallow minima usually have open structures which possess incipient instabilities. Such clusters resist only weakly certain types of deformations, i.e. they have low frequency vibrations. The entropy generating, low frequency vibrations tend to make the configuration more favorable as the temperature increases. In principal, that configuration with the lowest frequencies will always become the most favorable if the temperature is made high enough. We will now present a specific example to illustrate this point.

Consider again the case of the sulfer dimer $(S_8)_2$. The total energy surface of this cluster has four local mimima in addition to the global minimum, the "staggered stack". When describing the temperature dependence of the favorability of these four configurations, it is usual to formulate the discussion in terms of free energy rather than probability. However, the conversion is simple,

$$P_a/P_b = \exp[(F_b - F_a)/KT] \qquad (5)$$

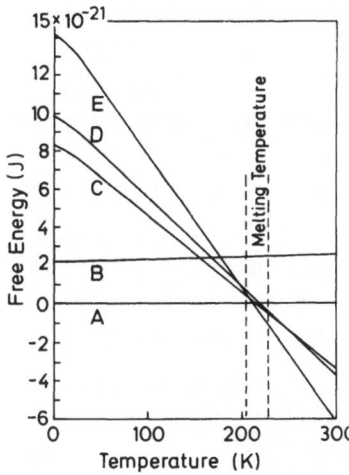

Fig. 9 The relative free energy of each type of dimer as a function of temperature. The letters correspond to those in Figs.1 and 2. Near 200 K configurations A,C,D and E become equally probable.

It should be remembered that this is not a usual definition of free energy because the integration in Eq. 2 is made over one catchment region only. The temperature dependence, in the harmonic approximation, of the free energy for each of the five stable configurations is shown in Fig. 9. Since normalization always presents a problem in this type of calculation, free energy differences have been plotted using the staggered-stack, configuration A, as reference. Although configuration A has the lowest free energy at low temperature, configurations C, D and E have several very low frequency vibrations drastically lower ing their free energy at high temperatures. All four free ener gy curves converge to the same value at a temperature just over 200 K, i.e. at this temperature all four configurations are equally probable. In this sense one might speak of a "melting" of the cluster at this temperature.

5. Concluding Remarks

The use of a classical interaction potential makes it possible to search the total energy surface of a large cluster relative ly thoroughly. However, such a search is meaningless if the surface has no basis in reality. Therefore, the classical approach should proceed hand-in-hand with quantum mechanical ab initio calculations. The classical potential should be fitted to realistic ab initio results for small cluster structures which sample a wide region of configuration space including many different interatomic distances and coordinations. The resulting interaction potential can then be used to pinpoint likely minima for large clusters. If possible, the classical results should then be further refined with a few selected ab initio calculations for the large clusters.

It is, without a doubt, convenient to think of a cluster of a given size as having a specific geometry. Unfortunately, this simplified way of thinking does not lead to an accurate de scription of a collection of clusters at finite temperature. The total energy surface of a cluster contains many local minima, most of which play an important role in determining cluster geometry. The calculational effort requred to characterize all of these minima is enormous because of the multidimensionality of configuration space. Numerical procedures which are simple in 3 dimensions become unthinkable in 100 dimensional space. However, some progress can be made through the use of two-body interaction potentials applied to clusters of small size.

References

1. The proceedings of three international conferences on clusters published in: J. Phys. (Paris), Suppl.C2 (1977); Surf. Sci. 106 (1981); 156 (1985)
2. M.R. Hoare: Adv. Chem. Phys. 40, 49 (1979)
3. T.P. Martin: Phys. Rep. 95, 167 (1983)
4. F.H. Stillinger and T.A. Weber: Science 225, 983 (1984); Phys. Rev. A 25, 978 (1982); 28, 2408 (1983)
5. J. Koutecky and P. Fantucci: Chem. Rev. (1986)
6. J. Flad, G. Igel-Mann, H. Preuss and H. Stoll: Surf. Sci. 156, 379 (1985)
7. P.S. Bagus, C.J. Nelin and C.W. Bauschlicher,jr.: Surf. Sci 156, 615 (1985)
8. C.L. Briant and J.J. Burton: J. Chem. Phys. 63, 2045 (1975)
9. R.D. Etters and J. Kaelberer: J. Chem. Phys. 66, 5112 (1977)
10. E.E. Polymeropoulos and J. Brickmann: Chem. Phys. Letters 92 59 (1982)
11. J. Farges, M.F. Feraudy, B. Raoult and G. Torchet, Surf. Sci. 106, 95 (1981)
12. E. Blaisten-Barojas and H.C. Andersen: Surf. Sci. 156, 548 (1985)
13. T.P. Martin: J. Chem. Phys. 67, 5207 (1977)
14. D.O. Welch, O.W. Lazareth, G.J. Dienes, R.D. Hatcher: J. Chem. Phys. 68, 2159 (1978)
15. B.I. Dunlap: J. Chem. Phys. 84, 5611 (1986)
16. F.H. Stillinger and A. Rahman: J. Chem. Phys. 60, 1545 (1974)
17. J.R. Reimers and R.O. Watts: Chem. Phys. 85, 83 (1984)
18. V. Carravetta and E. Clementi: J. Chem. Phys. 81, 2646 (1984)
19. I.P. Buffey and W. Byers Brown: Chem. Phys. Lett. 109, 59 (1984)
20. M.J. Ondrechen, Z. Berkovitch-Yellin and J. Jortner: J. Am. Chem. Soc. 103, 6586 (1981)
21. B.W. van de Waal: J. Chem. Phys. 79, 3948 (1983)
22. R. Steudel, R. Strauss and L. Koch: Angew. Chem. Int. Ed. Engl. 24, 59 (1985)
23. R. Steudel: Nova Acta Leopoldina 59, 231 (1985)
24. E. Blaisten-Barojas: to be published
25. T.P. Martin: J. Chem. Phys. 81, 4426 (1984)
26. R.P. Rinaldi and G.S. Pawley: J. Phys. C 8, 598 (1975)
27. J. Stoer: Einführung in die Numerische Mathematik I; 3rd ed. (Springer, Berlin, Heidelberg 1979)

Theoretical Study on Photodissociation Processes of Lithium Metal Microclusters

O. Sugino and H. Kamimura

Department of Physics, University of Tokyo, Bunkyo-ku, Tokyo 113, Japan

It is first shown by the MCSCF variational method that an optimized set of effective one-electron orbitals is obtained for the many-electron states of Li_4 microcluster and that its lower excited and ground states correspond to the multiplets consisting of mainly a single electron configuration. Then the mechanisms of photodissociation are investigated by calculating the adiabatic potentials for two dissociation processes of $Li_4 \rightarrow Li_3 + Li$ and $Li_4 \rightarrow 2Li_2$. It is concluded that photodissociation occurs with the loss of one atom rather than two atoms.

1. Introduction

In 1984, KNIGHT et al. [1] found the existence of magic numbers in alkali-metal microclusters in photoionization experiments using the supersonic cluster beam technique. For example, in sodium microclusters the magic number is 20, 40, 56 etc. These magic numbers suggest the existence of stable clusters in their ground states. In order to explain this fact they adopted a simple shell model of jellium droplets and showed that the magic number corresponds to the number of electrons in a closed shell.

On the other hand, another kind of magic numbers, magic fragments, at which a microcluster is dissociated efficiently has been suggested by BLOOMFIELD et al. [2], who worked on fission experiments on silicon and carbon microclusters. According to their experiments, carbon clusters dissociate in units of C_3 when a laser beam hits them, while in silicon clusters Si_4^+ is very stable against photodissociation. In order to explain the existence of magic fragments RAGHAVACHARI et al. [3] calculated the total energy of silicon microclusters, and then they discussed whether dissociation is possible or not by simply comparing the ground state energy at the optimized geometry of microclusters in an initial state with the sum of ground state energies of two dissociated clusters which appear in a final state. In their calculations the investigation on dissociation mechanisms in intermediate processes from an initial state to a final state has not been well taken into account. Such investigation is essential and important when one tries to clarify a mechanism which yields the magic fragments in photodissociation.

Recently BRUCAT et al. [4] carried out similar experiments on transition-metal microclusters. They found that Fe, Ni and Nb clusters always dissociate with a loss of one atom. But as far as we know, no clear-cut theoretical approach has been proposed as regards mechanisms of the photodissociation of transition-metal microclusters.

In this paper we investigate mechanisms by which magic numbers or magic fragments appear in excited as well as ground state s of metallic microclusters. For this purpose we choose the simplest metallic microcluster of Li_4 as an object of study, perform the ab initio total energy calculations of ground and excited many-electron states using a variational technique, and investigate whether the electronic shell model based on the one-electron picture holds or not in a microcluster system and how microclusters in their excited states dissociate by the laser illumination.

2. Theoretical Method

Since the electron-electron interactions play an essential role in microclusters, it is important in developing a theoretical formulation to take into account the effects of electron-electron interaction s as accurate as possible. From this standpoint we adopt the multiconfiguration self-consistent variational method [5] which we abbreviate MCSCF hereafter.

Let us describe this method briefly. First we assume that even in a many-electron system a set of effective one-electron orbitals $\{\phi_i(r_i)\}$ exists. Then by distributing electrons in these orbitals, a number of electron configurations appear. Each electron configuration can be expressed by a Slater determinant $\Phi_\alpha(\phi_i \chi_j, \ldots)$ consisting of one-electron spin-orbitals $\{\phi_i(r)\chi_i(\sigma)\}$, with spin function $\chi_i(\sigma)$. Then a many-electron eigen-function ϕ of a Schrödinger equation

$$H\Psi = E\Psi \tag{1}$$

is expanded in terms of the above Slater determinants as follows;

$$\Psi = \Sigma_\alpha C_\alpha \Phi_\alpha(\phi_i \chi_i). \tag{2}$$

We determine coefficients C_α and effective one-electron orbitals ϕ_i so as to minimize the ground state energy

$$E_0 = <\Psi \mid H \mid \Psi> / <\Psi \mid \Psi>. \tag{3}$$

Then, by using the set of effective one-electron orbitals thus determined, we vary coefficients C_α so as to optimize the energy of respective excited states. On doing so we obtain the following algebraic equation

$$\Sigma_\beta \{<\Phi_\alpha \mid H \mid \Phi_\beta> - E\} C_\beta = 0 \tag{4a}$$

and the secular equation

$$\det \mid <\Phi_\alpha \mid H \mid \Phi_\beta> - E \mid = 0. \tag{4b}$$

By solving equations (4a) and (4b) we obtain the eigenfunctions and eigenenergies of excited states.

3. Electronic States and Adiabatic Potentials of Li₄ along Its Dissociation Paths

In this section we apply the above-mentioned method to a dissociation process of Li_4. A Li_4 microcluster is the simplest metallic microcluster system for calculating the electronic structures of a metallic microcluster and for investigating its dissociation processes by light. The stable geometry of Li_4 molecule in its ground state was first calculated by BECKMANN et al. [6], and it is a rhombus shape, as shown in Fig.1. First we calculate the ground state energy for this stable shape and determine the set of effective one-electron orbitals by the above variational method. In doing so, we have adopted a LCAO form (the linear combination of atomic orbitals) as a trial functional form of an effective one-electron orbital, i.e.

$$\phi_i = \Sigma_{\eta,\rho} D^i_{\eta\rho} \phi_\eta(r - R_\rho) \tag{5}$$

where $\phi_\eta(r - R_\rho)$ is an η type atomic orbital of an atom at R_ρ. As regards ϕ_η, we adopt the Gaussian-type orbitals of 1s, 2s, 2p, 3s and 3p of Li atom [7].

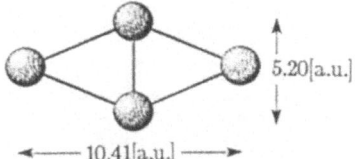

5.20[a.u.]

◄——— 10.41[a.u.] ———►

Fig.1 The stable geometry of Li_4 molecule

We consider the following two dissociation processes of Li_4 microcluster: (a) $Li_4 \rightarrow Li_3 + Li$ and (b) $Li_4 \rightarrow 2Li_2$. Then we investigate which of these two processes contributes to photodissociation. For this purpose we calculate the energies of the excited states of Li_4 clusters along the above two dissociation paths. These dissociation paths are shown in Fig.2, where in the path(a) Li atom is removed along a longer diagonal direction of the rhombus and in the path(b) two Li_2 molecules are separated so as to keep C_{2h} symmetry. In both cases the relaxation of bond lengths in a Li_3 triangle and in Li_2 molecules are not taken into account.

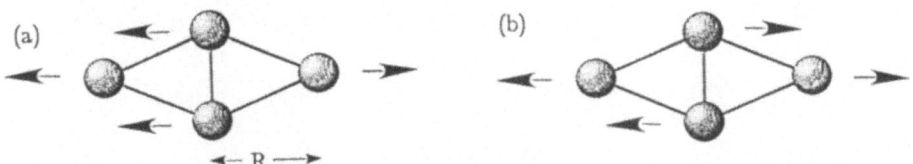

(a)

(b)

◄— R —►

Fig.2 The dissociation paths of (a) $Li_4 \rightarrow Li_3 + Li$ and (b) $Li_4 \rightarrow 2Li_2$

(a) $Li_4 \rightarrow Li_3 + Li$ dissociation path: First let us present the calculated results of an adiabatic potential of each eigenstate in a dissociating system of $Li_3 + Li$ as a function of R, where R is the distance between Li_3 triangle and Li atom, as shown in Fig.2(a). Before doing so, we first show some of the effective one-electron orbitals determined by the MCSCF method in Fig.3. In this figure ϕ_1 to ϕ_6 represent the wave functions of effective one-electron states with the increase of energy for two values of R, 6.4 a.u. and 9.4 a.u. Black and white parts represent the highest amplitudes of the wave function with plus and minus signs, respectively. As seen in this figure, ϕ_1 corresponds to the orbital which has bonding character in the left-side Li_3 triangle, while ϕ_2 has bonding character in the right-side one. ϕ_3 and ϕ_4 have a nodal line along the longer diagonal line of a rhombus.

Since the system of $Li_3 + Li$ has C_{2v} symmetry, each many-electron state of this system is classified according to the four irreducible representations of the C_{2v} point group, A_1, A_2, B_1 and B_2. The adiabatic potentials of the eigenstates with $^1A_1$, $^1B_2$ and $^1B_1$ symmetry are shown in Fig.4 as a function of R. The ground state is $^1A_1$ and the first excited state is $^1B_2$. In an equilibrium rhombus of Li_4 molecule the ground state energy is -29.819 a.u. and R=5.205 a.u. The first excited state $^1B_2$ lies at 0.064 a.u.(=1.25eV) above the ground state.

(a) 6.4 a.u.

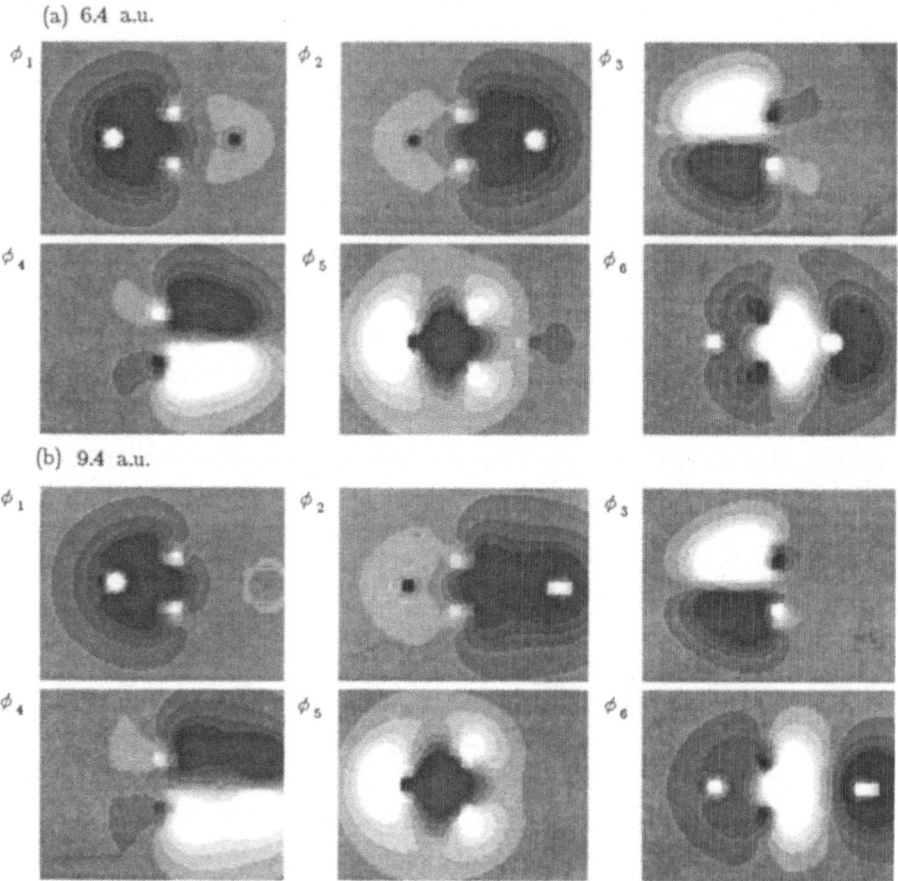

(b) 9.4 a.u.

Fig.3 The shape of the effective one-electron orbitals ϕ_1 to ϕ_6 for two values of R: (a) 6.4 a.u. and (b) 9.4 a.u.

The lowest four states of $^1A_1$, E_0, E_1, E_2, and E_3, correspond mainly to the electron configurations C_0, C_1, C_2, and C_3, where $C_0=(\phi_1)^2(\phi_2)^2$, $C_1=(\phi_1)^1(\phi_2)^1(\phi_3)^1(\phi_4)^1$, $C_2=(\phi_1)^2(\phi_2)^1(\phi_5+\phi_6)^1$, and $C_3=(\phi_1)^2(\phi_2)^1(\phi_5-\phi_6)^1$, respectively. The energies of these electron configurations are thought to vary with R, as shown in Fig.5. Because of interactions in the crossing regions of two energy curves, the energy curves are repelled with each other, and thus the adiabatic potential curves shown in Fig.4 are obtained.

From this behavior we can conclude that the effective one-electron picture holds very well not only for the ground state but also for several lower excited states and that only in the crossing region the configuration mixing becomes important. Although this conclusion has been obtained for a Li_4 microcluster, it may be considered that this feature appears also in larger microclusters, so that we can reasonably say that the electronic shell model is a good approximation for expressing the electronic structures of metallic microclusters if we adopt the effective one-electron orbitals ϕ_i obtained in the MCSCF method as the one-electron orbitals in the shell model.

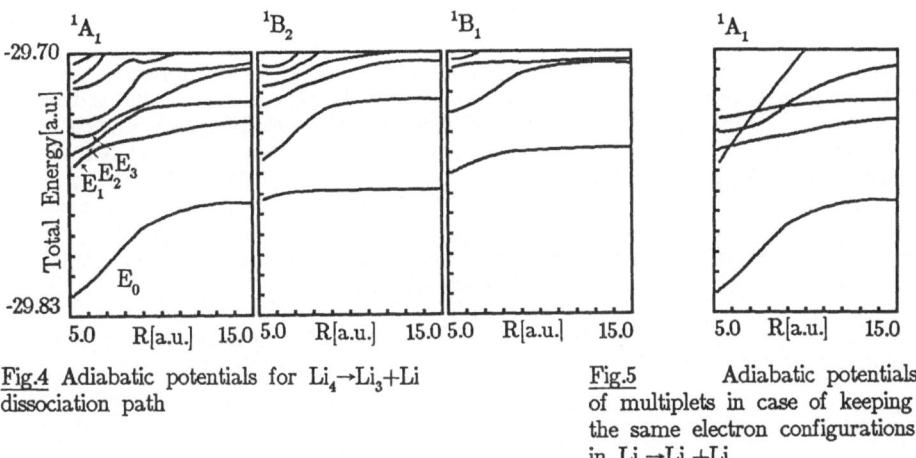

Fig.4 Adiabatic potentials for Li$_4$→Li$_3$+Li dissociation path

Fig.5 Adiabatic potentials of multiplets in case of keeping the same electron configurations in Li$_4$→Li$_3$+Li

Now we discuss possible photodissociation paths. Since the optical transitions are allowed to $^1A_1$, $^1B_1$ and $^1B_2$ from $^1A_1$ ground state, let us look at the adiabatic potentials for $^1A_1$, $^1B_1$ and $^1B_2$ states. Then we find that the first excited state of $^1B_2$ is almost constant as a function of R. If we allow the relaxation of Li atoms in the Li$_3$ triangle in the Li$_4$→Li$_3$+Li dissociation process, we can expect that the energy of the first excited $^1B_2$ state is lowered with increasing R so that Li$_4$ microcluster is easily dissociated into Li$_3$ and Li by light. This is also understood from the nature of orbitals involved in the electron configurations: The first excited $^1B_2$ state corresponds to the electron configuration $(\phi_1)^2(\phi_2)^1(\phi_3)^1$. Because the $^1B_2$ first excited state contains only one bonding ϕ_2 orbital while the $^1A_1$ ground state contains two bonding ϕ_2 orbitals, this $^1B_2$ state is easily dissociated while the $^1A_1$ ground state is not.

(b) Li$_4$→2Li$_2$ dissociation path: Now let us investigate another photodissociation process of Li$_4$→2Li$_2$. For this purpose we try to find a set of optimized effective one-electron orbitals and calculate the adiabatic potentials of various excited states and the ground state as a function of the distance between two Li$_2$ molecules which are being separated. The calculated adiabatic potentials are shown in Fig.6 as a function of R. As seen in this figure, all the adiabatic potentials of the excited states in this case increase their energy sharply with increasing R. Thus we can conclude that the photodissociation in the Li$_4$→2Li$_2$ path is difficult to occur. In this view the photodissociation of Li$_4$ microcluster occurs preferentially through the dissociation process from Li$_4$ to Li$_3$+Li , that is with the loss of one atom rather than two atoms. This situation is very similar to those found in photodissociation experiments of transition-metal microclusters.

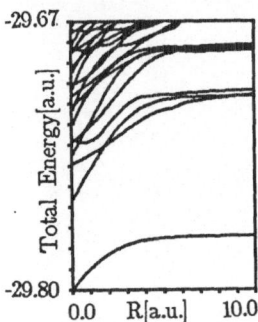

Fig.6 Adiabatic potentials for Li$_4$→2Li$_2$ dissociation : In this calculation a simpler basis set [8] is adopted.

4. Conclusions

We have found a set of effective one-electron orbitals in Li_4 microclusters by the MCSCF variational method and showed that lower excited states as well as the ground states correspond to the multiplets consisting of mainly a single electron configuration in this effective one-electron orbital scheme. From this fact we have concluded that the electronic shell model is a good approximation in metallic microclusters. Then we have investigated the mechanisms of photodissociation by calculating the adiabatic potentials for two dissociation processes of $Li_4 \rightarrow Li_3 + Li$ and $Li_4 \rightarrow 2Li_2$, and we have concluded that the photodissociation of Li_4 microcluster occurs preferentially in the process of $Li_3 + Li$, that is, with the loss of one atom rather than two atoms.

We would like to thank Prof. Shigeki Kato, Dr. Nobuyuki Shima and Mr. Shinji Tsuneyuki for valuable discussions.

References

1. W. Knight, K. Clemenger, W. de Heer, W. Saunders, M. Chous, and M.L. Cohen: Phys. Rev. Lett. **52**, 2141 (1984)
2. L.A. Bloomfield, R.R. Freeman, and W.L. Brown: Phys. Rev. Lett. **54**, 2246 M.E. Geusic, T.J. Mcllrath, M.F. Jarrold, L.A. Bloomfield, R.R. Freeman, and W.L. Brown: J. Chem. Phys. **84**, 2421 (1986)
3. K. Raghavachari and V. Logovinsky Phys. Rev. Lett. **55**, 2853 (1985)
4. P.J. Brucat, L.-S. Zheng, C.L. Pettiette, S. Yang, and R.E. Smalley: J. Chem. Phys. **84**, 3078 (1986)
5. S. Kato, K. Morokuma: Chem. Phys. Lett. **65**, 19 (1979)
6. H.-O. Beckmann, J. Koutecky, and V.B. Koutecky: J. Chem. Phys. **67**, 119 (1979) H.-O. Beckmann, J. Koutecky, P. Botochwina, and W. Meyer: Chem. Phys. Lett. **67**, 117 (1979)
7. S. Hujinaga : J. Chem. Phys. **53**, 2823 (1970)
8. J.D. Dill and J.A. Pople: J. Chem. Phys. **62**, 2921 (1975)

Ionization of Clusters in Collision with High-Rydberg Excited Rare Gas Atoms

T. Kondow

Department of Chemistry, Faculty of Science, The University of Tokyo, Bunkyo-ku, Tokyo 113, Japan

1. INTRODUCTION

In the studies of electron attachment processes to neutral clusters, nega-tive-ion mass spectrometry is used universally. One of the most fundamental issues is how efficiently electrons can be attached to the neutral clusters with a lesser extent of dissociation. Very slow electrons having a narrow energy spread are necessary to achieve this goal, since the cross sections for the attachment increase greatly with decrease in the incident electron energy, and consequently the excess energy due to the electron attachment can be minimized by reduction of the kinetic energy of the incident elec-trons. The basic principle of our ionization method is to use high-Rydberg rare gas atoms whose outermost electrons (Rydberg electrons) are trans-ferred to the weakly bound clusters.

In the collisional process involving a high-Rydberg rare gas atom, Rg^{**}, with a target molecule, M, its Rydberg electron is readily liberated as a result of rotational or vibrational de-excitation of M [1]. In particular, when M has a positive electron affinity, the Rydberg electron of Rg^{**} is attached gently to M with high efficiency and a transient negative ion state, M^{-*}, is formed. In such a collision, an effective distance for interaction between M and the Rydberg electron is much smaller than the radius of the Rydberg orbit, and consequently, the collision occurs almost exclusively with the Rydberg electron and the core ion simply behaves like a spectator. Therefore, the Rydberg electron can be treated as a free electron which has momentum and energy distribution identical with those of the Rydberg electron (essentially-free-electron model) [1,2].

The collisional electron transfer from Rg^{**} to van der Waals clusters has been examined and negative cluster ions have been detected by mass spectroscopy. When clusters having positive vertical electron affinities are used as targets, many-cluster ions have been produced efficiently and gently [3].

The design and performance of the apparatus used for this study are described by use of formation of cluster ions, $(CCl_4)_n Cl^-$ from $(CCl_4)_m$ in collision with Kr^{**}. Attachment of a Rydberg electron to a CO_2 cluster is shown as a test example, in comparison with attachment of free electrons having a kinetic energy of 1 eV to the same cluster.

2. IONOZATION PROCESSES

The electron-transfer collisions between electron-attracting molecules and Rg^{**} [1,2] indicate that these negative cluster ions are produced via the following processes:

$$(M)_m + Rg^{**} \rightarrow (M)_m^{-*} + Rg^+ \qquad (1)$$

$$(M)_m^{-*} \rightarrow (M)_n^- + (m-n)M \qquad (2)$$

$$\rightarrow \text{intracluster reactions} \qquad (3)$$

As discussed in previous papers [3,5], the vertical electron affinity of $(M)_m$ should not be negative, so that the electron attachment (process 1) proceeds with sufficiently high efficiency. On the other hand, when $(M)_m^{-*}$ is stabilized, the excess energy generated is transmitted to the internal degrees of freedom of the cluster, and the effective temperature of the cluster increases. If the temperature does not exceed the sublimation temperature of the cluster, no substantial evaporation is expected to occur (non-evaporative). Otherwise, a number of the component molecules are evaporated (evaporative) or the excess energy is released by the rupture of the chemical bonds of the component molecules (dissociative).

3. EXPERIMENTAL ASPECTS

3.1 Instrumentation

A schematic diagram of the apparatus is shown in Fig. 1. The apparatus consists of (1) a cluster beam source, (2) a triple-grid ion source, (3) a quadrupole mass spectrometer, and (4) a CAMAC system based on an LSI-11/23 microcomputer.

Formation of neutral clusters

Van der Waals clusters are produced from a sonic nozzle having an aperture of 50 μm diameter and a channel length of 0.2 mm. A sample gas is seeded in He, Ar or H_2 gas at room temperature with a stagnation pressure of 100 ∿

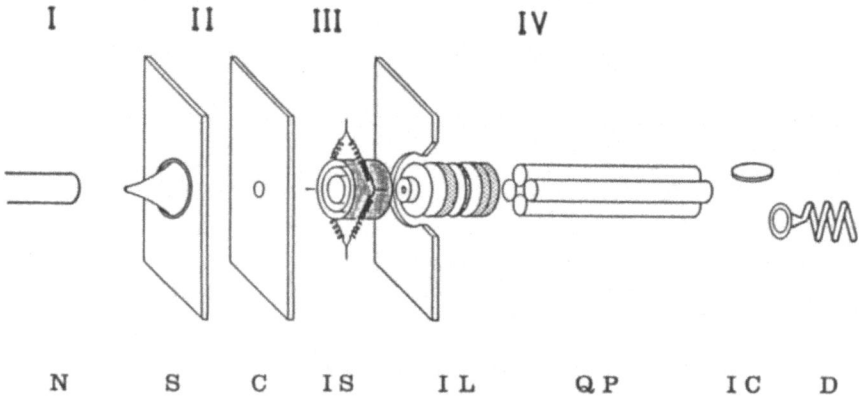

Fig. 1. A schematic diagram of the apparatus. It is composed of four chambers, I-IV, pumped separately, N: sonic nozzle, S: skimmer, C: collimator, IS: ion source, IL: ion lens, QP: quadrupole mass spectrometer, IC: conversion dynode, D: ion detector, P: oil diffusion pumps.

3000 torr, and the seeded gas is expanded through the nozzle into chamber I having a pressure of 10^{-4} torr. The central portion of the expansion is sampled by a conical skimmer (Beam Dynamics) and is further collimated by an aperture of 5 mm diameter into an ionization chamber II.

Ionization of clusters

The clusters are ionized in collision with Rg^{**} in a triple-grid ion source mounted in chamber III. The ion source has three concentric cylindrical grids and filaments (see Fig. 2). The grids, G_A, G_B and G_C, are made of stainless steel mesh. The four pieces of thoriated tungsten filaments form a rectangular square. The cluster beam passing through the collision region is ionized by impact of Rg^{**} atoms or electrons. Argon or Krypton gas with more than 99.95% purity is introduced to the ion source. The rare gas atoms are excited in the exterior of G_B by 50 eV-electrons. Ionic species and electrons are retarded by application of appropriate potentials to the three grids, and only neutral species including Rg^{**} are allowed to enter the central collision region. The pushing pressure of the rare gas introduced to the ion source is typically 0.05 - 0.2 torr. The pressure of chamber III, which is originally about 10^{-6} torr, is increased to about 10^{-5} torr with a rare gas in the ion source.

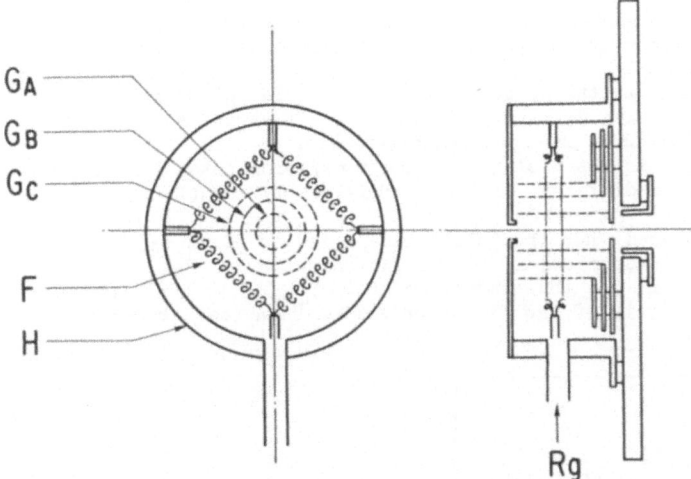

Fig. 2. A schematic diagram of the ion source. Three concentric grids, G_A, G_B, and G_C are installed for preventing charged species from penetrating into the central collision region. F and H denote filaments and a housing, respectively.

Electron impact ionization of the clusters can also be studied in the same ion source, where the potentials applied to the grids and the filaments are adjusted so as to allow free electrons to enter the collision region. The energy spread (fwhm) of the electrons is estimated to be about 2 eV at $\varepsilon = 4$ eV by comparison of the measured cross section curve for the 0^- production from CO_2 with that reported [6].

Detection of negative cluster ions

The negative ions thus produced are mass-analyzed by a quadrupole mass spectrometer (Extranuclear, 162-8) mounted coaxially with the cluster beam in chamber IV. The maximum mass-to-charge ratio of the mass-selected ions is m/z 1650 and a typical mass-resolution is about 300 at m/z 1460. The ions after the mass spectrometer are focused by a lens and converted to positive ions by use of an ion conversion dynode made of stainless steel at a voltage of about 5 kV. The positive ions ejected from the dynode are detected by a Ceratron (Murata, EMS-1081B). The use of this ion conversion dynode has improved the signal-to-noise ratio by more than three orders of magnitude, probably because the essential part of the stray electrons is eliminated. The measurement system is controlled by a CAMAC-crate-mounted LSI-11/23 computer. Fluctuation in the signal intensity is less than 10% for a period of several hours.

3.2 Performance

The following test experiments have been carried out and show that single collision-conditions are found to be fulfilled and that the observed negative ions originate from transfer of the Rydberg electrons to the neutral clusters in collision with Rg^{**}:

(a) The $(CCl_4)_n Cl^-$ ions produced from CCl_4 clusters are measured, as shown in Fig. 3; their intensities are proportional to the pushing pressure of the Kr gas, as shown in Fig. 4.

(b) The impact energy of electrons, given by the potential difference between the center pole of the filaments and grid G_C, is varied at a fixed pressure of the Kr gas in this pressure range (see Fig. 5). The ion intensities start to rise in the vicinity of the ionization potential of Kr. The cluster ions with different sizes, n, show nearly identical trends.

(c) The signals are reduced when the potential applied to G_B is increased; this observation can be interpreted as the field ionization of Rg^{**}. The relative cross sections for the production of $(CCl_4)_n Cl^-$ by Kr^{**} impact are estimated by the analysis of the field ionization data (see above).

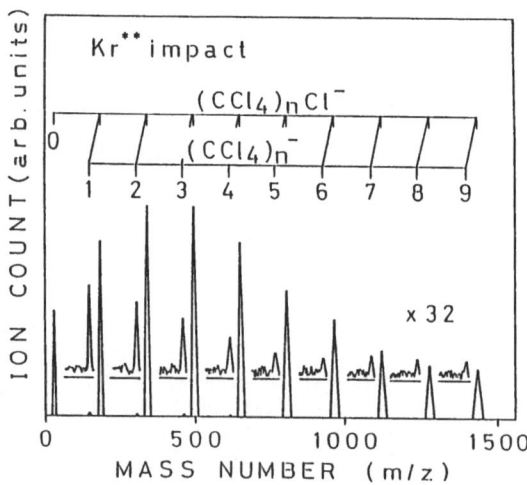

Fig. 3. Mass spectrum of negative cluster ions produced by impact of Kr^{**} atoms on $(CCl_4)_m$.

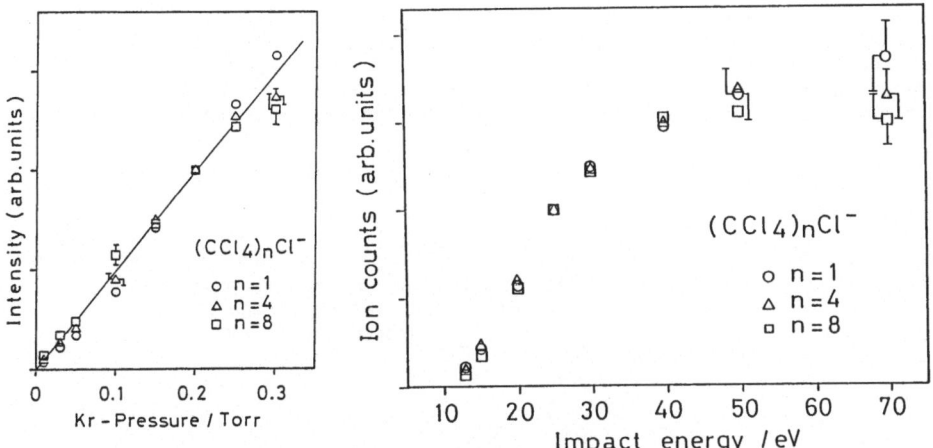

Fig. 4. Dependence of the intensities of $(CCl_4)_n Cl^-$ (n=1,4 and 8) on the pressure of the Kr gas used as a source of Kr^{**} atoms. Typical error bars are shown. The intensities for different ions are normalized at the Kr pressure of 0.2 torr.

Fig. 5. Dependence of the intensities of $(CCl_4)_n Cl^-$ (n=1,4 and 8) on the energy of the electrons used for excitation of Kr atoms to high-Rydberg states. The intensities for different ions are normalized at the electron impact energy of 25 eV.

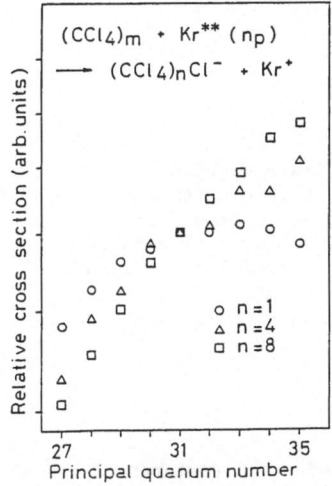

Fig. 6. Dependence of the relative cross sections for formation of $(CCl_4)_n Cl^-$ (n=1,4 and 8) from the neutral clusters of CCl_4 by impact of Kr^{**} on the principal quantum number of Kr^{**} atoms estimated by field ionization. The relative cross sections for different n are normalized at n_p=31.

Figure 6 shows the relative cross sections for the cluster ions with different sizes, n, plotted against the principal quantum number, n_p, of Kr^{**}. Since the observed trends are almost identical, the high-Rydberg atoms with different n_p are expected to give essentially the same size distribution. These n_p-dependences of the cross sections for the clusters agree with those for the monomer given by state-selected Rg^{**} [7].

4. EXAMPLE OF MASS SPECTRA

Negative ion mass spectrometry by use of Rg^{**} impact has been examined for various van der Waals clusters whose monomer molecules have positive and negative electron affinities [3-5]. For demonstration of the basic features of this technique, ionization of CO_2 clusters [3,7], $(CO_2)_m$, in collision with Kr^{**} is described.

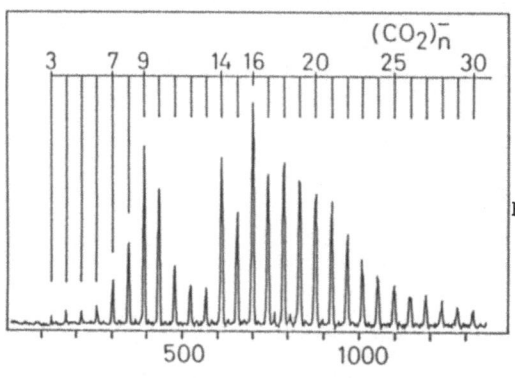

Fig. 7. Mass spectrum of $(CO_2)_n^-$ produced from $(CO_2)_m$ in collision with Kr^{**}. The stagnation pressure of the gaseous mixture of CO_2+He used was 2000 Torr.

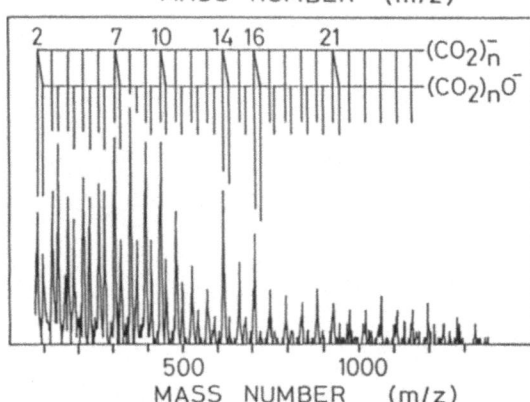

Fig. 8. Mass spectrum of negative cluster ions produced from $(CO_2)_m$ by impact of 1 eV-electrons. Two series of negative ions, $(CO_2)_n^-$ and $(CO_2)_nO^-$, are observed.

Figure 7 shows the spectrum of $(CO_2)_n^-$ thus produced. The features of the spectrum are as follows: (1) the threshold size is 3, (2) there is a region where the peak intensities are weak ($11 \leq n \leq 13$), and (3) n=14 and 16 are magic numbers. In the electron impact ionization, the peaks with n=14 and 16 are enhanced with increasing impact energy (see Fig. 8). This behavior indicates that n=14 and 16 are the magic numbers for $(CO_2)_n^-$ (n=14 and 16). It seems that a dimeric cluster ion is solvated by 12 CO_2 molecules, which make an icosahedron structure.

The CO_2 monomer cannot capture the Rydberg electron because the vertical electron affinity of CO_2 is negative. However, $(M)_m^-$ $(m \geq m_L)$ can capture the Rydberg electron with a large cross section because (1) the vertical electron affinity of $(CO_2)_m$ increases with m and becomes positive beyond m_L, (2) the state density of $(CO_2)_m^{-*}$ which takes part in the electron attachment increases with m, and (3) the lifetime for autodetachment from $(CO_2)_m^{-*}$ also increases with m. The attached electron is trapped in one of the CO_2 molecules (possibly with a bent structure in the cluster [3,7]. From feature 2, the number of molecules to be evaporated, m-n, is estimated to be at least 4, and hence, m_L is estimated to be at least 7. Detailed calculations on the cluster structures, the affinity levels and the cross sections for the CO_2 clusters have been carried out by Tsukada et al. [8].

5. CONCLUSION

The method for gentle and efficient production of the negative cluster ions by impact of high-Rydberg atoms developed in the present study provides an opportunity to investigate the dynamics of electron attachment processes to van der Waals clusters. The following information has been obtained from a systematic study of the mass spectra of the negative cluster ions produced by this method:
(1) the size dependence of the vertical electron affinity
(2) processes of relaxation in the cluster ions.

Acknowledgements

The author is grateful to Professors K. Kuchitsu and M. Tsukada, with whom the present study has been undertaken, and to Dr. K. Mitsuke, Messrs. F. Misaizu, H. Tada, and S. Yamamoto for their collaboration.

References

(1) M.Matsuzawa, In Rydberg States of Atoms and Molecules, ed. by R.F.Stebbings, F.B.Dunning, (Cambridge University Press, Cambridge, U.K. 1983) p.267
(2) B.G.Zollars, C.Higgs, F.Lu, C.W.Walter, L.G.Gray, K.A.Smith, F.Dunning, R.F.Stebbings: Phys. Rev. A. 32, 3330 (1985)
(3) T. Kondow: J. Phys. Chem. (in press)
(4) K.Mitsuke, T.Kondow, K.Kuchiutsu: J. Phys. Chem. 90, 1552 (1986)
(5) K. Mituske, T.Kondow, K.Kuchitsu: J. Phys. Chem. 90, 1505 (1985)
(6) P.J.Chantry: J. Chem. Phys. 57, 3180 (1972)
(7) T.Kondow, K.Mitsuke: J. Chem Phys. 83, 2612 (1985)
(8) M. Tsukada, N. Shima, S. Tsuneyuki, H, Kageshima, this volume.

Theory of Electron Attachment of Van der Waals Microclusters

M. Tsukada, N. Shima, S. Tsuneyuki, and H. Kageshima

Department of Physics, University of Tokyo, Hongo 7-3-1, Bunkyo-ku, Tokyo 113, Japan

1. INTRODUCTION

Recently, electron attachment phenomena have been observed on the molecular clusters such as $(CO_2)_N$, $(COS)_N$, $(H_2O)_N$ [1]. For some of these clusters, specific fine structures in mass spectra and critical size for the attachment have been found. The cross section of the electron attachment is found to markedly decrease with the increase of the electron energy. The purpose of this work is to clarify the mechanism of the electron attachment through a microscopic theoretical approach, and to estimate the energy and the cluster size dependences of the attachment cross section.

For the capture of an electron to the cluster, the electron kinetic energy must be transferred to some internal degrees of freedom of the cluster. If the electron impact energy is much less than the electronic excitation energies, this energy should be transferred to the intra- or intermolecular nucleus motions. Thus non-adiabatic processes are crucially involved in the attachment procedure.

In this paper, we mainly focus our attention on the $(CO_2)_N$ clusters, which show the typical threshold size for the attachment and characteristic fine structure in the mass spectrum. The crucial quantity which governs the attachment cross section is, as will be seen later, the vertical affinity energy of the cluster. If the energy position of the affinity levels is high, the attachment is hard to take place. This is because the deformation of the cluster to stabilize the affinity state, if any, should be large and requires long time, during which the electron escapes from the cluster.

For the larger clusters, the energy levels of the extended affinity states, i.e., not localised on any of the single molecules, are significantly lowered. And this causes the increase of the vertical affinity energy. Therefore the initial step of the electron attachment might be the transition from the free electron state to one of these extended affinity states. However, such a cluster state accomodating the electron in the extended affinity state is not the ground state of the negative cluster ion, but an excited state. This can be understood, since none of the molecules in the cluster are deformed significantly in such a case. The ground state is realized by the trapping of the electron to a single molecule, with the drastically bent O-C-O angle.

In this case the electronic energy is stabilized by about 3eV. The released stabilization energy is transferred to the inter- and intra-molecular vibrational energies, which results in the heating up of the cluster to cause evaporation of some molecules from the cluster.

2. MODEL OF THE ELECTRON ATTACHMENT

The Hamiltonian to describe the process discussed above is written as follows;

$$H = \sum_{k} \varepsilon(k) a_{k}^{\dagger} a_{k} + \sum_{i} \{ \varepsilon_{i} + \sum_{\lambda} \eta_{\lambda}^{i} (b_{\lambda}^{\dagger} + b_{\lambda}) \} a_{i}^{\dagger} a_{i} + \sum_{ik} (V^{i}(k) a_{i}^{\dagger} a_{k} + h.c.)$$
$$+ \sum_{\lambda} \hbar \omega_{\lambda} b_{\lambda}^{\dagger} b_{\lambda} \quad . \tag{2.1}$$

In the above a_{k}, a_{i} are the electron annihilation operator for the k-th free electron state and the i-th affinity state, with the energy $\varepsilon(k)$, ε_{i}, respectively. Inter- and intra-molecular vibration modes are treated by the harmonic model, and b_{λ}, $b_{\lambda}^{\dagger}$ are their boson operators. Since the affinity level positions are greatly influenced by these vibrational modes, this effect is taken into account by the linear coupling term in { }.

For more realistic discussions, we must distinguish the intra-molecular modes from the inter-molecular modes, as well as the extended cluster affinity states from the localized molecular affinity states. We postpone, however, this sophistication until a forthcoming paper, since the essential conclusion seems not much changed. It is convenient to introduce the interaction coordinate by

$$Q^{i} \equiv - \sum_{\lambda} \eta_{\lambda}^{i} (b_{\lambda}^{\dagger} + b_{\lambda}) \quad . \tag{2.2}$$

Then the potential energies V^{k}, V^{i} of the boson system for the neutral and negative charged state, respectively, are such as illustrated in Fig.1, as the function of Q^{i}.

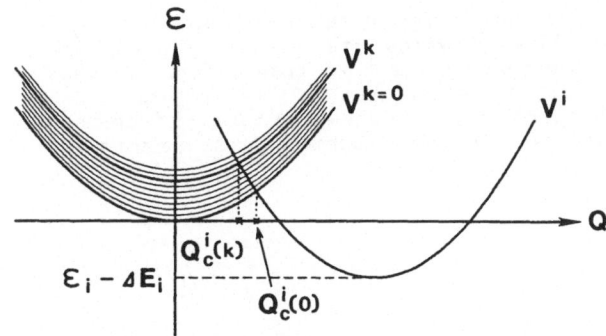

Fig.1 Potential surfaces of the neutral and negative charged state as the function of the interaction mode coordinate Q^{i}.

The transition from k to i is only possible at the cross over point $Q_{c}^{i}(k)$. The electron can be attached to the cluster, if it survives on the potential surface V^{i} until the critical point, $Q_{c}^{i}(k=0)$. Following the similar method by Sumi [2], we obtain the probability $P_{i}(t,t_{0})$ of the transition $k \to i$ at $t=t_{0}$, and remaining on the i-th affinity state during $t_{0} < t < t_{0} + \tau$ as follows;

$$P_{i}(\tau, t_{0}) = P_{i}(\tau)$$

175

$$= \frac{2\pi}{\hbar} |V^i(\mathbf{k})|^2$$

$$\times \langle\langle \delta(Q^i(t_0) - \varepsilon_{\mathbf{k}i}) \exp[-\frac{2\pi}{\hbar} \sum_{\mathbf{k'}} |V^i(\mathbf{k'})|^2 \int_{t_0}^{t_0+T} dt \, \delta(Q^i(t,t_0) - \varepsilon_{\mathbf{k'}i})] \rangle\rangle \quad , \quad (2.3)$$

with

$$\varepsilon_{\mathbf{k}i} = \varepsilon(\mathbf{k}) - \varepsilon_i \quad , \tag{2.4}$$

$$Q^i(t) \equiv -\sum_\lambda \eta_\lambda^i (Q_\lambda \cos\omega_\lambda t + P_\lambda \sin\omega_\lambda t) \quad , \tag{2.5}$$

$$Q^i(t+\tau, t) \equiv Q^i(t+\tau) + \sum_\lambda \frac{|\eta_\lambda^i|^2}{\hbar\omega_\lambda}(1 - \cos\omega_\lambda \tau) \quad . \tag{2.6}$$

The average $\langle\langle \quad \rangle\rangle$ is taken over the Gaussian distribution of the (complex) initial values of of the $\{Q_\lambda\}$, $\{P_\lambda\}$. Equation (2.3) is roughly estimated as

$$P_i(\tau) \sim \frac{1}{\sqrt{4\pi\Gamma_i^2}} \exp(-\frac{\varepsilon_{\mathbf{k}i}^2}{4\Gamma_i^2}) \exp[-\frac{2\pi}{\hbar} \sum_{\mathbf{k'}} \frac{|V^i(\mathbf{k'})|^2}{\sqrt{4\pi\Gamma_i^2}} \int_0^\tau \exp(-\frac{(\varepsilon_{\mathbf{k'}i} + u_i(\tau'))^2}{4\Gamma_i^2}) d\tau'] \quad , \tag{2.7}$$

with

$$\Gamma_i^2 \equiv \sum_\lambda (\eta_\lambda^i)^2 \coth(\hbar\omega_\lambda/kT) \quad , \tag{2.8}$$

$$u_i(\tau) \equiv \sum_\lambda (\eta_\lambda^i)^2 /\hbar\omega_\lambda (1 - \cos\omega_\lambda \tau) \quad . \tag{2.9}$$

It should be remarked that the potential energy of the interaction mode for the negative charged state relaxes towards its minimum $\varepsilon_i - \Delta E_i$ as $\varepsilon \sim \varepsilon_i - u_i(\tau)$ with the elapsed time τ after the $\mathbf{k} \to i$ transition. If the stabilization energy $\Delta E_i \sim \sum_\lambda |\eta_\lambda^i|^2/\hbar\omega_\lambda$ is larger than Γ_i, the integral over τ' in the r.h.s. of (2.7) is in proportion to the escaping time τ_c^i from the reemission zone. Thus, finally, the electron attachment cross section is roughly obtained as

$$\sigma \sim \frac{1}{v} \sum_i P_i(\tau \to \infty)$$

$$\sim \frac{2\pi}{\hbar v} \sum_i |V^i(\mathbf{k})|^2 \frac{\exp[-\frac{\varepsilon_{\mathbf{k}i}^2}{4\Gamma_i^2} - \frac{2\pi}{\hbar} \langle |V^i(\mathbf{k})|^2 \rangle_{\mathbf{k}} \rho_{\mathbf{k}} \tau_c^i]}{2\sqrt{\pi}\Gamma_i} \quad , \tag{2.10}$$

where $\rho_{\mathbf{k}}$ is the density of states of the free electron.

3. NUMERICAL ESTIMATION OF THE ATTACHMENT CROSS SECTION

To evaluate numerically the right-hand side of (2.10) the following quantities should be obtained:

1) affinity level distributions $\{\varepsilon_i\}$
2) coupling constants $\{v^i(k)\}$
3) width of the affinity level $\{\Gamma_i\}$ and
4) the escape time τ_c from the re-emission zone.

Since, of course, these quantities depend strongly on the cluster structure, we have determined by the potential energy minimization method several quasi-equilibrium structures of the clusters with the size up to N=13. In doing this, the inter-molecular force is calculated by the sum of the pairwise atomic potential used by van der Waal [3]. Examples of the cluster geometries are shown in Fig.2.

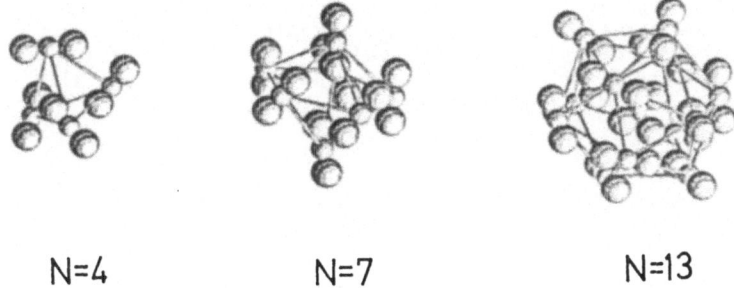

N=4   N=7   N=13

Fig.2   Structure of $(CO_2)_N$ clusters determined by the potential energy minimization method.

Affinity levels of these clusters are calculated as the transition state energy by the DV-Xα-cluster method [4]. The bands of the affinity levels of the clusters $(CO_2)_N$ (N=2,4,7,10,13) are shown in Fig.3. It should be remarked that it is rather difficult to determine accurately the affinity level positions relative to the vacuum level, since they somewhat depend on the choice of the basis set. However, the relative affinity level positions among different clusters are calculated with enough accuracy. We tentatively determined the vacuum level position E_{vac}, so that the vertical affinity energy of the monomer equals to -3.8 ev [5]. As seen in Fig.3, the band of affinity levels goes sharply down with the increase of the cluster size below N~7, and crosses with E_{vac} for N≥13. The size dependence of the

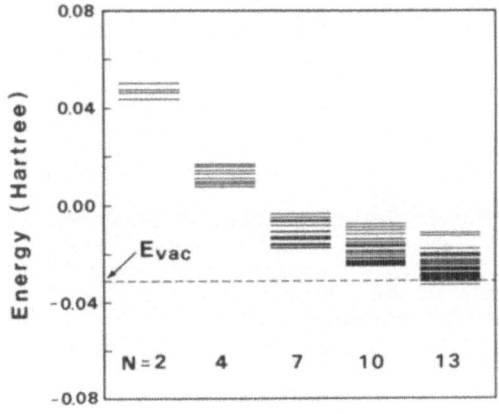

Fig.3   Affinity levels of $(CO_2)_N$ clusters. E_{vac} is the vacuum level (see text). Affinity level of the monomer is located at 0.108 Hartree, i.e. 3.8 eV above E_{vac}.

affinity band position is not significant for $N\gtrsim13$. As seen by (2.10) if the lowest affinity level is located at the higher position by $<\Gamma_i>_{av}$ than the vacuum level, the attachment cross section is drastically reduced. Thus, the existence of the threshold cluster size is explained qualitatively by the change of the affinity bands shown in Fig.3.

The squared coupling constant $|V^i(\mathbf{k})|^2$ can be numerically evaluated for respective clusters. For the rough estimate, however, we substitute for them an approximate expression like

$$<|V^i(\mathbf{k})|^2>_{av} \sim (N_s/N)(Z_B-Z_s)^2|V|^2\Omega \quad . \tag{3.1}$$

In the above N, Ns, V, Ω are the cluster size, number of the molecules on the cluster surface, the transfer integral of the affinity states between the nearest neighboring molecules and the volume per CO_2 molecule, respectively. Z_B (~12) and Z_s are the coordination number in the crystal and the cluster surface, respectively. Crude approximations,

$$\Gamma_i \sim \sqrt{<\hbar\omega_\lambda>_{av}<\Delta E_i>} \quad , \tag{3.2}$$

$$\tau_c^i \sim <\omega_\lambda>_{av}^{-1}\sqrt{\epsilon(\mathbf{k})/\Delta\epsilon} \quad , \tag{3.3}$$

are also used. In the above $\Delta\epsilon$ is the threshold energy measured from the potential minimum of the negative charged state (Fig.1).

Energy dependence of the cross sections is shown for a plausible set of parameters (Γ_i, V, Ω) (0.0184, 0.0033, 46 in atomic unit) in Fig.4.

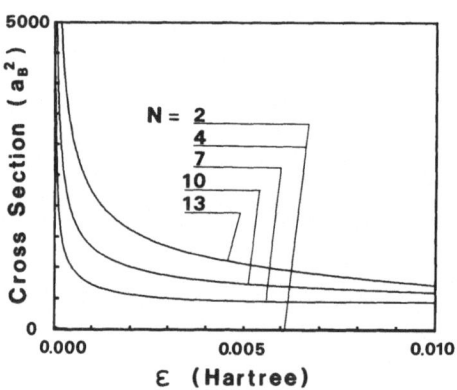

Fig.4 Electron attachment cross section of $(CO_2)_N$ clusters.

Experimentally estimated values of σ for $E\sim0.01eV$ is about $10^4(a_B)^2$ for $N > 11$, but σ is reduced by the factor of about 10^{-4} for the energy region, $\epsilon\sim3\sim5eV$. This sharp decrease of the cross section with ϵ is well explained by the present model, as well as the presence of the critical size for the electron attachment between N=4 and 7.

4. CONCLUSION

Mechanism of the electron attachment on CO_2 molecular clusters is proposed based on the strong coupled electron-cluster deformation model. The vertical affinity levels go down sharply with the increase of the cluster size, explaining the presence of the critical size for the attachment. Drastic decrease of the attachment cross section with the electron energy is also explained by this model.

For more quantitative arguments, a detailed model of the deformation modes and realistic calculation of $|V^i(\mathbf{k})|^2$ are necessary. Such refinement of the theory is now under way.

Acknowledgements

The authors would like to thank Prof. T. Kondow for stimulating discussions.

References

1. T.Kondo, K.Mitsuke: J. Chem. Phys. 83, 2612 (1983)
2. H.Sumi: J. Phys. Soc. Jpn. 49, 1701 (1980)
3. B.W.van de Waal: J. Chem. Phys. 79, 3948 (1983)
4. H.Adachi, M.Tsukada, C.Satoko: J. Phys. Soc. Jpn. 45, 875 (1978)
5. D.G.Hopper: Chem. Phys. 53, 85 (1980)

Time Evolution of Electronic States in Microclusters

N. Hamada

Fundamental Research Laboratories, NEC Corporation, Miyazaki 4-1-1, Miyamae-ku, Kawasaki 213, Japan

A microcluster can be utilized as a functional unit in a quantum computer. The time evolution of the electronic states in microclusters is studied on the basis of the tight-binding model. It is shown that a magnetic field can be used as a switch in quantum circuits.

1. Introduction

There have already been several proposals on the concept of a quantum computer[1]. Like conventional computers, a quantum computer needs functional units to perform computations. This functional unit is realized by an assembly of a small number of atoms, i.e. a microcluster. Recent technological developments are making it possible to design microclusters. In the near future, we expect to have such designed microclusters in our hands.

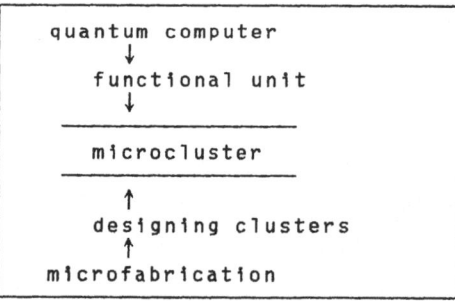

Such a microcluster may be constructed from ordinary atoms or from superatoms, a recently proposed spherical semiconductor heterostructure of mesoscopic dimensions[2].

To perform a calculation is to trace the time evolution of some physical system. In this paper, we study the time evolution of electronic states in a functional unit, i.e., a microcluster.

Let us, for simplicity, consider a square-shaped microcluster as shown in Fig. 1. We denote the side length of the square by a, and the electron transfer between the nearest-neighbor atoms as v. Typical values of these parameters are shown in Table 1.

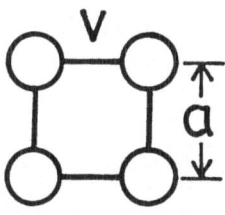

Fig.1. Cluster.

Table 1. Characteristic parameter values for microclusters

	atom cluster	superatom cluster
a	1 A	500 A
v	1 eV	1 meV
h/v	4.1 fs	4.1 ps
$\dfrac{hc/e}{a^2}$	4.1×10^5 T	4.1 T

For an ordinary atom cluster, a and v are on the order of 1 angstrom and 1 eV. For a superatom cluster, a~500 angstrom and v~1 meV. Therefore, the time scale characterizing the system, h/v~4.1 femto-second for the atom cluster and ~4.1 pico-second for the superatom cluster. The magnetic field corresponding to a unit quantum flux, hc/e, inside the square, ~4.1×10⁵ Tesla, and ~1.64 Tesla, respectively. 10⁵ Tesla is tremendously large and unrealistic, whereas 1 Tesla is easy to achieve today. A superatom cluster or an object somewhat smaller has suitable properties for the following discussion.

2. Tight-Binding Model

We adopt a tight-binding model for this system. The Hamiltonian is expressed as

$$H = \sum_i |i> \varepsilon_i <i| + \sum_{i \neq j} \sum |i> v_{ij} <j| , \qquad (1)$$

where i stands for the site with Cartesian coordinates, (x_i, y_i, z_i), ε_i the site energy, and v_{ij} the transfer matrix element between i and j sites.

Under a magnetic field, B, in the z direction, v_{ij} is expressed as

$$v_{ij} = u_{ij} \exp(i 2\pi A_{ij}) , \qquad (2)$$

where u_{ij} is a real parameter, and

$$A_{ij} = -B(x_i + x_j)(y_i - y_j)/2 , \qquad (3)$$

if we adopt the Landau gauge for the vector potential:

$$\vec{A} = (0, -Bx, 0). \qquad (4)$$

The time evolution of this system is described by a time-dependent Schrödinger equation:

$$H |\psi(t)> = i\hbar \frac{\partial}{\partial t} |\psi(t)> \qquad (5)$$

181

Using the eigenstate for H,

$$|\phi_n> = \sum_i |i> c_{in}: \quad H|\phi_n> = E_n |\phi_n> \, , \tag{6}$$

the solution of Eq.(5) is expressed as

$$|\psi(t)> = \sum_n |\phi_n> b_n \exp(-iE_n t/\hbar) \tag{7a}$$

$$= \sum_{i,n} |i> c_{in} b_n \exp(-iE_n t/\hbar) \, . \tag{7b}$$

The coefficient, b_n, is obtained by giving the initial state at t=0:

$$<i|\psi(0)> = \sum_n <i|\phi_n> b_n \quad (i=1,2,..,N) \tag{8}$$

for an N-site system.

For example, the probability of finding the electron at the i-th site is given by

$$\rho_i(t) = |<i|\psi(t)>|^2$$

$$= \sum_{n,m} c_{in} b_n \exp[-i(E_n - E_m)t/\hbar] \, b_m^* c_{in}^* \tag{9}$$

3. Numerical Results

The first example of time evolution is shown in Fig. 2. Here the site energies are all taken to be equal to zero and the transfer energies are all set to one energy unit. When the energy unit is 1 meV, the time unit is 4.1ps as shown for the superatom cluster in Table 1. It is assumed that the electron dose not transfer along the diagonal of the square. Initially, we assign an electron to site 1. The clockwise-travelling wave

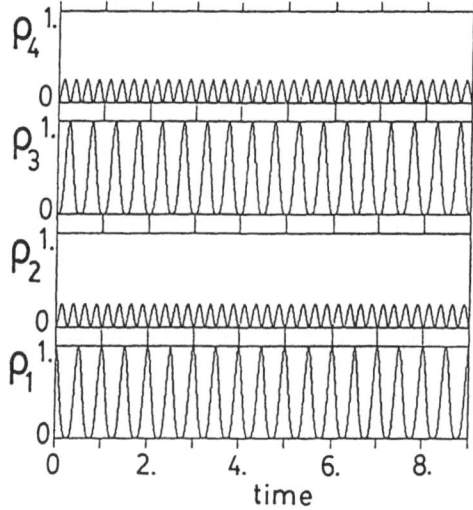

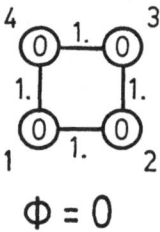

Fig.2. Time evolution in a square cluster.

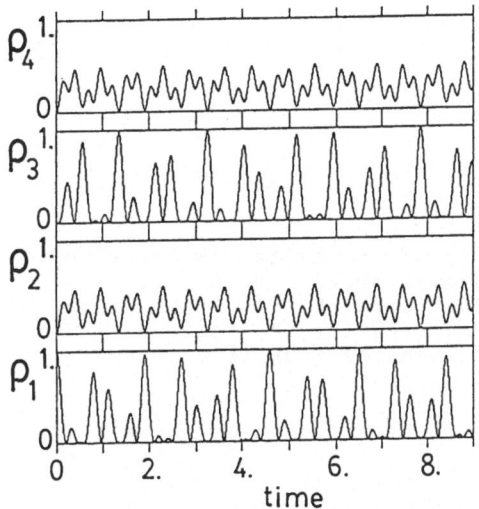
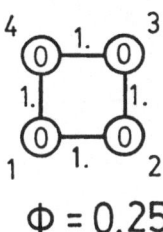

Fig.3. Time evolution under the magnetic flux, Φ=0.25(hc/e).

and the counterclockwise wave are produced from the site. The electron goes via sites 2 and 4 to site 3, and then comes back via sites 4 and 2 to site 1. The probability of finding the electron at each site changes periodically.

The second example is shown in Fig.3. We apply a magnetic field to the same cluster as above. The magnetic flux penetrating the square is 0.25 (hc/e). The oscillation of the probability at each site becomes complex and unperiodic.

We observe a remarkable phenomenon on applying the magnetic flux of 0.5(hc/e), as shown in Fig.4. The probability of finding the electron always vanishes at site 3. This is a typical illustration of the Aharonov-Bohm effect, which is a result of the interference between the clockwise and counterclockwise waves.

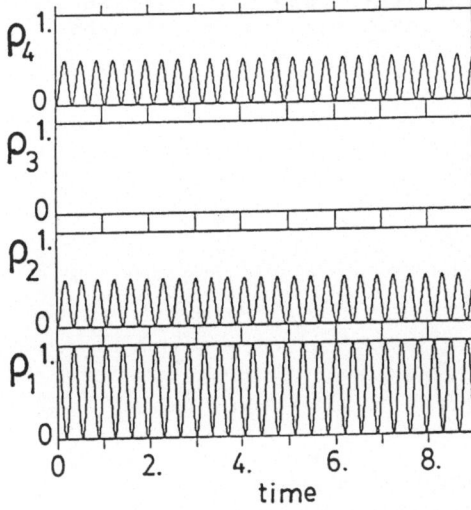
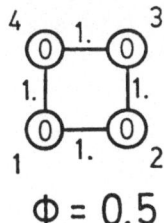

Fig.4. Time evolution under the magnetic flux, Φ = 0.5(hc/e).

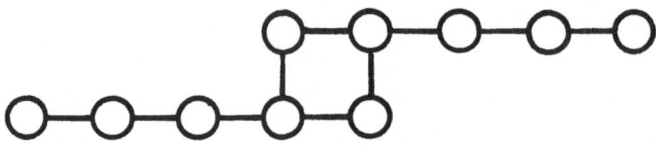

Fig.5. A part of a quantum circuit.

This phenomenon can be used as a switch in quantum circuits. For the array of atoms shown in Fig.5, the transmission probability from one terminal to the other changes with the magnetic flux penetrating the square. The transmission probability vanishes when the flux is 0.5(hc/e).

4. Conclusion
    I have shown some examples of time evolution of electronic states in a microcluster. A microcluster is utilized as a functional unit in a quantum computer. A magnetic field can be used as a switch in quantum circuits. These studies suggest the existence of many interesting phenomena with respect to the time evolution in microclusters. I hope that additional studies on these types of phenomena will lead to advancement in the field of microclusters, assemblies of microclusters, and quantum computers.

Acknowledgement ———— The author would like to express his thanks to Dr. T. Inoshita, Dr. S. Tanaka and Dr. S. Ohnishi for fruitful discussions.

References
[1] For example, P.A.Benioff, Int.J.theor.Phys.21(1982)177; R.P.Feynman, Int.J.theor.Phys.21(1982)467; D.Deutsch, Proc.R.Soc.Lond.A400(1985)97.
[2] H.Watanabe, 'The Physics and Fabrication of Microstructures and Microdevices', ed. M.J.Kelly and C.Weisbuch (Springer-Verlag, Heidelberg, 1986) pp.5; H.Watanabe and T.Inoshita, Optoelectronics ——— Devices and Technologies 1(1986)33; T.Inoshita, S.Ohnishi and A. Oshiyama, Phys.Rev.Lett.57(1986)2560; T.Inoshita and H.Watanabe, in this volume.

Part V

Structural Fluctuations

Some Experiments on Structural Instability of Small Particles of Metals

S. Iijima

Research Development Corporation of Japan, c/o Department of Physics, Meijo University, Tenpaku-ku, Nagoya 468, Japan

Fine particles of some metals up to 100 Å in size were examined at the level of the atomic resolution by an electron microscope equipped with a TV monitor system. As the size of the particles decreases to below 50 Å, the particle shapes were found to change continually through the internal transformation from a single crystal to a twinned crystal, including multiply-twinned crystals, and vice versa. The transformation, taking place in a cooperative motion of both internal and external atoms, occurs abruptly in less than a fraction of a second. The structural fluctuation of small clusters seems to be associated with electronic excitations due to irradiation of the electron beam.

1. INTRODUCTION

A decrease in the size of a solid often leads to anomalous properties: lowering of melting points, lattice parameter anomalies, multiple-twinning, modified crystal structures and so on. These properties have long been studied experimentally and theoretically by many workers [1]. Small clusters of multiply twinned particles (hereafter abbreviated MTP) [2,3,4], and thermal equilibrium forms of the particles are theoretically discussed within a framework of Curie-Wulff theorem of crystal growth by INO [5] and later by MARKS [6].

Recently structural anomaly was discovered in metallic clusters of less than 50 Å in diameter [7,8]. In that study, the dynamic behaviors of gold clusters were examined at the atomic resolution level by a modified electron microscope. Incidentally, the present state-of-art electron microscopy has already reached the level where microclusters and surface steps on crystals are able to be observed at the level of a single atom resolution by conventional transmission electron microscopes [9,10]. The gold clusters became structurally unstable under the electron beam irradiation (1.3×10^5 electrons per Å² per sec) and continually changed their shapes; among which were single, twinned and multiply-twinned crystals including icosahedron and decahedron.

In addition to the structural instability of small particles, anomalous behaviors of surface atoms on small clusters have also been reported [11, 12,13,14]. Some surface atoms appear to be bound loosely and they hang around near the surface as "atomic clouds". Atoms jump in and out of their atomic clouds, which extend 10-15 Å from the surface.

Two possibilities can be considered as causes of instability; a rise in specimen temperature during the electron beam irradiation [15] or surface charging effects due to the emission of secondary electrons [8]. The specimen temperature is determined by a balance between heat gain due to the inelastic scattering process (valence electron excitations and inner-shell excitations) and heat dissipation due to conduction through a substrate.

The present paper outlines the structural instability of small metallic clusters and then effects of specimen temperatures and charging up are examined experimentally. Through these experimental evidences, the origin of the instability is discussed in terms of charging up due to electronic excitation of the small clusters. As one of the possible mechanisms for structural reorganization of atoms in a cluster, a twin deformation is proposed.

2. EXPERIMENT

The specimen of metal clusters was prepared by the in-situ vacuum deposition onto various kinds of substrates such as fine and spherical particles of crystalline silicon, γ-Al$_2$O$_3$ and α-Fe$_2$O$_3$, partially graphitized carbon particles and a percolated amorphous carbon film. The small Si particles were made by a gas-evaporation method in argon under a reduced atmosphere, where the Si vapor was condensed into small particles as smoke. They were less than several hundred A in size and were usually covered with SiO$_2$ layers of 10-30 Å in thickness. The details of specimen preparation have been reported elsewhere [11].

The metal clusters deposited onto the substrates were initially smaller than 10 Å in size but they aggregated into larger clusters under the electron beam irradiation. The sizes of the cluster were controlled in the range from 10 Å to 100 Å by adjusting the electron beam intensity and its exposure time. Figure 1 shows an electron micrograph of spherical particles

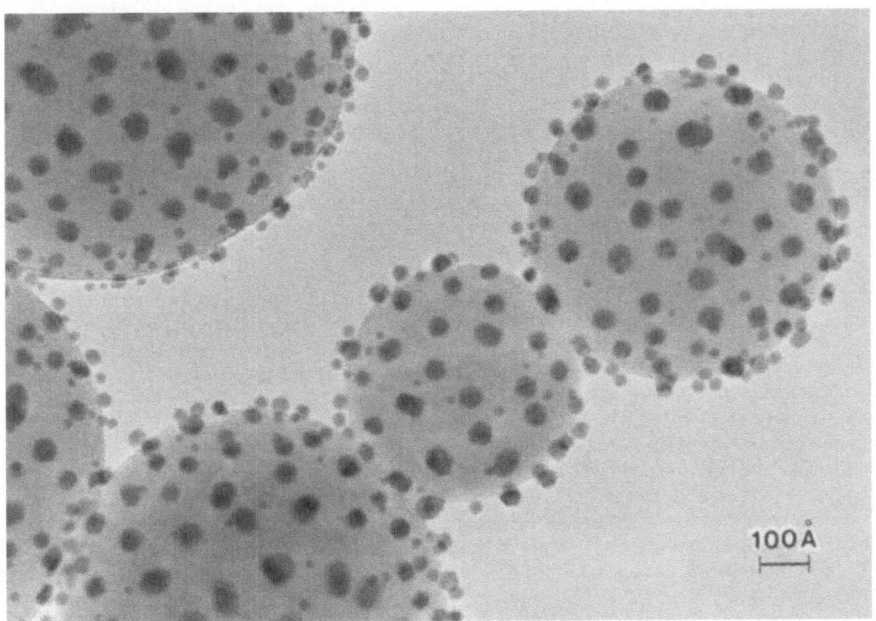

Figure 1 An electron micrograph of gold clusters vacuum-deposited on spherical particles of single crystal Si.

of Si onto which gold clusters have been deposited. The clusters (imaged as dark blobs) are 10-50 Å in diameter. The similar cluster specimens are prepared by impregnating various metal carbonyl clusters onto the Si or γ-Al_2O_3 particles [16]. For electron microscopic observations, the metal clusters attached on the periphery of the spherical particles were chosen and observed as the profile images along the electron beam.

The employed electron microscope, equipped with a real-time TV monitor system, has been specially improved for the present study. The resolution of the microscope was 2.3 Å at 120 keV which allowed us to image individual rows of atom columns in gold crystals. To reduce the problems of specimen contamination, the instrument was operated under ultra-high vacuum of around 3×10^{-8} Torr by a dry vacuum pumping system with turbo-molecular pumps and ion sorption pumps.

All micrographs presented in this paper were reproduced from single frames of the VTR tape. A sequence of evolution of a particular gold cluster was analyzed by examining a set of individual frame pictures. The image-recording pick-up system (SIT TV-camera from Hamamatsu Photonics and Sony BVU-820) had a time resolution of 1/60 of a second.

3. RESULTS

3.1. Observation of Structural Instability

A typical series of electron micrographs of the same cluster of gold are reproduced in Fig. 2. These were selected from a VTR over a 5-minute span. The cluster resting on a SiO_2-covered Si substrate is about 20 Å in size. According to the real-time observation on a TV screen, the shape of the cluster itself constantly changed approximately every fraction of a second. The change was often accompanied with rotational and translational motions of the cluster. Its center of gravity moved over distances of 30 Å to 60 Å on the substrate. The moving clusters often collided with each other to cause coalescence or grain growth. As mentioned below, the internal structure also changed from a single crystal to a twinned crystal and vice versa. With the increase of the particle size, the movement became slow and no rapid change was observed for the cluster larger than about 100 Å in size. It is emphasized, however, that the surface atoms of such a large cluster move around actively [11].

The momentary change of the cluster was promoted by the intense electron beam irradiation (about 1.3×10^5 electrons per $Å^2$ per sec at the specimen position) and was not observed when the beam intensity was below an order of 10^3 electrons per $Å^2$ per sec. The rate of movement increased with the decrease of the area of the cluster in contact with the substrate, which is located at the bottom portion of each micrograph. The evolution of the clusters, however, became sluggish when electrically good conductors such as graphite or amorphous carbon were used as a substrate. Alumina substrates gave the same results as the SiO_2-covered Si but almost no activities of the clusters were observed when a substrate of α-Fe_2O_3 was used. These observations suggest that the instability of the clusters is related partly to the charge fluctuation on the clusters or in their vicinity.

3.2. Morphologies of Small Clusters

The gold clusters in Figs. 2c, e, f and k are well-faceted cuboctahedra, as illustrated in Fig. 3a. The two sets of lattice fringes in Fig. 2k corres-

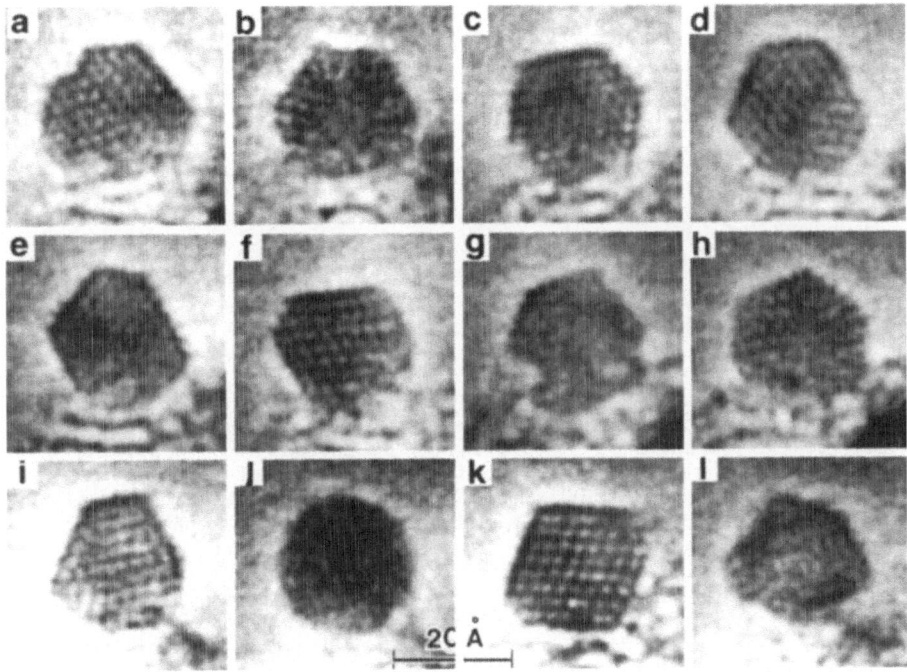

Figure 2 A series of the electron micrographs showing structural changes of a gold cluster consisting of about 460 atoms. The changes occurred under the electron beam irradiation. Single crystals of cuboctahedra (c,e,f,k), simple twins (a,d,i) and icosahedral MTPs (b,h,i) are observed but no decahedral ones appear.

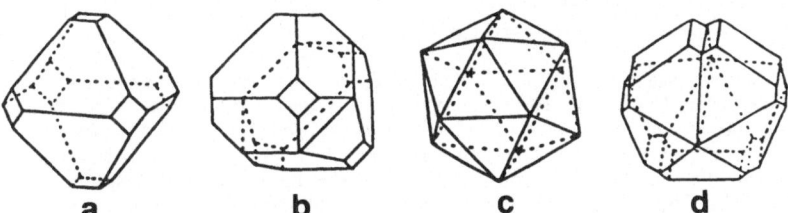

Figure 3 Clinographic views of various shapes of metal clusters. (a) cuboctahedron, (b) twinned cuboctahedron, (c) icosahedral MTP, and (d) truncated decahedral MTP.

pond to the net-planes of Au d_{111}(=2.35 Å). The number of the fringes gives the size of the cuboctahedron, which is 8x2.35 Å=18.8 Å in vertical thickness. Such an ideal cuboctahedron contains theoretically 459 atoms (see Fig. 4). Its (100) facet, indicated by hatched balls, has a square arrangement of only 3x3 gold atoms. A similar but somewhat larger cuboctahedron consisting of about 940 gold atoms has been reported in the previous paper [11].

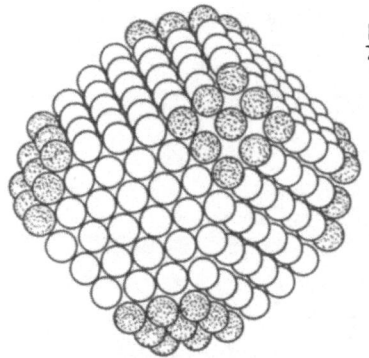

Figure 4 A model for a cuboctahedral cluster of 459 atoms which appears in Fig. 2k.

A VTR analysis disclosed that the cluster was not always a single crystal but frequently transformed into a twin. A re-entrant angle appearing in the left top of the cluster of Fig. 2a is characteristic of twinning for fcc metal with facetings of {100} and {111} surfaces. The twinning can also be recognized by noticing the {111} lattice fringe systems that have changed in their orientations by 140 degrees (Figs. 2a, d and i). These twinned clusters are oriented with their common [110] axes vertical to the plane of the page. In many cases, twin planes and stacking faults initiated with the transformation of a cluster which took place with cooperative motion of both external and internal atoms. The detailed account of the twinning will be described later.

In some instances, the cluster transformed into an icosahedral MTP, which was identified as the closest packing of 20 tetrahedral units. The cluster with a hexagonal shape (Figs. 2b and h) is the icosahedral MTP viewed down one of its three-fold axes (Fig. 3c). The cluster in Fig. 2h appeared 20 seconds after the one in Fig. 2g had been observed. According to INO's calculation [5], the icosahedral MTP is the most stable form of a gold cluster smaller than 100 Å in size. For the ideal icosahedral MTPs, the number of atoms in the clusters N is given by a relation $N=3^{-1}(10n^3-15n^2+11n-3)$, where n=integer and gives N=1, 13, 55, 147, 309, 561,... [17]. The cluster concerned here is close to the N=561.

In contrast to the smaller particle of Fig. 2, the larger cluster shows often a decahedral MTP (Fig. 3d) as well as other shapes already mentioned above. This is shown in another series of electron micrographs of a slightly larger gold cluster of about 30 Å in size which consists of about 1000 atoms (Fig. 5). The decahedral MTP consists of 5 tetrahedra with a common <110> edge (Fig. 5d). The imperfect shape of the decahedron is due to the contact effect between the substrate and the cluster. For fcc metal particles of less than 100 Å in diameter, their stabilities decrease in the order of icosahedral MTP, single and decahedral MTP [5]. A lower occurrence of the decahedral MTP accords with the present observations. In these series of micrographs, however, a particular preference among the cluster shapes for a given number of atoms has not been confirmed.

The decahedral particles were found to be always truncated with {100} surfaces and to form re-entrant angles [8]. The truncation seems energetically favorable in reducing surface energy over the regular pentagonal decahedral MTP, which was predicted by MARKS [6]. It is mentioned that almost the same structural instability described above has been confirmed in other fcc metals, such as Pt, Rh, Ni and Ag.

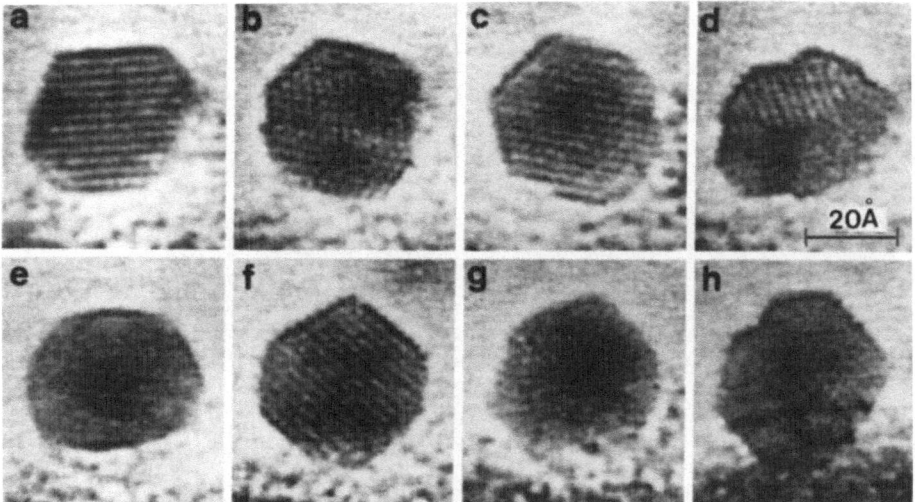

Figure 5 A series of the electron micrographs of a gold cluster of about 1000 atoms which is slightly larger than one in Fig. 2. The cluster changes into various shapes including a decahedral MTP which is not found in smaller clusters. Cuboctahedra (a,c,f), twinned cuboctahedra (b,h), decahedral MTP (d) and icosahedral MTP (g).

3.3. Featureless Clusters

In very few instances a cluster takes a round or spherical form, as seen in Figs. 2j and 5e. Since these clusters usually do not exhibit lattice fringe images, the forms might be liquid droplets of gold. A typical example is seen in a pair of photographs of two Pt clusters on a SiO_2-covered Si particle (Fig. 6). The size of the clusters is about 6 Å in diameter, which is calibrated from the lattice images of Si d_{111} (=3.1 Å) appearing in the vertical direction. These clusters should be composed of approximately 20 atoms. The two clusters moving on the SiO_2 surface collide and coalesce to become a larger cluster of 12 Å in diameter, as mentioned in sect. 2 (Fig. 6b). It is worth mentioning that the cluster activities are enhanced under a H_2 atmosphere of 10^{-5} Torr [18]. A granular object lying between the Pt clusters and the crystalline Si particle is a SiO_2 surface layer of about 10 Å thickness. The reasons why these clusters do not show lattice images are: occurrence of non-fcc structures, liquid droplets, or very rapid motion of the clusters. A possibility of melting of the clusters is discussed further below.

3.4. Experiment on Cluster Temperature

On specimen temperatures, two experiments are worth mentioning. Firstly, Bi clusters were prepared and observed in exactly the same procedure as the Au clusters. The results confirmed a solid state of the Bi clusters, since they exhibited lattice fringe images. This means that the substrate temperature during the observation in the microscope would not exceed the melting temperature of Bi which is 271°C. The featureless clusters mentioned above, therefore, are not likely to be caused by melting, although a question

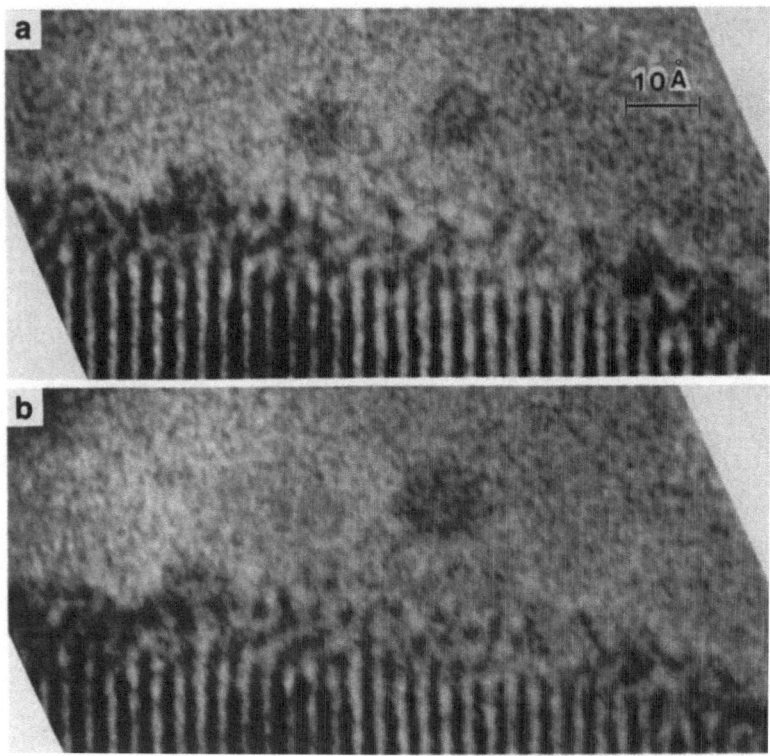

Figure 6 A pair of the electron micrographs showing Pt clusters moving on a SiO_2-covered Si substrate. (a) The two 6 Å clusters don't show lattice images. (b) They collide with each other to coalesce.

remains to be answered as to how much the melting point of small Pt clusters (1769°C)is lowered due to the size effect [19].

Secondly, to examine an effect of the electron beam heating, two modes in electron microscopic observations were compared in terms of the probe size. In the conventional mode, which was employed for taking pictures as shown in Figs. 2, 5 and 6, the specimens were illuminated by an electron beam of several thousand Å in diameter. Thus, the beam irradiates inevitably the whole particle of a Si substrate (see Fig. 7a) and some of the energies of the incident electrons will be lost in the substrate as a result of inelastic scattering events. This will lead to a temperature rise in the specimen depending on the size (thickness) of the particles. To avoid the electron beam irradiation of the substrate during the observation of the clusters, a micro-beam technique was employed [20]. The technique allowed us to irradiate only a particular cluster with a finely focused electron beam (Fig. 7b).

Figure 8 demonstrates an effect of sizes of the electron beam probe on the cluster stability. The gold clusters (Figs. 8a and b) were recorded on a conventional mode. The one in Fig. 8a appears to be a single crystal but a moment later it has changed to a decahedral MTP (Fig. 8b). After confirming the instability of the cluster through the TV monitor, the microscope

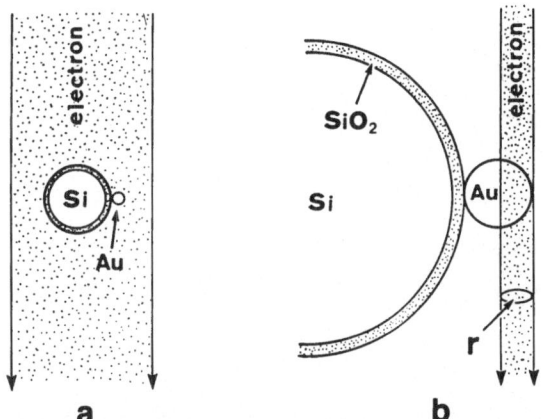

Figure 7 Two models for the electron-microscopic observation of clusters supported on a spherical Si particle. (a) conventional mode where the electron beam size is 2000 Å in diameter. (b) micro-beam mode where the beam size (indicated by r) is 55 Å.

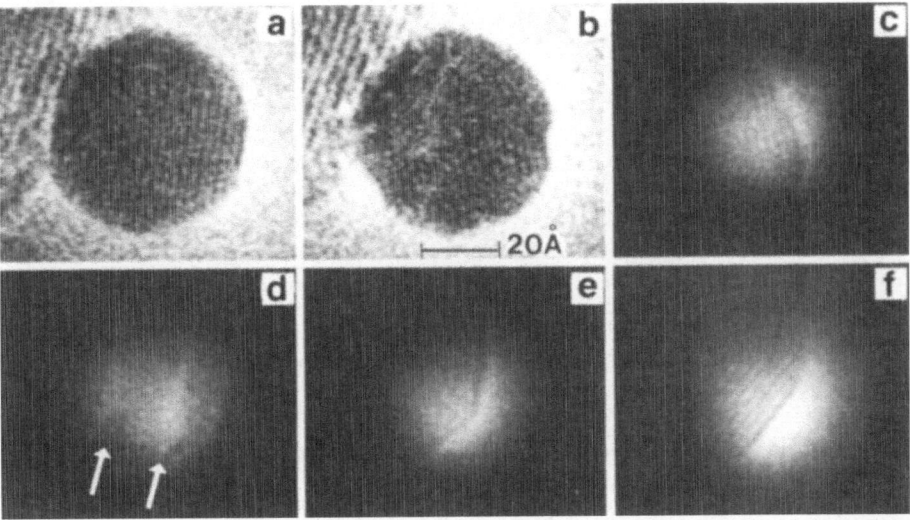

Figure 8 A series of the electron micrographs of a gold cluster of 40 Å demonstrating the structural instability. Conventional HREM-VTR images (a) showing a single crystal and (b) a decahedron. Microbeam-HREM images of the cluster (c,d,e,f) disclosing changes in the outer shapes as well as arrangements of the internal atoms. The fringes appearing in the clusters correspond to d_{111}=2.35 Å. Black and white bands in (d) show twin bands.

was switched to the microbeam imaging mode. The micrographs (Figs. 8c, d, e and f) are some of the instant shots reproduced from the VTR tape. The bright areas indicate a size of the electron probe which is 55 Å in diameter. It should be emphasized that the beam illuminates not a whole cluster but only a portion of the cluster. The outlines of the clusters are different in the four images, and also the lattice fringes change their direction. The alternating black and white bands (indicated by arrows in Fig. 8d) are caused by twinning. These results demonstrate that the clusters are still moving and changing the internal arrangement of atoms. A smaller probe of 35 Å in diameter makes the cluster movement slightly slower but gives no major changes in the cluster activities.

4. DISCUSSION

4.1. Estimation of Specimen Temperatures

As we described above, the specimen temperature appears not to be high enough for the Au clusters to melt during the electron microscope observation. A cluster temperature will be determined by a balance of heat dissipation through conduction and heat gain through inelastic scattering events of the incident electron beam in the cluster. Since the direct measurement of the specimen is difficult, it seems sensible to estimate the order of magnitude to an appropriate model.

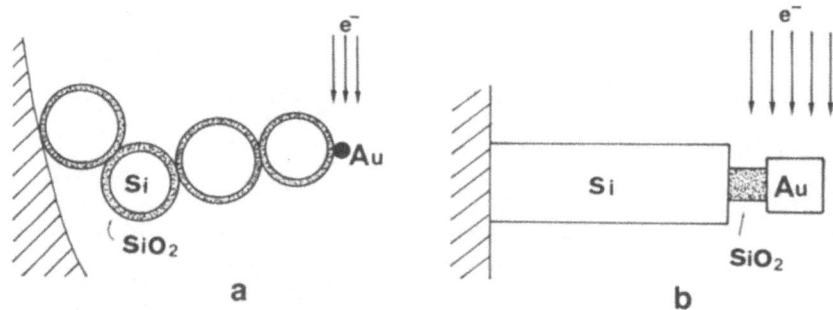

<u>Figure 9</u> Illustrations of an experiment on observation of the clusters. (a) A schematic representation of a small gold cluster on SiO_2-covered Si balls. (b) A simplified model for estimation of a specimen temperature. In the microbeam mode, only the cluster is illuminated by a focused electron-beam.

Our experimental situation of the microbeam observation is illustrated schematically in Fig. 9a. A gold cluster lying on the end of a chain of SiO_2-covered Si particles is irradiated by a microbeam. The model is simplified, (as shown in Fig. 9b), where a gold cluster is attached to a Si rod via a rod of SiO_2. The temperature difference between the cluster and substrate, δT, is given by an equation $\delta T = K^{-1} e^{-1} J_b t (dE/dZ)_0$, where t is the thickness of a cluster, J_b is the total electric current, and K is given by the specimen geometry and thermal conductivities of the specimen and the substrate material. $(dE/dZ)_0$ is a Bethe's stopping power for the fast electron. Assuming a 20 Å gold cluster and putting experimental parameters to the equation, the cluster temperature turns out to be almost the same as that of the substrate, which is room temperature. The result supports our assumption that the instability of the clusters may not be caused by a rise in specimen temperature.

4.2. Charging Effect on the Structural Instability

Under the electron beam irradiation, the small clusters suffer various electronic excitations such as plasmons and inner shell excitations. These excitations will decay radiatively to result eventually in thermal energy in the cluster. The excitations cause the metal clusters to be charged positively due to emission of the secondary electrons if the clusters were not well grounded. A feasibility of the multiple charging of small metallic clusters under the intense electron beam irradiation has been suggested by HOWIE [21]. He thought that the charging could take place in the Auger cascade process during relaxation of the inner shell excitations.

A few experimental evidences indicate that either a local area of the substrate or a gold cluster is deviated temporarily from electrical neutrality. If a metal cluster is sitting on the substrate which is an electrical insulator and is positively charged, the charge will impose a coulomb repulsive force on another charge appearing on the substrate. The translational motion of the cluster will be explained by this force (Fig. 10a). If the coulomb force becomes larger than a total adhesion force between the cluster and the substrate, the cluster will be detached from the substrate. This has been observed occasionally during the experiment.

Figure 10b illustrates another example of the surface charging effect. Two clusters of about 30 Å in diameter which are separated by a distance of 40 Å are changing structurally but their movements are quite independent. At certain occasions they all of a sudden become still upon thoroughly touching the substrate. It happens simultaneously. The sudden stop of the movements could be caused by a temporal holding of surface charging up of both clusters. The observation suggests that there has been a coulomb repulsive force interacting between the two clusters. It should be empha-

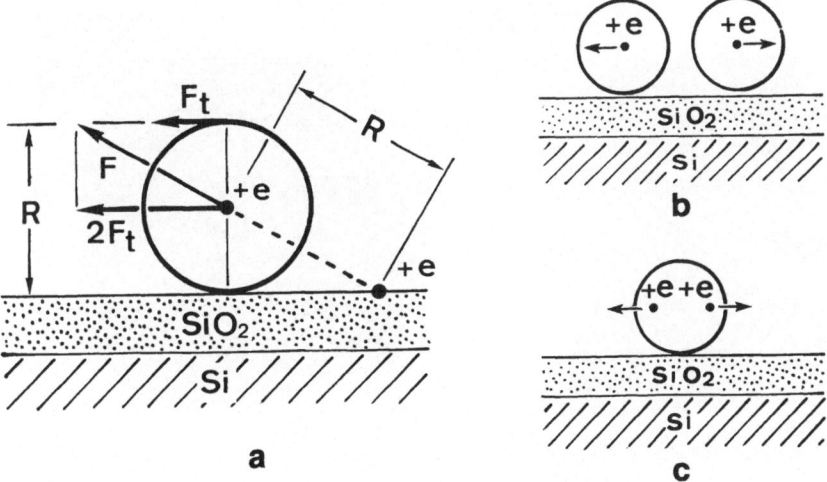

Figure 10 Coulomb forces imposing on the clusters due to the charging up. (a) Coulomb repulsive forces on a singly charged cluster acting to another charge on a nearby substrate. (b) A repulsion force between two singly charged clusters on the substrate. (c) Two positive charges, which might be generated temporarily on a cluster, which could cause a shear deformation of the cluster.

sized that the charging up is not perpetual but fluctuating, so that the coulomb force is also temporary.

4.3. A Possible Mechanism for Rearrangement of Atoms

Figure 11 shows a sequence of micrographs of a gold cluster similar to the one shown in Fig. 5 [18]. The last micrograph (Fig. 11d) was recorded one second later after the first one (Fig. 11a) was taken. The cluster is viewed along the exact [110] direction, and thus dark spots correspond approximately to individual columns of gold atoms. The arrangements of the atoms for the clusters of Figs. 11a, c and d are illustrated in Fig. 12a, b

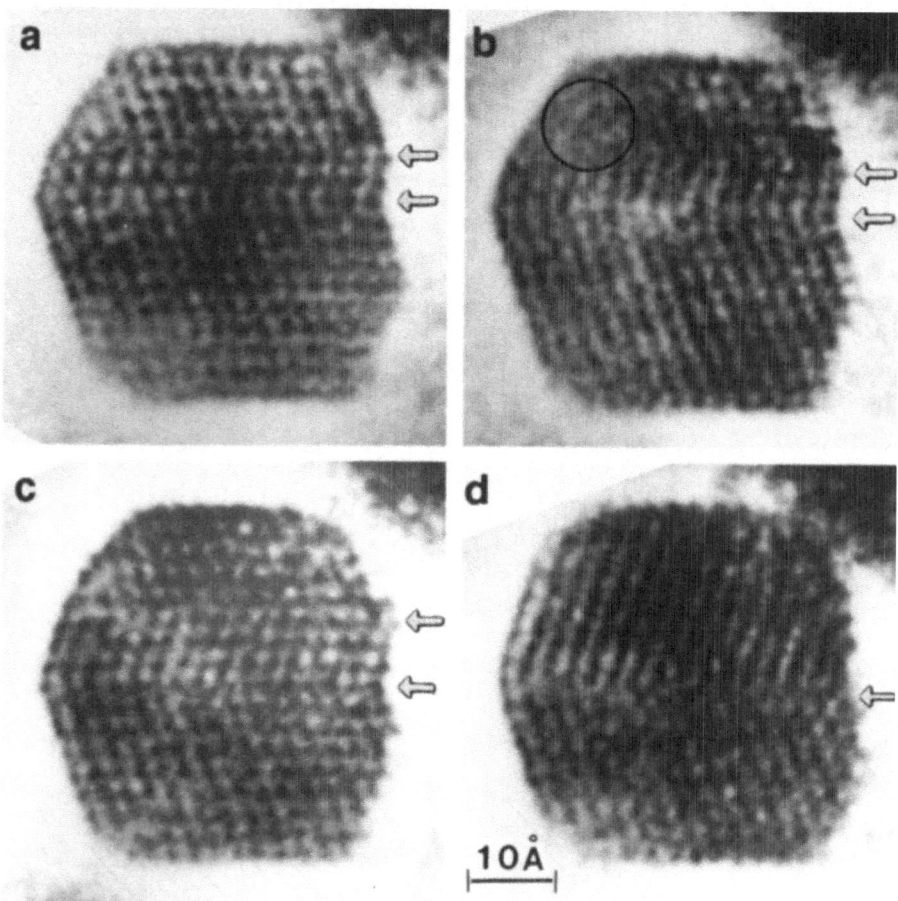

Figure 11   A sequence of the electron micrographs of a gold cluster in [110] zone axis. (a) Two twin planes separated by a distance of $3Xd_{111}$. (b) A half second after (a), the distance became $4Xd_{111}$. (c) After a another half second from (b), one of the twin planes has been removed to cause a shear deformation in the upper half of the cluster. A lattice anomaly (encircled region in Fig. 11c) may be caused by a twin dislocation.

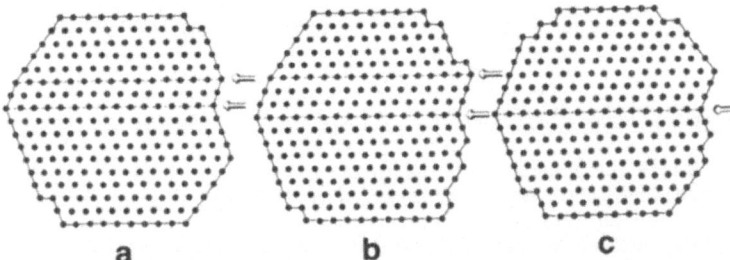

Figure 12 Three models for atomic arrangement of the clusters which are directly derived from the micrographs shown in Figs. 11a,c and d.

and c respectively. The cluster has a couple of twin planes (arrowed) which traverse the whole particle. One of the twin planes in Fig. 11c has shifted by one atomic plane and another one has disappeared to leave a single-twinned cluster in Fig.11d. In addition to the major reorganization of the atoms, small modifications of the surface atoms are also noticed, particularly near the edges of the cluster.

It is noted that a crystal orientation in the upper part of the cluster has been swung into the other direction after the removal of the twin plane (see Figs. 11d and 12c). In order to remove a twin, the cluster should have undergone a shear deformation. It took place possibly by a twin deformation. The deformation is accomplished by moving a twin dislocation through the cluster. A region where the lattice image becomes blurred as indicated by a circle in Fig. 11b appears to show this twin dislocation. A similar lattice modification is also found in a cluster shown in Fig. 2f which is enlarged in Fig. 13. The encircled region shows a somewhat hexagonal arrangement of atoms, which can be explained by the twin deformation as well. An

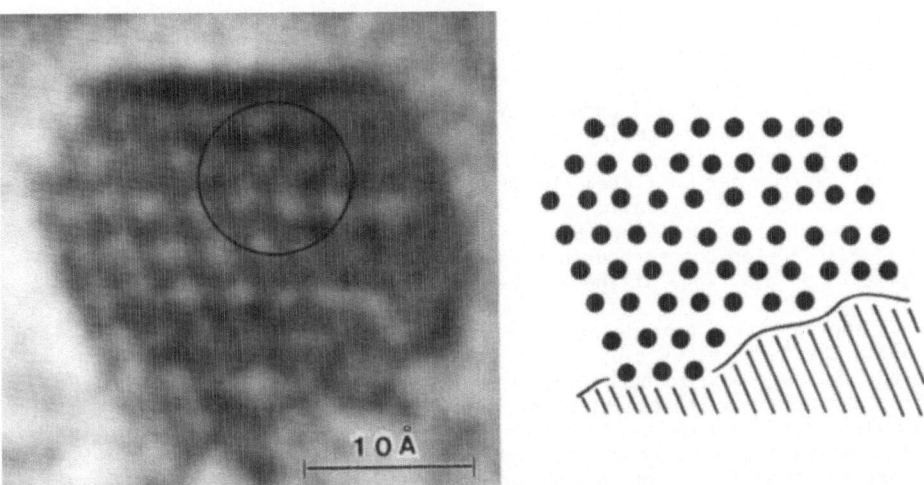

Figure 13 Another micrograph showing a lattice anomaly in a gold cluster appearing in Fig. 2f. A circled region shows a hexagonal packing of the atoms (See the insert).

origin of the shear force to cause the twin deformation will be discussed below.

4.4. Shear Deformation of the Clusters

Let us consider the shear deformation of fcc metal clusters in terms of the coulomb force. Suppose that a spherical cluster of a diameter R carries a single positive charge which repels another positive charge located at a distance R on a nearby substrate. Its coulomb force F has a tangential component F_t as shown in Fig. 10a. Furthermore, if the cluster is adhered firmly to the substrate, it can not roll away from the positive charge on the substrate. Therefore, the tangential force F_t imposes a shear on the cluster. If a distance R is 20 Å, a magnitude of the coulomb force F becomes 6×10^{-6} dyn and the shear force F_t becomes $\frac{1}{2}(3/4)^{\frac{1}{2}}F$, which is 2.6×10^{-6} dyn. A maximum stress which is required to move a dislocation in a metal crystal is generally an order of 1/10 of Young's modulus, and 3×10^{-5} dyn for the case of gold. The maximum stress therefore becomes comparable to the shear force and the 20 Å cluster can be deformed more easily than a larger one.

The shear deformation of the clusters which creates or annihilates stacking faults and twin planes can explain the cooperative atomic arrangement in the small clusters. To deform complicated MTP clusters which contain non-parallel twin planes (see Figs. 3c and d), shear forces which act in radial directions can be required. Such forces can be generated by multiple charging of the clusters. According to time-of-flight mass spectrum experiments [22], Multiply charged clusters could have been exploded as their sizes decrease. The coulomb explosion was observed for very small metal clusters of less than 30 atoms. The clusters which were dealt with in the present study could be too large for the coulomb explosion but small enough to cause the cluster deformation. The nature of such multiple charging in metallic clusters has not been explored yet.

5. CONCLUSION

A morphology of a crystal under a thermal-equilibrium condition is discussed by the Curie-Wulff theory which is based on minimization of surface energy ($\int \gamma(n)dS$), where $\gamma(n)$ is the surface energy of a particular crystal surface per unit area. In this context, the present experiment is to give an idea on the validity domain of the classical consideration of the crystal morphology. The fact that even a 20 Å gold cluster often has a very well-faceted cuboctahedron suggests the validity of the classical argument to this range of the cluster sizes.

On the contrary, considerations along the Curie-Wulff theory can not explain the featureless clusters of less than 10 Å in size described in sect. 3.3. They tend to become round and do not show the fcc crystalline structure. The results can be understood as one of the consequences of the anomalous surface tension. As a cluster size decreases, a surface tension will have its large component towards a center of the cluster. This tension compresses the cluster so that it would become spherical. In this consideration,physical roles of the corner or edge atoms whose numbers are no longer trivial to the total system are not clear. Theoretical studies on cluster structures based on quantum mechanics should be awaited for larger clusters of several hundred atoms,as presented in this study.

Fluctuations in morphologies of small clusters are hardly explained by considerations along the Curie-Wulff theory, which predicts a stable crystal shape for a given number of atoms. Obviously there are various saddle points in the total energy surface which are energetically subtle under a given temperature. The structural instability is regarded as a special phenomenon which takes place only under the intense electron beam irradiation. The present experiment favors to say that the instability is most likely caused by the electronic excitations rather than the electron beam heating.

Temporary charging up of the clusters appears to play an important role in the transformation of the clusters. The positively charged clusters are sheared by the coulomb repulsive force. The shear deformation might be initiated by the strain induced through the contact plane with the substrate.

The essence of our observations lies in the unstable behavior of metallic clusters. It is not straightforward at the present stage to interpret the observed phenomena in terms of macroscopic concepts such as the averaged temperature, because its fluctuation would be large in such a small cluster. It has been proposed that the state of the fluctuating cluster structures, which is neither solid nor liquid according to the conventional concepts of matter, is called a "quasi-solid" state [8].

6. LITERATURE

1. M. R. Hoare : Advances in Chem. Phys., 40, 49 (1979)
2. K. Mihama and Y. Yasuda : J/ Phys. Soc. Jpn., 21, 1166 (1966)
3. J. G. Allpress and J. V. Sanders : Surf. Sci., 7, 1 (1967)
4. S. Ogawa, S. Ino, T. Kato, and H. Ota : J. Phys. Soc. Jpn., 21, 1963 (1966)
5. S. Ino : J. Phys. Soc. Jpn. 27, 941 (1967)
6. L. Marks : Phil. Mag. A 49 81 (1984)
7. S. Iijima : Nature 315, 628 (1985)
8. S. Iijima and T.Ichihashi : Phys. Rev. Lett. 56, 616 (1986)
9. S. Iijima : Optik 47 437 (1977)
10. S. Iijima : Optik 48, 193 (1977)
11. S. Iijima and T. Ichihashi : Jpn. J. Appl. Phys. 24, L125 (1985)
12. J/-O. Bovin, R. Wallenberg and D. J. Smith : Nature 317, 47 (1985)
13. S. Iijima : J. Electron microscopy 34, 249 (1985)
14. D. J. Smith, A. K. Petford-Long, L. R. Wallenberg and J.-O. Bovin : Science (1986), in press.
15. L. D. Marks : Phys. Rev. Lett. 51, 1000 (1983).
16. S. Iijima and M. Ichikawa : J. Catal. 94, 313 (1985).
17. A. L. Mackay : Acta. Crystallogr. 15, 916 (1962)
18. S. Iijima and T. Ichihashi : Proc. XIth Int.Cong.on Electron Microscopy, Kyoto, 1986, p.1439
19. Ph. Buffat and J. -P. Borel : Phys. Rev. A13, 2287 (1976)
20. S. Iijima : Proc.XIth Int.Cong.on Electron Microscopy, Kyoto, 1986, P.87
21. A. Howie : Nature 320, 684 (1986)
22. K. Sattler : Surf. Sci, 156, 292 (1985)

Melting and Freezing of Microclusters from Analytics and Simulations

R.S. Berry

Department of Chemistry and the James Franck Institute,
The University of Chicago, Chicago, IL 60637, USA

Analytic theory and numerical simulations indicate that microclusters may
exhibit solid-like and liquid-like phases which can coexist over a non-zero
range of temperature. The lower bound for existence of a stable liquid-like
phase is a sharp freezing temperature T_f and the upper bound for existence
of a stable solid-like phase is a sharp melting temperature T_m. The temperatures T_f and T_m are unequal for small systems; between them is the temperature T_{eq} at which the free energies of the solid-like and liquid-like
clusters of a specified composition are equal. Observable phase-like behavior also requires that a dynamical criterion be satisfied, i.e. that the
cluster persist in each phase long enough to develop well-defined equilibrium
properties of that phase. This criterion is met by small clusters of argon
of most but not all sizes. Taking averages inadvertently over both phases
can hide the coexistence phenomenon; averaging properties separately for the
two phases exhibits the double-valued character of the equation of state and
other properties in the coexistence range.

1. Introduction and Theoretical Background

The phase equilibrium of clusters can be studied in a detailed manner that
cannot be approached if one begins with bulk matter. Because small clusters
can be treated as molecules, solid clusters are ordinary molecules with small
amplitude vibrations and rigid-body rotations, and liquid clusters are very
non-rigid molecules with large-amplitude motions. With this conceptual starting point, a short sequence of steps takes us to a theory of the equilibrium
between solid-like and liquid-like forms of a cluster. First, postulating
that the two forms may exist, we select a model for each form from which the
corresponding energy levels and degeneracies - i.e. the densities of states -
can be estimated. These models are taken as limiting cases along a scale of
non-rigidity, and the conservation of parity, angular momentum and permutational symmetry among identical particles permits the connection of the limits in an energy level correlation diagram [1]; Fig. 1 is an example for Ar_5.
The models for the limiting cases may be chosen at whatever level of accuracy
is appropriate. It is useful to take for the solid limit a model that represents fairly realistically the equilibrium geometry and harmonic normal
modes of the cluster in its ground state [2], although simpler models have
been used [3]. For the liquid form of a cluster of N identical atoms, it has
been adequate [2,3] to use the density of states based on the model of 3N-3
identical harmonic oscillators, which is just the model of Gartenhaus and
Schwartz [4] for liquid-like nuclei. Greater accuracy could be achieved for
the liquid from molecular dynamics simulations, by constructing the velocity
autocorrelation function and taking its Fourier transformation.

From the densities of states, one can readily construct the partition
functions and from them the thermodynamic functions of the solid-like and

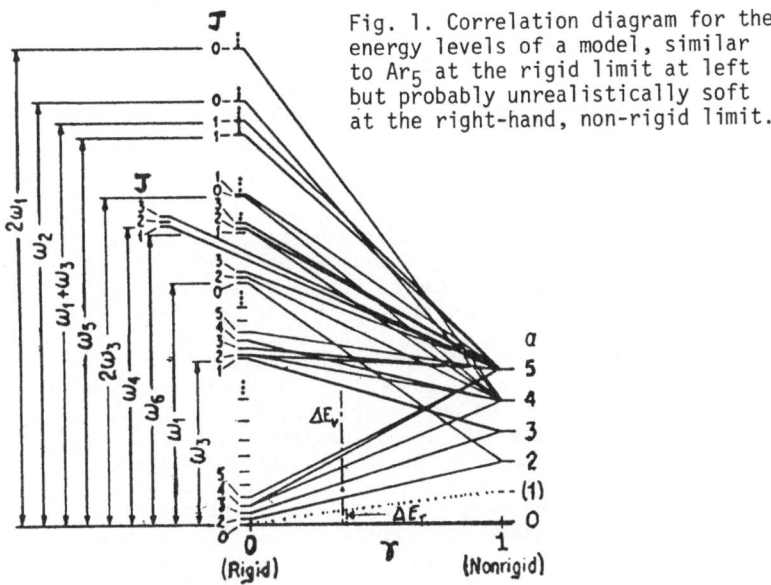

Fig. 1. Correlation diagram for the energy levels of a model, similar to Ar_5 at the rigid limit at left but probably unrealistically soft at the right-hand, non-rigid limit.

liquid-like forms of the cluster of any specific number of atoms. The essential point at this stage is that the density of states for the solid-like form is larger than that of the liquid-like form at low energies, but the liquid-like form gains states considerably faster than the solid-like form at higher energies. The consequence of this is that at low temperatures, for which only low-lying states are well populated, the solid form has the larger partition function and the lower free energy. At high enough energies, the reverse holds, and the liquid form has the larger partition function and the lower free energy. Fig. 2 illustrates this with the cumulative densities of states for a model case of Ar_5 but with an unrealistically weak Ar-Ar attraction for the liquid form, a necessity of the particular computational procedure. The outcome of this part of the analysis is simply that if the two forms coexist, then properties of each form are calculable from the corresponding thermodynamic functions. In fact, the results of prior simulations [5] were reasonably well fit by this simple quantum statistical model [3], properties such as the N dependence of T_{eq}, the temperature range ΔT over

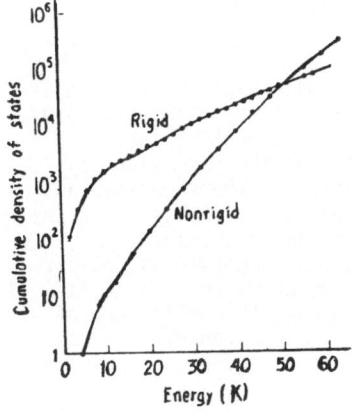

Fig. 2. Cumulative densities of states for the two limiting cases of Fig. 1.

which the solid-like and liquid-like forms might be seen in equilibrium, and the surface free energy and surface tension [3,6].

The next stage of the theory is the search for necessary and sufficient conditions for the validity of the assumption of coexistence of the two phase-like forms. This section concludes with the background for that search; its results and consequences are the basis for the next section.

Stability is always associated with a minimum in a generalized potential. For liquids and solids, the minimum is that of a free energy, and for clusters as we examine them here, of the free energy of a cluster of fixed number of constituent particles. The subtle point is the selection of the quantity with respect to which the free energy is minimized. In this instance, the quantity is a parameter measuring the degree of non-rigidity of the cluster; the non-rigidity parameter γ which quantifies the abscissa of the correlation diagrams is much like a Landau order parameter. We define γ as twice the ratio of two excitation energies. The numerator, ΔE_r, is the lowest excitation energy to a state which becomes the lowest excited rotational state in the rigid limit; the denominator, ΔE_v, is the lowest excitation energy to a state that becomes the lowest rotationless, vibrationally excited state in the rigid limit. The reason we define $\gamma=2\Delta E_r/\Delta E_v$ is that if the non-rigid limit is taken to be the Gartenhaus-Schwartz harmonic model, then at that limit, $\Delta E_v=2\Delta E_r$, so at that limit, $\gamma=1$. If the vibrations were infinitely stiff at the rigid limit, γ would be zero there. Hence, as defined, $0<\gamma<1$. The choice of γ is a direct analogy to the parameter defined by Yamada and Winnewisser [7] for triatomics and linear molecules.

With γ defined, the Helmholtz free energy F can now be treated as a function of the thermodynamic variable T and the parameter γ:$F=F(T;\gamma)$, rather than as a function of temperature alone. If the solid-like form of a cluster is thermodynamically stable at a temperature T, then at some value of γ near zero, $[\partial F(T;\gamma)/\partial\gamma]_T = 0$ and $[\partial^2 F(T;\gamma)/\partial\gamma^2]_T>0$ unless, of course, the minimum in $F(T;\gamma)$ occurs at the lower limit of the range of γ. Similarly, if the liquid-like form is stable, the same conditions must hold at some value of γ near 1, again unless the minimum is a boundary minimum. Thus one minimum in $F(T;\gamma)$ as a function of γ is the necessary and sufficient condition for thermodynamic stability of a single phase; two minima in $F(T;\gamma)$ for different values of γ at the same temperature is a necessary and sufficient condition for the thermodynamic stability of two "phases" of the cluster. The next step is therefore to examine whether the correlation diagram provides enough information about the densities of states to tell us what to expect of the minima in $F(T;\gamma)$.

2. The Prediction of Coexistence

To inquire into the form of $F(T;\gamma)$ we need only recognize that every state at the left-hand side of Fig. 1 must connect with a state at the right-hand side and conversely; also, we assume that most of the lines connecting the two limits have slopes over most of the range of γ that have the same sign as $E(\gamma=1)-E(\gamma=0)$, that is, the slopes of the connections have the same sign as a straight line connecting the two limits. This immediately implies that the energy levels populated at very low temperatures have positive slope, $[\partial F/\partial\gamma]_T>0$, for all γ. This in turn implies that at low temperatures the partition function is a monotonically decreasing function of γ, and that $F(T;\gamma)$ is a monotonically increasing function of γ, so that only the solid form of the cluster is stable. Exceptions to this are the rare substances such as Li_3 and Na_3, and probably small clusters of helium whose zero-point amplitudes extend throughout all their potential minima.

At higher temperatures, levels are occupied which for the most part have negative slopes, i.e. for which $\partial E(\gamma)/\partial\gamma<0$. Fig. 2 illustrates for one case how the density of states near the non-rigid limit overtakes that of the rigid limit so much that the cumulative density of the non-rigid limit becomes the larger. This means that the partition function for γ near 1 receives significant contributions from energy levels with negative slope at temperatures at which the partition function near $\gamma=0$ is dominated by levels with positive slopes. In turn, this means that $F(T;\gamma\approx1)$ has a falling off in slope as T increases, and, as T increases still more, a change from positive to negative slope. The temperature at which this change in slope occurs is T_f, the lower limit of stability of the liquid-like form of the cluster, to which we give the obvious name of "freezing temperature".

As the temperature is increased further, the negatively-sloping energy levels dominate a larger part of the range of γ until at high enough T, the partition function becomes a monotonically increasing function of γ and the free energy, a monotonically decreasing function of γ, except perhaps for a small region near $\gamma=1$. At such temperatures, only the liquid-like form of the cluster is stable. Because the energy levels are smooth functions of γ, $F(T;\gamma)$ is a smooth function of both T and γ. Hence there must be a value of T for which $[\partial F(T;\gamma)/\partial\gamma]_T=0$ at some value of γ near $\gamma=0$. This is precisely the upper limit for the thermodynamic stability of the solid-like form of the cluster, which we call T_m, the "melting temperature". Thus, from only general qualities of the density of states as a function of non-rigidity, we conclude that finite clusters of most substances should exhibit sharp freezing and melting temperatures but that these temperatures, T_f and T_m, should be different. This is behavior unlike any phase transition of any order for bulk matter; it is a property specific to finite systems. And of course T_f and T_m are functions of N; we can expect $T_m(N)-T_f(N)$ to go to zero as N becomes large, but not necessarily monotonically.

Another aspect of the solid-liquid equilibrium enriches the picture further. In the temperature range where the two forms may coexist in equilibrium, each has its own free energy. Conventional thermodynamics tells us immediately that the relative concentration is determined by the temperature-dependent "equilibrium constant" $K_{eq}\equiv\exp[\Delta F(T)/kT]$, where $\Delta F(T)=F_{solid}-F_{liquid}$ and the concentration ratio [liquid]/[solid]=$K_{eq}(T)$ if the system is in equilibrium. The implication of the foregoing argument is that K_{eq} has non-zero, finite values throughout the range $T_f<T<T_m$ but for $T<T_f$, K_{eq} is zero and for $T>T_m$, K_{eq} is infinite. Whether the amount of liquid present at temperatures just above T_f, or of solid present just below T_m is large enough to observe cannot be answered on the basis of any general arguments. We must turn to specific examples to estimate whether a discontinuity in K_{eq} would be observable. The possibility of the coexistence of two phases of a cluster, e.g. two solid forms, had been recognized by Nauchitel and Pertsin [8] on the basis that the free energy difference between the two forms is finite for a cluster.

Another general question can be raised here but cannot be answered without information about specific systems. Put most broadly, the question is simply "can we expect to see the two forms, solid-like and liquid-like, coexisting in a dynamic equilibrium?" For a phase to be detectable, it must persist long enough in that form for its properties to be seen and measured. If the intervals spent in each form are too short for the cluster to develop the averaged equilibrium characteristics of that form, we can only observe the averages over both forms, and therefore must see a sort of slush, rather than two distinct phases. Which of these is the correct description for any particular cluster is only answerable from dynamics, not from the statistical thermodynamics of equilibrium. The next section, on simulations, treats this matter for clusters of argon atoms.

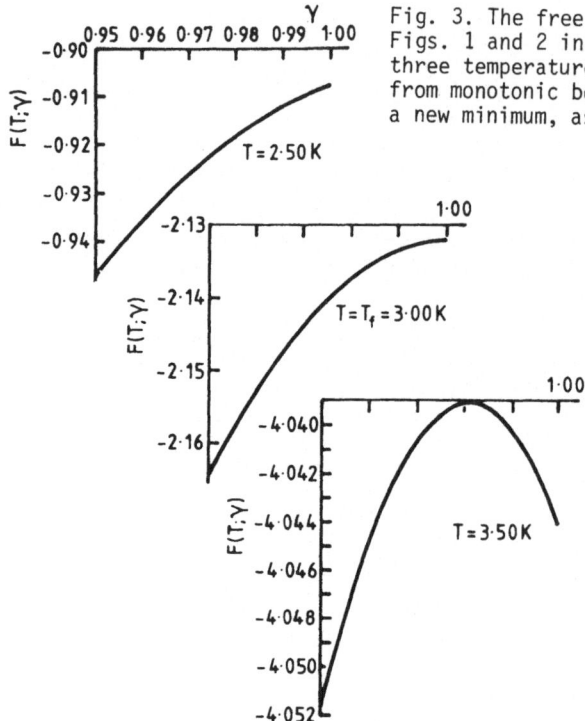

Fig. 3. The free energy for the example of Figs. 1 and 2 in the region near $\gamma=1$, for three temperatures, illustrating the change from monotonic behavior to a zero-slope to a new minimum, as T increases.

To conclude this section, we cite the results of calculations of the free energy as a function of γ, in the range $0.95<\gamma<1$ and for T in the neighborhood of T_f, for the cluster pseudo-Ar_5 used to generate Figs. 1 and 2. Fig. 3 shows $F(T;\gamma)$ in this liquid-like end of the range, and for three temperatures. To do these calculations, it was assumed that the connections in the correlation diagram are all straight lines. The three curves demonstrate that indeed, for this model $F(T;\gamma)$ does develop a minimum in the liquid-like region. For this specific model system, the liquid-like minimum occurs at $\gamma=1$. However this is not necessary in general, either conceptually or logically. Fig. 4 is a sketch of how the curves of $F(T;\gamma)$ vs. γ might look at various temperatures for a real substance.

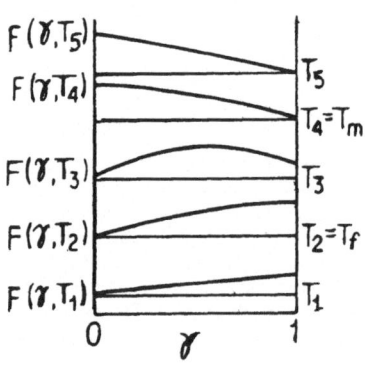

Fig. 4. A sketch of the simplest sort of behaviour $F(T;\gamma)$ might show in order for the cluster to exhibit a region of coexisting phases. For other possibilities, see Ref. 2.

3. Information from Simulations

Until we have considerably more information about the range of temperatures within which solid-like and liquid-like forms may coexist, it is difficult to plan laboratory experiments to test the theory just outlined. Other difficulties may have to be overcome also, such as the problem of dealing with the polydispersity of clusters prepared by most techniques now available. Furthermore, any laboratory experiment that is designed to study a system at equilibrium is always vulnerable to the charge that it is only displaying the results of a metastable condition. In the case of trying to find an equilibrium between solid and liquid, this charge touches the issue at its heart.

Consequently investigating the freezing and melting of clusters via simulations is not only useful; it is necessary, if not sufficient, to establish or destroy the validity of the theory. Furthermore, dynamic simulations are necessary to indicate to us whether there are really any systems in which the solid and liquid forms may be seen together and yet be distinguished.

For these reasons, we undertook an extensive set of simulations of argon clusters, testing results by others [5,9] and extending the calculations in several ways. Here we review the results of the isoergic molecular dynamics (MD) simulations of Ar_7 and Ar_{13} in some detail and treat more briefly the isoergic MD simulations of other argon clusters and the MD and Monte Carlo (MC) simulations of Ar_{13} done at constant temperature.

All the simulations described here were done with pairwise Lennard-Jones potentials. We took the internuclear distance for zero interaction potential $\sigma = 3.4 \times 10^{-10}$ m, and well depth $\varepsilon = 1.67 \times 10^{-21}$ j, values chosen because they are reasonably accurate and have been used in all the other simulations with which we wish to compare results.

We begin with Ar_{13}, the clearest, least ambiguous case [10]. The first, most important diagnostic here is the mean kinetic energy, i.e. the temperature, as a function of energy. At low energies such as -5×10^{-21} j per atom, the temperature, defined by 500-step averages, is essentially constant and the cluster is a tight icosahedron. At higher energies, the temperature shows small fluctuations; at an energy as high as -4.36 (the same units will be used throughout), occasional permutational isomerizations occur but the time required to accomplish one of these is brief, of order 1000 steps or two averaging periods, and the cluster is essentially always an icosahedral solid. At energies from about -4.42 to about -3.80 (probably ±0.02) the temperature, as a function of time, shows a bimodal distribution, with the temperature persisting in the vicinity of one mode or the other for long intervals, tens of averaging periods or more. An example of the time history and bimodal temperature distribution for one energy in the coexistence range is shown in Fig. 5.

Clearly the two temperatures of the bimodal distribution correspond to two kinds of regions on the potential surface. Examination of the structure of the cluster in these regions confirms our preconceptions, that the high-temperature form corresponds to the cluster residing in a deep, narrow potential well, with the form and structure of a hot solid; the low-temperature form corresponds to residence in a region of high potential and many weakly separated minima so that the cluster moves throughout a large "area" of the potential surface. Frequently in this region the geometry of Ar_{13} is recognizable as an icosahedron that has had one of the 12 equivalent atoms promoted from the closed shell to the "surface", leaving a vacancy. Both the promoted

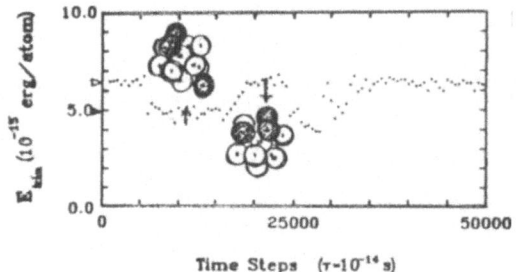

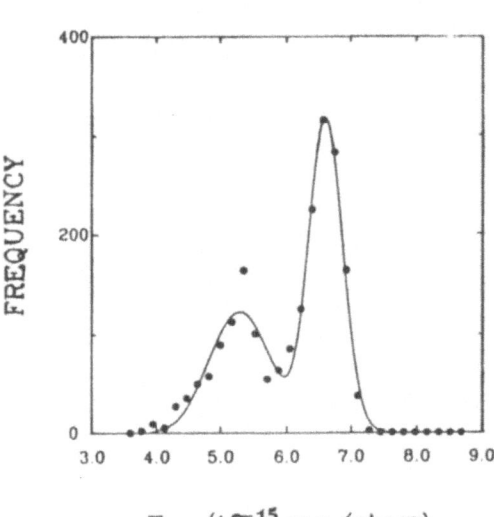

Fig. 5. (a) The time history and (b) the bimodal temperature distribution of an Ar_{13} cluster at a temperature in the coexistence region.

particle and the hole are mobile so that the cluster looks like a liquid to someone following its configuration. Hence we are justified in calling the energy range of the bimodal temperature distribution a coexistence range for solid-like and liquid-like clusters.

More quantitative measures confirm the solid-like behavior of the low-energy form and the high-temperature form in the coexistence range of energies, and the liquid-like character of the high-energy form and its extension as the low-temperature form in the coexistence region. The standard diagnostics here are the mean square displacement as a function of time and its slope which is the diffusion coefficient, the relative rms bond length fluctuation, the velocity autocorrelation function and its Fourier transform, which is the spectral density for vibrational states. The cold solid shows essentially no diffusion; the warm liquid has a high diffusion coefficient (but the curve of mean square displacement, which is linear in time to approximately the square of the cluster radius, flattens in that region), and in the coexistence region the warm solid shows a small but non-zero slope, while the cool liquid shows a considerably larger slope but smaller than that of the liquid in its single-phase region.

The velocity autocorrelation functions show a characteristic "bounce" at short times for the solid but a much damped return of velocity for the liquid. The spectral densities provide even sharper distinction between the two forms: the liquid has a significant density of states close to zero frequency, the so-called "soft modes", while the solid has a very low density of states there in the coexistence range of energies and essentially no soft modes at lower energies.

The evidence is thus rather strong that Ar_{13} has a two-phase region of energies. Before concluding the discussion of this cluster with an examination of its caloric function - temperature vs. energy - we anticipate a point that has been studied by MC rather than MD, but which is relevant here; exactly the same observations and reasoning could have been done with MD simulations. One objection [9] was raised to the idea of distinct phase-like forms on the ground that the angular distribution function determined by Monte Carlo simulation shows a significant density at 90°, at 33 K, a temperature in the coexistence range for which solid clusters should predominate. (The concept of the co-existence range had not yet been introduced.) Furthermore the mean potential energy as a function of the temperature of the simulated ensemble varied smoothly, showing no characteristics of a first-order phase transition. We return shortly to the variation of energy or potential energy with temperature. Regarding the angular distribution function, more recent and more extensive isothermal MC simulations [11] of Ar_{13} have shown a bimodal distribution of potential energies. If the two parts of this distribution are averaged separately, so that each has its own angular distribution function, then the solid-like, low-potential form has a very small density at 90° while the liquid-like form has a very significant density there. The average of the two for 33 K, weighted by the frequency factors, is just the distribution reported in Ref. 9. The variations with temperature of the average and individual angular distributions are smooth and unexceptional.

The culminating point for this case is the behavior of the two average temperatures as functions of energy in the isoergic MD simulation. One is the short-time average temperature (500 step average); the other is the average taken over the entire time of the simulation at that energy. The short-time averages give the bimodal distributions mentioned previously. If they are used separately to specify points in a graph of <T> vs. E, we obtain the points indicated by triangles in Fig. 6; if the full-time averages are used, we obtain the points indicated by circles. What this figure shows is now very clear indeed: the equation of state is single-valued at low temperatures and low energies where only the solid exists, and again at high energies and temperatures where only the liquid exists; between these two is a region in which the equation of state is double-valued. We do not know what that equation is in any analytic sense but we know its form from Fig. 6. The double-valued region can of course be treated as single-valued by taking the weighted average of the two branches. That is just what the black circles represent and what was done in Ref. 10. The implication of the analytic theory outlined in the previous section is that because the equilibrium constant is discontinuous, the average curve connecting the single-phase regions should also be discontinuous. However if the fraction of the total material in the minority phase is smaller than about 10 or 15%, it is unlikely that the simulations to date could exhibit that discontinuity. The prediction of the temperature bounds of the coexistence region and the calculation of accurate free energies for the two forms at the boundary temperatures seems now to be a formidable challenge.

MD simulations by Briant and Burton [5] had indicated loops, like van der Waals loops, in the fully averaged curves of <T> vs. E in what we now recog-

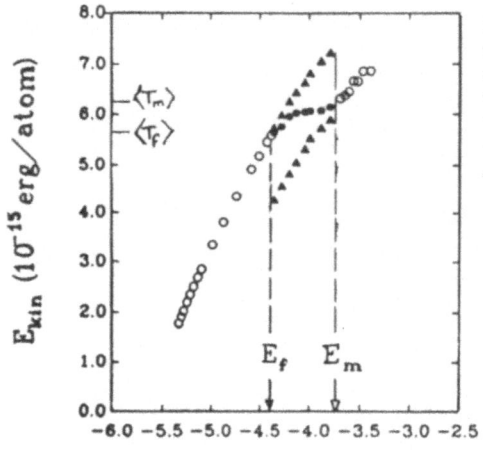

Fig. 6. The mean temperature or kinetic energy as a function of the energy for Ar_{13}, showing the single-valued regions (open circles), the two branches in the double-valued region (triangles) and long-time average in the double-valued region (filled circles).

nize as the coexistence region. Those loops are artifacts of runs too short to give proper averaging. By running for 10^4 steps as Briant and Burton did, instead of 5×10^5 or 10^6, and changing the energy by heating, we were able to reproduce the loops. Just how long is required to establish equilibrium averages is of course a deep question connected with the observability of ergodic behavior, which brings us to the MD simulation of the Ar_7 cluster.

The motivation for simulating Ar_7 by isoergic MD was initially quite different from that for Ar_{13}. The reason for studying Ar_7 was the testing of a conjecture [3] that melting can be characterized usefully by the breakdown in the separability of two time scales. The shorter time scale is that of the vibrations of the system about a minimum on its potential surface; the longer time scale is that of passage from one such minimum to another. This question has been studied in one way, by following the history of the energies of the potential minima about which the system vibrates [12]. This work indicates that the time scale for changing the energy of the underlying potential minimum is very long for solid-like systems and becomes shorter and shorter as the energy or temperature is raised, particularly as the system becomes liquid. By inference, from the supposition that long intervals above the same unchanging value of the potential minimum mean long intervals in the same potential, the conjecture is well borne out by these simulations.

The simulations of Ar_7 [13] were complementary in the sense that every permutation of the system and every passage from well to well was followed explicitly up to the point at which the jumping rate between wells exceeded the sampling rate. Sampling consisted of stopping the dynamics and quenching the system down to the bottom of the potential well above which the system moved at the moment of sampling. This was the procedure used previously in Ref. 12.

Ar_7 has four distinct equilibrium geometries, the most stable of which is the pentagonal bipyramid. The statistical weights for the structures, if the component atoms are classical, are 504, 1080, 1080 and 5040, in order of increasing energy. As with Ar_{13}, The Ar_7 cluster exhibits a single-form region at low energies where the solid is stable, another single-form region at high energies where the liquid is stable, and between them a two-phase region, from about 16.2 K to about 20 K. As with Ar_{13}, the mean bond length fluctuation

increases very gradually with energy up to T_f, where it is about 0.1, and then
jumps sharply upward. That is, the Lindemann criterion for formation of a
liquid, that the bond length fluctuations be about 10%, is supported by these
simulations.

The rate of isomerization of Ar_7 is a very rapidly rising function of the
energy of the cluster. At 15 K, the rate is about 0.2 ns^{-1}; at 15.8 K, it is
1.0; at 16.38 K it is 2.0; at 18.28 K it is 8.36; at 19.63 K it is 26.5 ns^{-1}
and at 21.22 K, it is 63.7. Very long trajectories, approaching 10^7 steps,
exhibit distributions of total residence times among the four structures
that mimic the statistical weights of the structures. Shorter runs could be
used for high energies in the pure liquid region but only the longest runs
gave even semiquantitative agreement with statistical predictions for the
lowest part of the coexistence region.

Clusters of other sizes were also investigated [14], more to study sys-
tematics of size dependence than to examine each cluster in the detail used
for A_7 and Ar_{13}. For all cluster sizes studied, N of 7, 8, 9, 11, 13, 14,
15, 17, 19, 20, 22, 26 and 33, a rather sharp increase occurs in the mean
bond length fluctuation as a function of energy, and at slightly higher energies, the diffusion coefficient increases. The former seems to be associated
more with the onset of isomerization of the solid form and the latter, with
the beginning of stability of the liquid.

Bimodal temperature distributions like that of Ar_{13} occur for N of 7, 9,
11, 13, 15 and 19, within limits of energy that depend on N. No such bimodal
distributions were found for N of 8, 14 or 17 or for N larger than 19. Ar_{19}
is most like Ar_7 and especially Ar_{13} in having well-defined two-phase regions
and a particularly stable ground state structure. The freezing temperatures
of these three are somewhat higher than those of clusters of adjacent sizes.
Ar_{17} is unusual in passing between solid-like and liquid-like structures too
frequently to allow equilibrium properties of each form to develop; one would
expect observations of Ar_{17} to show a sort of slush. Possibly Ar_8 and Ar_{14}
are also "slush-balls". Clusters larger than Ar_{19} seem to have several structures with energies only slightly above their ground state energies and consequently instead of bimodal temperature distributions they show wide unimodal
distributions. We have not yet studied clusters as large as Ar_{55}, the next
size expected to show exceptional stability. It appears that if Ar_{33} has a
bimodal distribution, it occurs over a temperature range too narrow for us to
have resolved. Thus, for weakly bound clusters, it appears that the phenomenon of a discrete temperature band of solid-liquid coexistence is a property
of small clusters. One may imagine that with more strongly bound clusters
the phenomenon would extend to larger sizes, but this is presently only
speculation.

Monte Carlo and Nosé-type MD simulations [15] of isothermal conditions for
Ar_{13} confirm that the results obtained for this species by isoergic MD can be
used to predict properties of canonical distributions. The full curve of
<E> vs. T from the isothermal MC calculations is superimposable on the single-
valued curve of Fig. 6 of <T> vs. E obtained by MD simulation under isoergic
conditions, and if the MC results are broken into two distributions, the two
branches are recovered.

4. Concluding Remarks

The behavior of clusters under conditions where solid-like and liquid-like
forms may exist has given us new insights into the nature of the melting
process. This insight also may open new approaches to the use of clusters

for preparation of materials in forms that cannot be achieved any other way. Not only are the structures of solid clusters unlike the structures of bulk materials; the temperature and size of cluster can, as we now see, very much affect its form. The melting and freezing behavior dictate conditions required if one wishes to prepare annealed or unannealed depositions: a jet of clusters prepared at any temperature below T_f is presumably not annealed, but if the clusters have internal, vibrational temperatures above T_f, then at least a fraction will have experienced a period in liquid form and therefore presumably anneal. It may be possible to exploit differences in T_f's and rates of passage between liquid and solid forms to prepare streams of well-controlled partially annealed clusters. And of course the possibilities are fascinating indeed for using the melting and freezing characteristics of clusters for the controlled preparation of heterogeneous clusters [16].

5. Acknowledgments

The author wishes to express his appreciation to his collaborators and students Francois Amar, Thomas L. Beck, Heidi L. Davis, Julius Jellinek and Grigory Natanson, without whose efforts this work could not have been done. The author also acknowledges the hospitality of the Physical Chemistry Laboratory, Oxford University where this manuscript was written. The research by the University of Chicago group reported here was supported by a Grant from the National Science Foundation.

References

1. M.E. Kellman, R.S. Berry: Chem. Phys. Lett. 42, 327 (1976); F. Amar, M.E. Kellman, R.S. Berry: J. Chem. Phys. 70, 1973 (1979); M.E. Kellman, F. Amar, R.S. Berry: J. Chem. Phys. 73, 2387 (1980); G.S. Ezra, R.S. Berry: J. Chem. Phys. 76, 3679 (1982).
2. R.S. Berry, J. Jellinek, G. Natanson: Chem. Phys. Lett. 107, 227 (1984); Phys. Rev. A30, 919 (1984).
3. G. Natanson, F. Amar, R.S. Berry: J. Chem. Phys. 78, 399 (1983).
4. S. Gartenhaus, C. Schwartz: Phys. Rev. 108, 482 (1957).
5. C.L. Briant, J.J. Burton: J. Chem. Phys. 63, 2045 (1975); R.D. Etters, J.B. Kaelberer: Phys. Rev. A11, 1068 (1975); R.D. Etters, J.B. Kaelberer: J. Chem. Phys. 66, 5112 (1977); R.D. Etters, R. Danilowicz, J. Dugan: J. Chem. Phys. 67, 1570 (1977); R.D. Etters, R. Danilowicz, J.B. Kaelberer: J. Chem. Phys. 67, 4145 (1977); J.B. Kaelberer, R.D. Etters: J. Chem. Phys. 66, 3233 (1977).
6. N. Nishioka: Phys. Rev. A16, 2143 (1977).
7. K. Yamada, M. Winnewisser: Z. Naturforsch. A31, 134 (1976).
8. V.V. Nauchitel, A.J. Pertsin: Mol. Phys. 40, 1341 (1980).
9. N. Quirke, P. Sheng: Chem. Phys. Lett. 110, 63 (1984).
10. J. Jellinek, T.L. Beck, R.S. Berry: J. Chem. Phys. 84, 2783 (1986).
11. H.L. Davis, J. Jellinek, R.S. Berry (in preparation).
12. F.H. Stillinger, T.A. Weber: Phys. Rev. A25, 978 (1982); ibid. 28, 2408 (1983); J. Phys. Chem. 87, 2833 (1983); R.A. LaViolette, F.H. Stillinger: J. Chem. Phys. 83, 4079 (1985).
13. F. Amar, R.S. Berry: J. Chem. Phys. (in press).
14. T.L. Beck, J. Jellinek, R.S. Berry (in preparation).
15. S. Nosé: Mol. Phys. 52, 255 (1984); J. Chem. Phys. 81, 511 (1984).
16. T.E. Gough, D.G. Knight, G. Scoles: Chem. Phys. Lett. 97, 155 (1983).

Dynamics of Transition-Metal Clusters

S. Sawada

Fundamental Research Laboratories, NEC Corporation, Miyazaki 4-1-1, Miyamae-ku, Kawasaki 213, Japan

The structure and thermodynamic properties of transition-metal clusters containing N atoms are investigated for N=6 or 7 using molecular dynamics. The melting transition and phase diagram (energy-temperature curve) are studied. The "quasi-solid state" is found in the region of the temperature near and below the melting point, where clusters make structural transition from one isomer to others without topological change. The rate of transition is found to increase as temperature increases. This result is proposed as an interpretation for the structural instability of gold metal clusters observed experimentally by S.Iijima and T.Ichihashi (Phys. Rev. Lett. 56, 616 (1986)).

1. Introduction

It is known that the physical properties ,e.g. electronic structure and melting temperature of microclusters are different from those of macroscopic condensed matter. This paper compares dynamic properties of microclusters with those of bulk.

Recently, dynamic behavior of gold clusters containing several hundred atoms was examined by S.Iijima and T.Ichihashi [1] with an electron microscope. It was observed that the shapes of the clusters changed continually through an internal transformation from a single crystal to a twin crystal, and vice versa. They considered that the transformations were induced by an electron-beam irradiation , i.e. by charge fluctuation, and examined this experimentally to some extent. In this paper, it is shown that such kinds of structural instability take place even for isolated microclusters in one temperature region.

Microclusters in both stable and metastable states were observed experimentally as static structures. Those stable and metastable states correspond to local minima of the potential energy and there are potential barriers or activation energies among them [2]. With this in mind, the dynamic aspects of microclusters are considered. As long as the cluster excited in the neighborhood of any local minimum has kinetic energy lower than the activation energies, its motion is bounded to the region around the minimum and is close to harmonic. If kinetic energy increases and exceeds activation energy, and if the system is ergodic,the cluster moves into the region around another minimum and is temporarily trapped there. After vibrating for some period, it returns to the previous region or moves into the region around another minimum. This state, which appears between solid and liquid states, is called the "quasi-solid state". This conjecture will be demonstrated below using molecular dynamics for clusters containing 6 or 7 atoms.

2. Molecular dynamics of 6- or 7- atom cluster

To simulate cluster motion , the following model potential, which was used to calculate the properties of the transition-metal surface [3] and clusters [4,5] by other authors, is used intead of calculating the electronic cohesive energy:

$$V = 1/2 \cdot U \sum_j \{ A \sum_i \exp(-p(r_{ij} - r_0)) - [\sum_i \exp(-2q(r_{ij} - r_0))]^{1/2} \}, \quad (2.1)$$

where r_{ij} is the distance between the i-th and j-th atom, and r_0 is the lattice constant of the perfect crystal (FCC). Constant A is determined by minimizing the cohesive energy of the perfect crystal for lattice constant r_0. The values of U, p and q are determined such that the bulk cohesive energy and the bulk modulus calculated by (2.1) using the experimental value of r_0 are in good agreement with corresponding experimental results. Thus, values $p = 9 \cdot r_0^{-1}$ and $q = 3 \cdot r_0^{-1}$ are appropriate for transition metals [3]. After calculation A= 0.101035 and the lattice sum in (2.1) determines that the bulk cohesive energy $E_{bulk} = 1.17674 \cdot U$.
 Hereafter, we use r_0, E_{bulk} and $r_0 \cdot (m/E_{bulk})^{-1/2}$ (the order of the period of the atomic vibration) as the units of distance, energy and time, respectively, where m is the atomic mass.
 Searching minima of the potential energy surface of the 6-atom cluster gives two kinds of isomers, octahedron (OCT) and tripyramid (TP) corresponding to the global minimum and the local minimum, respectively. These are illustrated in Fig.1. Molecular dynamics is started with the initial condition that the atomic coordinates are arranged in OCT form and the atomic velocities are chosen randomly to establish some definite kinetic energy with the translational and rotational degrees of freedom frozen.
 The phase diagram (temperature as a function of total energy) given by molecular dynamics for the 6-atom cluster is shown in Fig.2. The temperature is defined by the time avarage of the kinetic energy [6,7,8],

$$T = (3N/3N-6) \cdot (2/3) \cdot \langle N^{-1} \sum_i m v_i^2 / 2 \rangle. \quad (2.2)$$

It is clear from the figure that the transition from solid state to liquid state is not sharp and the intermediate quasi-solid state appears.
 To get information about the temporary structure of the cluster from the atomic coordinates given by molecular dynamics, nearest neighbor distance r_n is introduced and the number of each atom's nearest neighbors is calqulated. A value $r_n = 1.2$, was found to be appropriate for this purpose. The number sets of the nearest neighbors are (4,4,4,4,4,4) for OCT and (4,4,3,3,5,5) or (3,3,4,4,5,5) e.t.c. for TP. Another

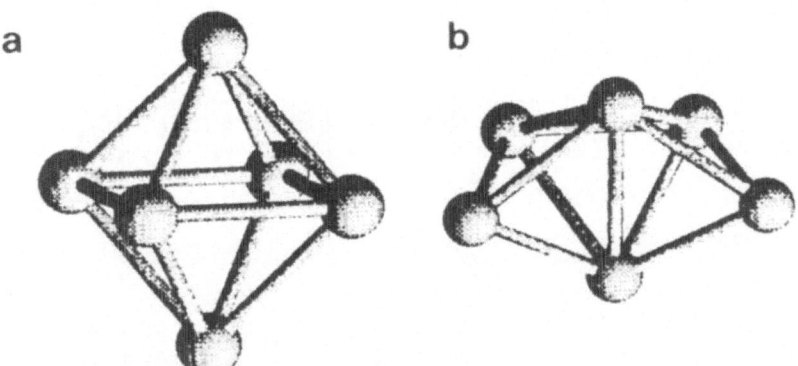

a   **b**

Fig.1. 6-atom clusters: (a) Octahedron (OCT), (b) tripyramid
(TP).

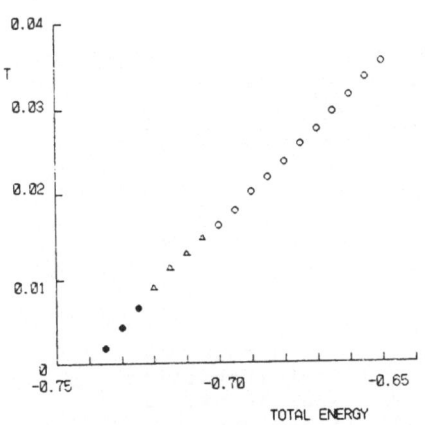

Fig.2. The phase diagram (temperature as a function of total energy) for 6-atom cluster:
● solid state, △ quasi-solid state, ○ liquid state.

quantity is introduced to reduce the information further. Let D
be a 6×6 matrix the value of whose elements is 1 for $|r_i-r_j| <$
r_n and 0 otherwise. A distance in matrix space δ is defined by

$$\delta = ||D-D_{oct}||^2/2, \qquad (2.3)$$

where D_{oct} is D for OCT and the norm of matrix A is defined by

$$||A|| = (\sum_i \sum_j A_{ij})^{1/2}. \qquad (2.4)$$

It should be noted that there are different OCT isomers for
the different atomic arrangements and, therefore, different
D_{oct}'s. At first, (2.3) is calculated using D_{oct} for the
initial OCT.
  The time evolution of δ is shown for various total energies
in Fig.3, where a - b and c - f corresponds to the solid
and quasi-solid states, respectively. Values $\delta=0$ and 2
correspond to OCT and TP, respectively, and $\delta=1$ to
(3,4,4,4,3,4) or (5,4,4,5,4,4) e.t.c. ,i.e. the intermediate
structures considered as fluctuation between OCT and TP. Value
$\delta=2$ also corresponds to (3,3,3,4,5,4) or (4,3,4,2,4,5) e t c.,
being considered a fluctuation from TP. It is seen that in the

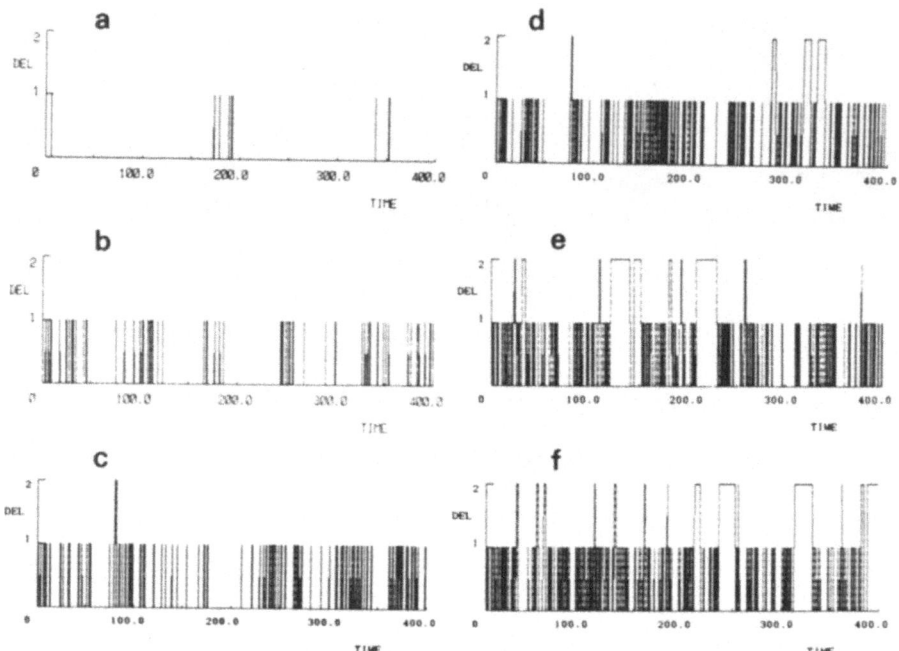

Fig.3. The time evolution of δ for various total energies:
E= (a) -0.730, (b) -0.725, (c) -0.720, (d) -0.715, (e) -0.710
and (f) -0.705.

solid state the motion is limited to vibration around OCT and,
on the other hand, in the quasi-solid state the transition
between OCT and TP takes place continually and the system
vibrates in some period around OCT or TP. It is "breathing"
motion, which is distinct from melting; there is no exchange
of the positions of the atoms.
  Next, motion in the liquid state is examined. The time
evolution of δ for total energy E=-0.700 is shown in Fig.4.a.

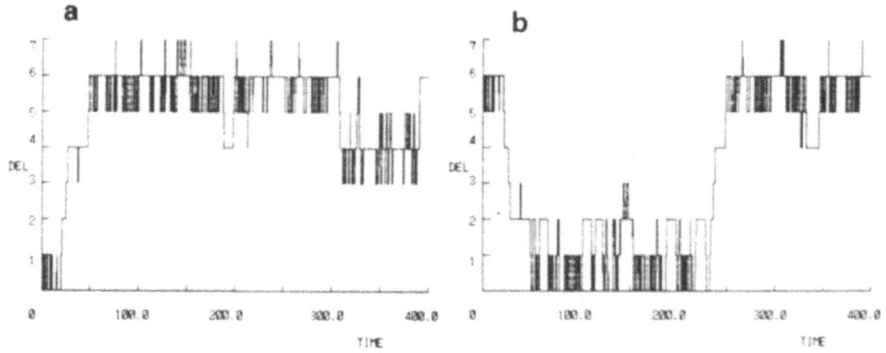

Fig.4. The time evolution of δ for total energy E= -0.700.

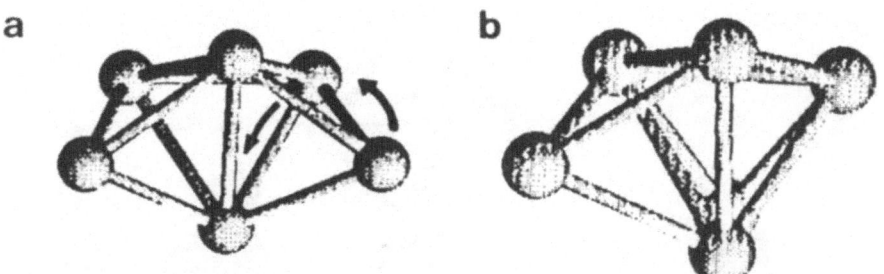

a   **b**

Fig.5. Illustration of cooperative motion of two atoms, which
makes the transition from one TP isomer (a) to another TP
isomer (b).

It is seen that δ takes values greater than 2. It was found
by tracing the time evolution for the numbers of each atom's
nearest neighbors that the change in the relative positions of
the atoms takes place as shown in Fig.5. The cooperative
motion of the two atoms makes the transition from one TP
isomer to another TP isomer ($\delta=4$) around t=25.0. The
following breathing motion makes the transition from TP to OCT
($\delta=6$), which is another OCT isomer different from the initial
one, i.e. the two atoms in it have exchanged. Using D_{oct} for
the new OCT in (2.3), time evolution of the new δ is calculated. The result is shown in Fig.4.b. It is seen that the
system exhibits breathing motion around the new OCT for a while
and then is suddenly interrupted by exchange motion. It was
found that as total energy (temperature) increases the
period of the breathing motion decreases and the exchange
motion takes place more frequently.

 Below, the qualitative features of the 7-atom cluster
dynamics, which is very similar to that for the 6-atom
cluster, are shown. There are four kinds of isomers;
pentagonal bipyramid (PBP), octahedron plus one (OCT+1),
skewed arrangement (Skew) and incomplete stellated tetrahedron
(IST) in increasing order of potential energy. They are
illustrated in Fig.6. The quasi-solid state appears in a
temperature region between the solid state and the liquid
state. In the quasi-solid state, the breathing motion occurs
only between PBP and OCT+1 at lower temperature and among all
four kinds of isomers at higher temperature.

3. Discussion

 Similar situations can be expected for clusters containing
any number of atoms. First assume that there is the steepest
descent path, which corresponds to cooperative motion of the
atoms in general, between two neighboring minima on the
potential energy surface [2]. This motion is, so to speak, the
collective mode of the cluster motion. In fact, such a path can
be found between icosahedron (IC) and cubooctahedron (COCT)
for N=13,55,137,309 and so on. For example, the coordinates
of the atoms for N=13 are given by

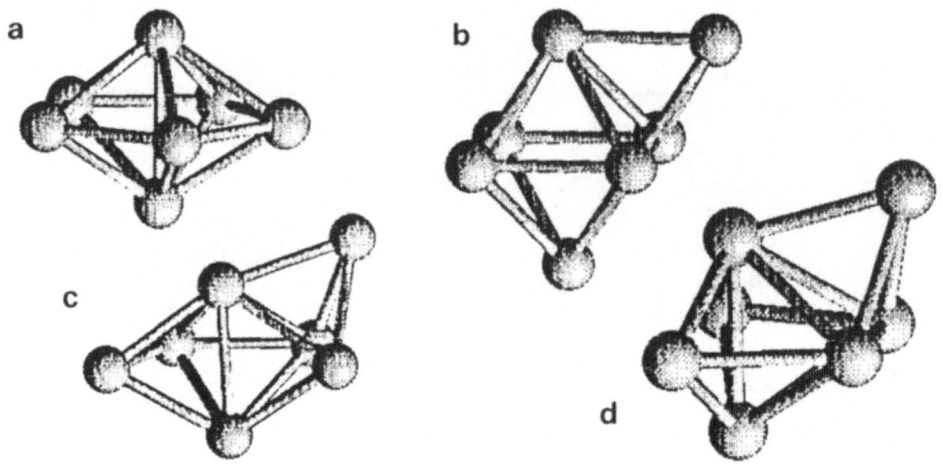

Fig.6. 7-atom clusters: (a) pentagonal bipyramid (PBP), (b) octahedron plus one (OCT+1), (c) skewed arrangement (Skew) and (d) incomplete stellated tetrahedron (IST).

$$r_0=(0,0,0), \qquad\qquad r_1=d(0,\cos\theta,\sin\theta),$$
$$r_2=d(\sin\theta,0,\cos\theta), \qquad r_3=d(\cos\theta,\sin\theta,0),$$
$$r_4=d(\cos\theta,-\sin\theta,0), \qquad r_5=d(0,\cos\theta,-\sin\theta),$$
$$r_6=d(-\sin\theta,0,\cos\theta),$$
$$r_{6+i}=-r_{7-i} \quad (i=1,2,\ldots,6), \tag{3.1}$$

where $\theta=\pi/4$ and $\cot^{-1}((1+\sqrt{5})/2)$ correspond to COCT and IC, respectively [9]. Minimizing the potential energy for these values and the intermediate values of θ by varying only d, a path between two structures is obtained. The coordinates of the atoms for N=55,137,309 and so on can be written with the

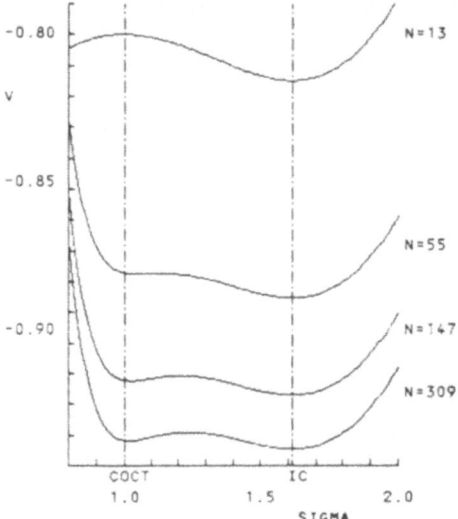

Fig.7. Potential energies as functions of $\sigma=\cot\theta$.

variables θ, d_1, d_2, d_3 and so on in the manner similar to
(3.1) and the potential energy is minimized for each fixed
value of θ. In this way potential energies are obtained as
functions of θ for various N. The results are shown in Fig.7,
where $\sigma = \cot \theta$. These paths are not necessarily equal to the
steepest descent paths but may be very close.
 There are other modes corresponding to individual atomic
vibration. (Those and the collective mode are not necessarily
equal to the modes of the harmonic vibration around any of the
minima.) Only when the collective mode gains energy larger
than the activation energy from the other modes, does the
transition from motion around one minimum to that around
another take place; otherwise the motion remains the vibration
around one of the minima. The probability of this transition
occurring is proportional to exp(-N) and also depends on the
temperature. As N increases, the solid-liquid transition gets
sharper and the region of the quasi-solid state narrower. The
quasi-solid state can be observed experimentally for appropriate N in some region of the temperature.
 Dynamics of larger clusters is now being worked on.

Acknowledgments

The author are very grateful to Prof. S. Sugano for his
suggestion and comments on this work and his continuous
encouragement. The author also acknowledges Dr. S. Saito and
Dr. S. Ohnishi for their valuable discussion and comments.

References

1. S. Iijima and T. Ichihashi, Phys. Rev. Lett. $\underline{56}$, 616
   (1986).
2. M. R. Hoare, Adv.·Chem. Phy. $\underline{40}$, 47 (1979).
3. R. P. Gupta, Phys. Rev. $\underline{B23}$, 6265 (1981).
4. M. B. Gordon, F. Cyrot-Lackmann and M. C. Desjonquéres,
   Surf. Sci. $\underline{80}$, 159 (1979).
5. D. Tománek, S. Mukherjee and K. H. Bennemann, Phys. Rev.
   $\underline{B28}$, 665 (1983).
6. D. J. McGinty, J. Chem. Phys. $\underline{58}$, 4733 (1973).
7. W. D. Kristensen, E. J. Jensen and R. M. J. Cotterill,
   J. Chem. Phys. $\underline{60}$, 4161 (1974).
8. C. L. Briant and J. J. Burton, J. Chem. Phys. $\underline{63}$, 2045
   (1975).
9. T. Ogawa and S. Nara, Z. Phsik $\underline{B33}$, 69 (1979).

Part VI

Larger Microclusters

Energy-Level Statistics of Metal Particles

S. Tanaka

Fundamental Research Laboratories, NEC Corporation, Miyazaki 4-1-1,
Miyamae-ku, Kawasaki 213, Japan

The relation between the shape irregularities of metal particles and
energy-level statistics is discussed here according to the results of
computer simulations /1/ and theoretical consideration /2/. We obtain the
relation $\omega \sim 1 - c/R^\eta$, where ω, R and c represent the so-called level-
repulsion exponent, the degree of shape irregularities and a certain
positive constant, respectively. Here η is given as $\eta \approx 2$ by means of both
the simulations and the theoretical consideration.

1. Introduction

It is widely accepted that metal particles have various kinds of
microscopic shape irregularities. KUBO /3/ has pointed out that these
irregularities could lead to fluctuations of electronic energy levels, and
he discussed the thermodynamics of metal particles on the basis of his
energy-level statistics. After that, GOR'KOV and ELIASHBERG /4/ have
discussed the optical properties of the metal particles in terms of the
energy-level statistics predicted by the Dyson's random matrix theory
(RMT) /5/. To my knowledge, only a few authors /6,7/, however, have
attempted to show how the shape irregularities have effects on the energy-
level statistics. It is important to examine the applicability of RMT to
metal particles before discussing their physical properties. Thus, I deal
with the statistical properties of ideal particles which will be referred
to as clusters.

2. Methods of Simulations

The level statistics of metal particles can be reduced to that of
artificial clusters which lose their shape. These clusters are created by
the following algorithm: when the cluster of size (N'+1) is grown out of
the N' cluster, one atom is added at one of the unoccupied sites
neighboring the occupied sites with the probability $Q(\xi)$ given by

$$Q(\xi) = \frac{\exp(\alpha\xi)}{\sum\limits_{\xi=1}^{z} \exp(\alpha\xi)} . \tag{1}$$

Here ξ, z and α represent, respectively, the number of occupied sites
around the site under consideration, the number of nearest neighbor
lattice sites and the parameter which governs the irregularities in
cluster shape.

From now on, we confine ourselves to clusters created on a 2-
dimensional triangular lattice. Another parameter is introduced, which
represents the degree of the shape irregularities. It is defined by

$$R = \frac{L^2}{48N} - 1 . \tag{2}$$

Here L stands for the perimeter of the cluster and integer 48 comes from such normalization that R=0 if L takes the minimum value. It is evident that the cluster becomes more irregular as R increases. Let us suppose that the ensemble denoted by (N, R) contains only clusters composed of N atoms and with the same value of irregularity parameter R.

Our goal is to understand statistical properties of electronic energy levels of the clusters in an ensemble (N, R). Simulations are performed according to three steps: (i) Creating a cluster belonging to the ensemble (N, R). Clusters are grown by using a fixed value of the parameter α. Then the clusters with the same value of R are collected. (ii) Diagonalizing the following tight-binding Hamiltonian of the electron in the cluster.

$$H = - \sum_{<i,j>} C_i^\dagger C_j , \tag{3}$$

where <i,j> represents the i-site at the nearest neighbor of the j-site, $C_i^\dagger$ and C_i are the creation and the annihilation operators, respectively. For simplicity, the spin is neglected and energy scale is chosen so that the transfer integral is 1. (iii) Repeating the procedures (i) and (ii) 600 times and examining the distribution of the level spacings.

We are especially interested in the fluctuations of energy levels. For the purpose of eliminating systematic behavior of levels which relates to the sensity of states, we define unfolded level spacing by the relation

$$x_\nu = \frac{E_{\nu+1} - E_\nu}{\delta_\nu} , \tag{4}$$

where E_ν is the νth energy level and δ_ν an ensemble average of level spacing $(E_{\nu+1} - E_\nu)$. We examine closely the properties of distributions of these unfolded level spacings around the middle of the energy band. When it is not confusing we refer to both the unfolded level spacing and the original as simply the level spacing.

3. Results of Simulations
In analysis, as is usually done /8/, we compare the level-spacing distributions obtained from our simulations with the Brody distribution involving parameter as follows:

$$P_B(x;\omega) = (\omega+1) \kappa(\omega)x^\omega \exp[-\kappa(\omega)x^{\omega+1}], \tag{5a}$$

$$\kappa(\omega) = [\Gamma(\frac{\omega+2}{\omega+1})]^{\omega+1}, \tag{5b}$$

where $\Gamma(x)$ is Gamma function of x. Here ω evidently corresponds to the exponent representing the degree of level repulsion, which we call briefly the level-repulsion exponent. This distribution is devised as a formula interpolating between the Poisson distribution (ω=0) and the Wigner distribution (ω=1). The latter is known to be an excellent approximate formula for the results of RMT in the orthogonal case.

We take into account about 100 level spacings around the middle of the band. Histograms of Figs.1(a) and 1(b) show the level-spacing distributions in ensembles (300, 0.068) and (300, 0.210), respectively. Solid curves show the Brody distributions with respective values of ω. These are given by the use of the relation between ω and a variance of spacings

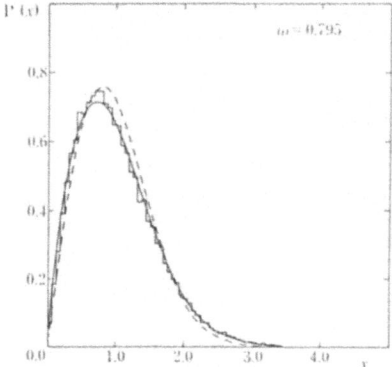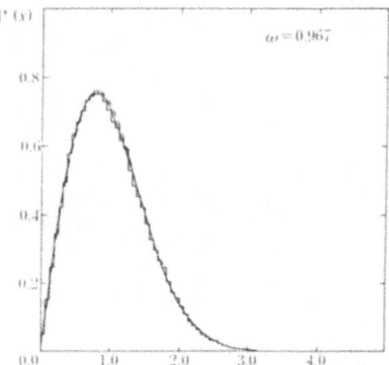

Fig.1 Level spacing distributions P(x) of the ensembles (300, 0.068) and (300, 0.210) in (a) and (b), respectively. A good agreement of $P_B(x;\omega)$ with P(x) is found in each case.

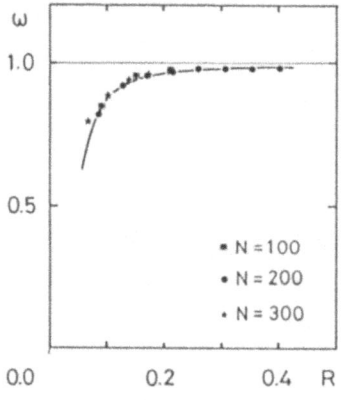

Fig.2 Level-repulsion exponent v.s. shape-irregularity parameter R for N=100, 200, 300. The solid curve represents $\omega = 1 - c/R^\eta$, where $\eta=1.7$ and $c=2.6 \times 10^{-3}$.

which is obtained directly by the simulations. For comparison,the Wigner distributions (dashed curves) are also shown.

Next we exhibit the dependence of ω on R in the three cases of N, that is, N=100, 200, 300 in Fig.2. This indicates that if the degree of shape irregularities is small, the level repulsion is weak,in disagreement with the prediction of RMT. The theory, however, successfully describes the fluctuations of energy levels as long as the cluster shape is irregular enough, since ω approaches 1. The ω seems to have no dependence on the cluster size. From the linear behavior of log(1-ω) vs. log R, we can get the value of the slope (-η) easily. These facts lead to the following relation:

$$\omega \sim 1 - c/R^\eta, \tag{6}$$

where c is a positive constant and $\eta=1.7 \pm 0.1$.

4. Equation of Motion of Hamiltonian

Now we consider analytically the relation between ω and R. Let us assume again that clusters consisted of N atoms are well described by the tight-binding Hamiltonian with a single band. The $N \times N$ Hamiltonian matrix can be written as

$$\underline{H} = \underline{H}^\circ + \underline{V}, \qquad <\underline{V}> = 0. \tag{7}$$

Here $\underline{H}^\circ$ is the ensemble average of $\underline{H}$. $\underline{V}$ is a random matrix written in the partitioned form as

$$[\ \underline{V}\ ] = \begin{bmatrix} 0 & 0 \\ 0 & v \end{bmatrix}, \tag{8}$$

where v is an $n \times n$ random matrix, supposing that n is the number of atom sites involving microscopic irregularities at the surface of the cluster. For simplicity, the matrix $\underline{V}$ is assumed to be described by the use of the $N \times N$ Gaussian random matrix $\underline{M}$ and the projection operator $\wp$ which takes only the n irregular sites. Thus we get a model Hamiltonian

$$\underline{H} = \underline{H}^\circ + 1/\sqrt{n}\,\wp \underline{M} \wp. \tag{9}$$

Here variances of the matrix elements of $\underline{M}$ have the same value 1. The coefficient $1/\sqrt{n}$ is determined by n-dependence of $\underline{V}$. Unfortunately it is impossible to derive the joint probability density of energy levels of the Hamiltonian (9) in the conventional manner /9/. Nevertheless we want to have some knowledge about the level repulsion of this system. We need a special technique to break through this situation. Now as we are dealing with the stochastic differential equation of $\underline{H}$ with respect to the virtual time t, we can discuss the level-repulsion exponent, although we cannot derive the joint probability density. The equation of $\underline{H}(t)$ is easily obtained so that the corresponding Fokker-Planck equation has a probability density of the Hamiltonian (9) as a stationary solution /10/. It is given as

$$d\underline{H}(t) = - [\ \underline{H}(t) - \underline{H}^\circ\ ]dt + 1/\sqrt{n}\,\wp d\underline{B}(t)\wp, \tag{10a}$$

$$< dB_{ij}(t) > = 0, \qquad < dB_{ij}(t)dB_{k\ell}(t) > = (\ \delta_{ik}\delta_{j\ell} + \delta_{i\ell}\delta_{jk}\ )dt, \tag{10b}$$

where $<...>$ means averaging over the Brownian-motion matrices $\underline{B}(t)$. We can derive equations of motion of energy levels $\{E_\nu(t)\}$ and wavefunctions $\{\psi_\nu(t)\}$ in the way of perturbation theory. If we define $P_{\mu\nu}(t)$ and $H^\circ_{\mu\nu}(t)$ as

$$P_{\mu\nu}(t) = < \psi_\mu(t)|\wp|\psi_\nu(t) > = \sum_{j \in (\text{irreg.sites})} \psi^j_\mu(t)\psi^j_\nu(t), \tag{11a}$$

$$H^\circ_{\mu\nu}(t) = < \psi_\mu(t)|\ H\ |\psi_\nu(t) >, \tag{11b}$$

respectively, we obtain the equations which couple to each others and look very complicated. Now we adopt the following assumptions; (i) wavefunctions fluctuate so fast that

$$< \psi^j_\nu > = 0, \qquad < \psi^j_\nu\psi^j_\mu > \simeq 1/N \cdot \delta_{\nu\mu}, \tag{12}$$

and (ii) $P(t)$ and $H(t)$ can be replaced by their ensemble averages which are roughly estimated as

$$< P_{\mu\nu} > \simeq n/N \cdot \delta_{\mu\nu}, \tag{13a}$$

$$< H^\circ_{\nu\mu} > \simeq \overline{E}_\nu. \tag{13b}$$

Here $\overline{E}_\nu$ denotes the ensemble-averaged νth level. We obtain simpler

223

equations

$$dE_\nu(t) = -[\ E_\nu(t) - \overline{E}_\nu]dt + \frac{1}{n}\sum_{\mu\neq\nu} \frac{<P_{\nu\nu}><P_{\mu\mu}>}{E_\nu(t) - E_\mu(t)}\ dt + \frac{1}{\sqrt{n}}\ dW_{\nu\nu}(t) \qquad (14a)$$

$$d\psi_\nu(t) = \sum_{\mu\neq\nu} \frac{H^\circ_{\mu\nu}(t)}{E_\nu(t) - E_\mu(t)}\psi_\mu(t)dt - \frac{1}{2n}\sum_{\mu\neq\nu} \frac{<P_{\nu\nu}><P_{\mu\mu}>}{[E_\nu(t) - E_\mu(t)]^2}\psi_\nu(t)dt$$

$$+ \frac{1}{\sqrt{n}}\sum_{\mu\neq\nu} \frac{dW_{\mu\nu}(t)}{E_\nu(t) - E_\mu(t)}\ \psi_\mu(t), \qquad (14b)$$

$$< dW_{\mu\nu}(t) > = 0,$$

$$< dW_{\mu\nu}(t)dW_{\mu'\nu'}(t) > = <P_{\mu\mu}><P_{\nu\nu}>(\delta_{\mu\mu'}\delta_{\nu\nu'} + \delta_{\mu\nu'}\delta_{\nu\mu'})dt \qquad (14c)$$

Furthermore we can get equations of motion of unfolded-level spacing x_ν from (14).

$$dx_\nu(\tau) = -1/n\ [x_\nu(\tau) - 1]d\tau + \omega_\nu/x_\nu\ d\tau + dW_{\nu\nu}(\tau) + \text{(residual terms)}, \qquad (15a)$$

$$\omega_\nu = \frac{2<P_{\nu\nu}><P_{\nu+1,\nu+1}>}{<P_{\nu\nu}>^2 + <P_{\nu+1,\nu+1}>^2} \qquad (15b)$$

where τ is the scale-transformed time, so that we have $<dW_{\nu\nu}(\tau)dW_{\nu\nu}(\tau)>=2d\tau$. This leads to the fact the stationary solution of the corresponding Fokker-Planck equation for small x_ν is given by

$$P(x_\nu) \sim x_\nu^{\omega_\nu}. \qquad (16)$$

Therefore we acquire the formula of the level-repulsion exponent ω_ν as (15b). From the definition of ω_ν, we have $0 < \omega_\nu \leq 1$, where $\omega_\nu = 1$ if and only if $<P_{\nu\nu}> = <P_{\nu+1,\nu+1}>$. In other words, $\omega_\nu = 1$ suggests that the specificity of each wavefunction vanishes because of its stochasticity.

If we define r as $r = n/N$, we can give the scale-transformed equations of motion of $\{\psi_\nu(\tau)\}$ as

$$d\psi_\nu(\tau) = -\frac{1}{2}\sum_{\mu\neq\nu}\frac{\psi_\nu(\tau)}{(e_\nu - e_\mu)^2}d\tau + \sum_{\mu\neq\nu}\frac{\psi_\mu(\tau)dW_{\mu\nu}(\tau)}{e_\nu - e_\mu} + \frac{1}{r}\sum_{\mu\neq\nu}\frac{\psi_\mu(\tau)}{e_\nu - e_\mu}\cdot\frac{H^\circ_{\mu\nu}}{N\delta_\nu}d\tau, \qquad (17)$$

where e_ν is given as E_ν/δ_ν. The first two terms show that ψ_ν behaves as that of the Gaussian random matrices. The last term describes the regular behavior of the wavefunctions. In addition, it is of the order of $1/r$ because $H^\circ_{\mu\nu}/N\delta_\nu$ is of order 1. Therefore $\psi_\nu(\tau)$ can be expanded by $1/r$, and we get $<P_{\nu\nu}>$ as

$$<P_{\nu\nu}> \sim <<\psi_\nu^{(0)} + 1/r\cdot\psi_\nu^{(1)}+\ldots|\wp|\psi_\nu^{(0)} + 1/r\cdot\psi_\nu^{(1)} +\ldots>>$$

$$\sim r(1 + \alpha_\nu^{(1)}/r + \alpha_\nu^{(2)}/r^2 +\ldots). \qquad (18)$$

$\alpha_\nu^{(1)}$ and $\alpha_\nu^{(2)}$ are certain constants. Replacing $<P_{\nu\nu}>$ of (15b) by the above formula, we obtain $\omega\sim 1 - c/r^2$, where c is a positive constant. It is natural to suppose $r\sim R^\rho$, $(\rho \approx 1)$, and we find a formula similar to (6):

$$\omega \sim 1 - c'/R^{2\rho}. \qquad (19)$$

5. Conclusion

We understand that if the degree of shape irregularities become small, that is, r or R is small, ω decreases from 1 which is the value predicted by RMT. I think this tendency is seen in the temperature dependence of the spin-susceptibility of Mg particles at low temperature in a recent experiment by KIMURA/11/. His result shows a rapid increase of the susceptibility near zero temperature. The susceptibility χ is known to depend on ω as $\chi \sim (kT/\delta)^\omega$. Therefore the experiment can be explained, supposing that the shape of the Mg particles is so regular that ω is less than 1.

Acknowledgment

I would like to thank Dr. K. Kimura for useful discussions.

References

1. S.Tanaka and S.Sugano, Phys. Rev. B34 740 (1986).
2. S.Tanaka and S.Sugano, Phys. Rev. B, in press.
3. R.Kubo, J. Phys. Soc. Jpn. 17 975 (1962).
4. L.P.Gor'kov and G.M.Eliashberg, Sov. Phys. JETP 21 940 (1965).
5. F.J.Dyson, J. Math. Phys. 3 140, 157, 1191, 1199 (1962).
6. K.F.Ratcliffe, in Multivariate Analysis V (North-Holland, New York) (1980) edited by P.R.Krishnaiah.
7. J.Barojas, E.Cota, E.Blaisten-Barojas, J.Flores and P.A.Mello, Ann.Phys. (N.Y.) 107 95 (1977), J.Phys.(Paris)38 C2-129 (1977).
8. T.A.Brody, J.Flores, J.B.French, P.A.Mello, A.Pandey and S.S.M.Wong, Rev. Mod. Phys. 53 385 (1981).
9. C.E.Porter, in Statistical Theories of Spectra: Fluctuations (Academic New York), (1965) edited by C.E.Porter.
10. F.J.Dyson, J. Math. Phys. 3 166 (1962), ibid 13 90 (1972).
11. K.Kimura, private communication.

Cluster Reaction from the Viewpoint of Nuclear Reaction

S. Sugano

The Institute for Solid State Physics, The University of Tokyo, Roppongi, Tokyo 106, Japan

1. Introduction

In the present paper, we bear in mind metal clusters containing from a few tens through a few hundreds atoms, which are called microclusters. Since both microclusters and nuclei are systems of the finite number of particles, we may expect some similarities in their physical properties, in spite of the great difference in the interaction properties of their component particles. The purpose of the present paper is to point out first these similarities and then discuss possible directions of the cluster reaction research by following well-developed ways of the nuclear reaction research.

We may easily find similarities in the following properties;
(i) Saturation property, which is seen in constancy of the particle density giving the radius of the system as $R \propto A^{1/3}$ (R:radius, A:the number of particles) and in the total energy proportional to A.
(ii) Shell structure, which arises from one-particle motion in an effective potential field.
(iii) Deformation, which arises from collective motion of particles.
(iv) Compound nucleus (cluster) processes, consisting of dispersion of the kinetic energy of an incident particle into all the modes of internal motion of the compound and re-concentration of the dispersed energy into a localized mode inducing particle emission.

2. Saturation Property

In nuclear physics[1], Bethe-Weizsäcker mass-formula, given by the liquid-drop model based on the saturation property, is very useful for discussing nuclear fission following a large nuclear deformation. Let us denote the numbers of neutrons and protons as N and Z and the number of total particles as A=N+Z. The mass-formula states that the mass of a nucleus (A,Z) is given as

$$M(A,Z) = NM_n + ZM_p - E(A,Z)/C^2, \qquad (1)$$

where M_n and M_p are the masses of a neutron and a proton and the total binding energy $E(A,Z)$ is given for nuclei of N=Z as

$$E(A,Z) = u_v A - u_s A^{2/3} - u_c A^{5/3}. \qquad (2)$$

The first, second and third terms represent the volume, the surface and the Coulomb energies, respectively. Constants u_v, u_s and u_c are of the order of 10,10 and 0.1 MeV, respectively.

For ^{4}He clusters, calculation of the total binding energies has been performed by Pandharipande et al[2] by employing the Green's function Monte Carlo method. The result shows that the total binding energy is given in kelvins as

$$E(N) = 7.02N - 18.8N^{2/3} + 11.2N^{1/3}, \tag{3}$$

where the number of ^{4}He atoms, N, is in the range $20 \leq N \leq 112$. The presence of the volume and the surface energies in eq.(3) is quite similar to that in eq.(2), although the energy unit is entirely different.

The liquid-drop model based on the saturation property has been used for studying surface oscillations of nuclei and clusters. For clusters, the calculated frequency spectrum[3] has been compared with that of gallium droplets of 70 - 100Å diameter impregnated in porous glass measured by the neutron inelastic scattering method[4]. The observed small hump in the low-frequency region has been attributed to the contribution of the surface oscillation.

For alkali metal clusters,the saturation property is assumed from the beginning[5,6]. This assumption determines the radius of a jellium sphere representing the distribution of the ion cores. Then, the distribution of valence electrons in this jellium sphere is determined self-consistently. The total energy per atom thus calculated is shown in Fig.1. The total energy behaves like $-u_V N + u_s N^{2/3} (u_V, u_s > 0)$ as expected, if the superposed oscillation due to the shell structure effect is ignored. By using the calculated total energy curve, fragmentation of the metal clusters has been discussed[6], which will be presented in the next section.

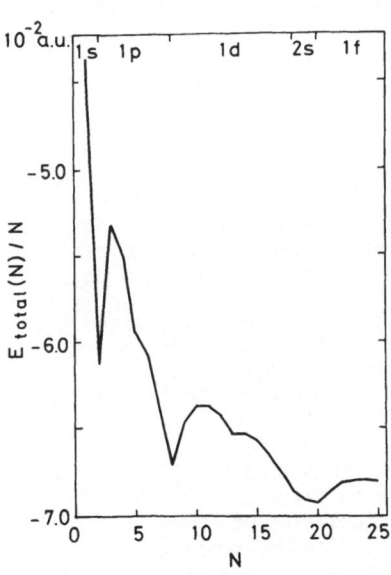

Fig.1. The total energy (minus the total binding energy) per atom for a jellium sphere of R= $3.93N^{1/3}$a.u.

3. Shell Structure Effects

The magic numbers observed in the mass spectra of alkali[5] and noble[7] metal clusters have been ascribed to shell structure effects arising from one-electron motion in an effective potential field. As already mentioned at the end of the previous section, the shell effect is seen as an oscillation superposed on the total energy curve as shown in Fig.1. The presence of this oscillation is very important in discussing fragmentation and coagulation of clusters.

Let us first consider the case where the shell effect is absent. Since the total energy per atom is monotonously decreasing function of N, it is easy to see that fragmentation of a cluster is an endothermic reaction and forbidden if the initial cluster is not internally excited, while coagulation of clusters is an exothermic reaction and allowed from an energetical point of view. Now we introduce the shell effect, i.e., an oscillation on the total energy curve. Then, fragmentation of a specific cluster of N atoms into specific clusters of $N_i (i=1,2,\cdots)$ becomes allowed. This has been carefully examined by using the total energy curve of better quality for sodium cluster[6]. It has been found that the following fragmentation processes, $X_N \rightarrow X_{N-\nu} + X_\nu$, for X=Na are allowed by the shell effect;

$$X_{11} \rightarrow X_9 + X_2 \rightarrow X_8 + X_2 + X_1,$$

$$X_{12} \rightarrow X_{10} + X_2 \rightarrow X_8 + 2X_2,$$

$$X_{13} \rightarrow X_8 + X_5,$$

$$X_{14} \rightarrow X_8 + X_6,$$

$$X_{15} \rightarrow X_8 + X_7,$$

$$X_{16} \rightarrow 2X_8,$$

$$X_{23} \rightarrow X_{21} + X_2 \rightarrow X_{19} + 2X_2,$$

$$X_{24} \rightarrow X_{22} + X_2 \rightarrow X_{20} + 2X_2,$$

$$X_{25} \rightarrow X_{17} + X_8. \tag{4}$$

The presence of these allowed fragmentation processes is consistent with the experimental results[5] showing the pronounced abundance of clusters N=8 and 20 relative to those of N=9-11 and 21-24. It should be noticed that our argument of the forbidden fragmentation processes is valid only if the initial cluster is not internally excited and the allowed processes may be forbidden by the presence of high potential barriers in the fragmentation processes. Actually, the allowed fragmentation processes mentioned above are inconsistent with a relatively large abundance of the N=12 and 14 sodium clusters observed in the mass spectrum.

In nuclear theory, one-particle motion in a deformed potential has been discussed by assuming constancy of the particle density, and the observed quadrupole moments and quadrupole transition intensities of heavy nuclei have been found to be in agreement with the calculated deformation[8]. A similar argument has also been applied to alkali metal clusters to explain the observed fine structure of mass spectra[9]. It has been pointed out that the equilibrium shape is ellipsoidal and a strong correlation is observed between the energy sequence of the ellipsoidal subshells and the observed fine structure.

4. Unification of the Liquid-Drop and the Shell Models

The liquid-drop and the shell models shed light on two different aspects of a small many-body system, collective motion of the component particles and one-particle motion in an effective potential. A unified treatment of these two models has been given by Bohr and Mottelson[10]. According to their unified model, Hamiltonian of the system is given by

$$H = H_{coll} + H_p + H_{int}, \tag{5}$$

where H_{coll}, H_p, and H_{int} are the Hamiltonians for the collective motion, the one-particle motion in a spherical effective potential $V_0(r)$ plus the pair interaction, and the interaction between these two types of motion, respectively. The interaction Hamiltonian is given as

$$H_{int} = \sum_i \{V(\vec{r}_i : \alpha_{\lambda\mu}) - V_0(r_i)\}, \tag{6}$$

where $V(\vec{r}_i : \alpha_{\lambda\mu})$ is an effective potential in the deformed nucleus whose coordinates of deformation are given by $\alpha_{\lambda\mu}$: $\alpha_{\lambda\mu}$ is the μ-th component of the λ-th rank irreducible tensor. By expanding $V(\vec{r}_i : \alpha_{\lambda\mu})$ in terms of $\alpha_{\lambda\mu}$, we have

$$H_{int} = -\sum_i K(Y_i) \sum_{\lambda\mu} \alpha_{\lambda\mu} Y_{\lambda\mu}^*(\theta_i \phi_i), \tag{7}$$

$$K(r) = r \{ dV_0(r)/dr \}. \tag{8}$$

Since the radial change of $V_0(r)$ occurs mainly at the nuclear surface, the interaction may be considered to be induced by the collision of a particle undergoing one-particle orbital motion with the potential wall undergoing collective oscillatory motion. Bohr-Mottelson's unified model, however, is not adapted to dealing with large deformations such as those important for nuclear fission etc..

The method of treating large deformations, as well as shell effects of heavy nuclei, has been developed by Strutinsky[11]. In this theory, the main part of the total energy of a deformed nucleus is treated by using the liquid-drop model and the shell effect is taken into account as a small correction to it. This idea is based on the fact that the total energy is approximately given by λN while the shell-energy correction by $\lambda N^{1/3}$ where λ is the Fermi energy of the order of 40 - 50 MeV. Accordingly, the total energy of a deformed nucleus is given as

$$E = \tilde{E} + \delta U + \delta P, \tag{9}$$

where $\tilde{E}$ is the total energy of the liquid-drop model which may be calculated by using empirical parameters. The second term, δU, is defined as

$$\delta U \equiv U - \tilde{U}, \tag{10}$$

where U is the sum of one-particle energies of the shell model with a given deformation and $\tilde{U}$ that of underlined uniformly distributed one-particle energies. A similar definition is given for the pairing energy,

$$\delta P \equiv P - \tilde{P}. \tag{11}$$

The calculation of $\tilde{U}$ and $\tilde{P}$ has been made by using an appropriate "uniform" level distribution function. The order of magnitude of δU is a few MeV while that of U a few GeV for heavy nuclei of $N \sim 100$. The contour map of the total shell-energy correction, $\delta U + \delta P$, has been calculated in the N-η plane where η measures the magnitude of deformation. We note that the valley of the contour map at certain N elongates almost parallel to the η axis near η=0, but at some value of η the valley suddenly shifts to a different point of N elongating again parallel to the η-axis. This sort of theory seems to be very powerful in discussing fragmentation processes of microclusters of $N \sim 100$.

5. Reactive Scattering

The present situation of microcluster research provides us with good reason to hope that the following research of reactive scattering of micro-clusters would greatly be developed in the near future by use of cross-beam experiments, as already seen in nuclear physics. The reactive scattering may be divided into two, direct processes and compound cluster processes.

The direct processes involve (i) inelastic scattering with excitation of collective modes and shell structure, and (ii) transfer reactions such as stripping, pick-up, knock-on reactions and so on as follows;

stripping: $A + a(b + x) \rightarrow b + B(A + x)$,

pick-up: $A(B + x) + a \rightarrow B + b(a + x)$,

knock-on: $A(c + b) + a \rightarrow b + B(C + a)$.   (12)

where A is a target cluster, $a(b + x)$ an incident particle containing component particles b and x, B a residual cluster, b an outgoing particle, and so on.

More interesting are the compound cluster processes described as

$$A + a \rightarrow C^* \rightarrow B + b, \qquad\qquad\qquad\qquad (13)$$

where C^* is the compound cluster in the metastable state with life time τ. The concept of compound nucleus processes was introduced by Bohr[12], and Breit and Wigner[13], and a great development has been achieved on this subject. To obtain the cross section of the process, the dispersion formula has been derived without adopting any approximation[1]. The formula contains resonant terms expressed in terms of resonance para-meters: in a simple case, in terms of parameters such as the energies and the life times of the metastable states of the compound nucleus. These parameters are considered to be quite sensitive to the size of the system and the forces between the component particles. Thus, characteristic features of a small many-body system are well represented in the compound cluster processes.

In nuclear physics, the colliding time of a 1 eV neutron with a nucleus of a medium size is $\sim 10^{-18}$s and the life time of a compound nucleus up to $\sim 10^{-15}$s. On the other hand, the colliding time of a 1 eV H^+ with a cluster of 10 Å diameter is estimated to be $\sim 10^{-12}$s and the life time of a cluster may be much longer than the colliding time, as seen in the results of the dynamical studies of transition-metal clusters of N=6 and 7 by Sawada presented in this volume. Therefore, we have great possibility to measure the life time directly. Furthermore, for clusters, we have a possibility to calculate non-empirically the values of the resonance para-meters and compare them with the experimental ones.

230

References

1. J. M. Blatt and V. F. Weisskopf: <u>Theoretical Nuclear Physics</u>, (John Wiley and Sons, 1952)
2. V. R. Pandharipande, J. G. Zabolitzky, Steven C. Pieper, R. B. Wiringa and U. Helmbrecht: Phys. Rev. Lett. <u>50</u>, 1676 (1983)
3. A. Tamura and T. Ichinokawa: Surf. Sci. <u>136</u>, 437 (1984)
4. V. N. Bogomolov, N. A. Klushin, N. M. Okuneva, E. L. Plachenova, V. I. Pogrebnoi and F. A. Chudnovskii: Soviet Phys. — Solid State <u>13</u>, 1256 (1971)
5. W. D. Knight, K. Clemenger, W. A. de Heer, W. A. Saunders, G. Y. Chou and M. L. Cohen: Phys. Rev. Lett. <u>52</u>, 2141 (1984)
6. Y. Ishii, S. Ohnishi and S. Sugano: Phys. Rev. B<u>33</u>, 5271 (1986)
7. I. Katakuse, T. Ichihara, Y. Fujita, T. Matsuo, T. Sakurai and H. Matsuda: Int. J. Mass Spectrom. Ion Processes <u>67</u>, 229 (1985)
8. S. G. Nilsson: Mat. Fys. Medd. Dan. Vid. Selsk. <u>29</u>, No.16 (1955)
9. K. Clemenger: Phys. Rev. B<u>32</u>, 1359 (1985)
10. A. Bohr and B. R. Mottelson: Mat. Fys. Medd. Dan. Vid. Selsk. <u>27</u> No.16 (1953)
11. V. M. Strutinsky: Nuclear Phys. <u>A95</u>, 420 (1967); ibid. <u>A122</u>, 1 (1968)
12. N. Bohr: Nature <u>137</u>, 344 (1936)
13. G. Breit and E. P. Wigner: Phys. Rev. <u>49</u>, 519 (1936)

Stability of a Charged Small Particle Against Surface Shape Fluctuations

A. Tamura

Department of Applied Physics, School of Science and Engineering, Waseda University, 3-4-1 Ohkubo, Tokyo 160, Japan

On the basis of a local density functional theory (LDF), surface shape effect on the total energy of a charged small particle is discussed. The electron density profile is assumed to be of Fermi's distribution function type and the positive background is assumed to be a uniform jellium. The total energy is written by the normal coordinate of a surface shape deformation parameter. Thus the stability of a charged small particle against surface shape fluctuations is treated.and the mechanism of the Coulomb explosion is shown from this viewpoint.

1. Introduction

Many kinds of aspects of small particles have been investigated, not only on the static but the dynamical properties. Among them, the stability of a charged small particle is an essential subject. Sattler et al [1] measured a mass spectrum of charged Pb clusters and observed the existence of the critical number of composing atoms below which charged clusters cannot be observed. This instability is inferred to be due to the Coulomb explosion which occurs by excessive Coulomb repulsion. Sattler et al [1] and Bennemann [2] explained this phenomenon by comparing the binding energy per composing atom with the repulsive Coulomb energy between two holes created by ionization. Iakubov et al [3] estimated the critical number of atoms by considering the energy difference between Z and (Z-1) charged clusters calculated from the ion work function and the self-interaction energy of surplus charge. On the other hand, Iijima and Ichihashi [4] observed the real-time images of an Au small particle on a Si large particle. In their movie, the Au small particle changes transiently its shape. This phenomenon is thought to be closely correlated with the fluctuations of the number of charges. In the present paper, we treat these phenomena by considering the effect of surface shape fluctuations on the stability of a charged small particle.

2. Surface Shape Deformation

We define the surface shape deformation by the following equation:

$$\zeta(\theta) = R(\theta) - R_0 = R_0 \sum \alpha_L P_L(\cos\theta), \qquad (1)$$

where $R(\theta)$ represents a distance from the origin to the point of the deformed surface in the direction θ from z-axis and R_0 the radius of an equilibrium sphere. For simplicity, it is assumed that the system has an axial symmetry around z-axis. To describe the total energy as a function of surface shape deformations, we define the normal coordinate of the surface shape deformation as the expansion coefficient of $\zeta(\theta)$ by Legendre polynomials. Several examples of shape deformation modes are shown in Fig.1.

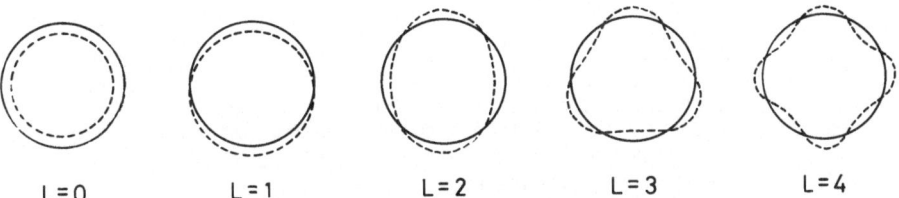

L = 0   L = 1   L = 2   L = 3   L = 4

<u>Fig.1</u> Schematic diagram of surface shape deformation modes.

The mode L=0 shows a compression in a radial direction. The mode L=1 shows a translation of the whole system. These modes conserve a spherical shape and do not have a shape deformation. The angular momentum L specifies the surface shape.

For the electron density profile, we assume it to be of Fermi's distribution function type:

$$\rho_e(r) = \rho_{eo} \frac{1}{1 + \exp[\beta\{r - R(\theta)\}]} .$$ (2)

The $1/\beta$ means a surface diffuseness of the electron density. For the positive background, we assume its distribution to be uniform:

$$\rho_p(R) = \rho_{po} \ \theta(R(\theta) - R),$$ (3)

where $\theta(x)$ means the Heaviside's function.

Since the surface shape deformation does not change the total number of electrons,

$$N_e = \int d\mathbf{r} \ \rho_e(\mathbf{r}) .$$ (4)

After expanding the distribution function around $\beta\zeta(\theta) = 0$, we obtain

$$\alpha_o = - \eta \{ \sum \frac{1}{2L + 1} \ \alpha_L^2 + \sum\sum\sum T(L,L',L'') \ \alpha_L \ \alpha_{L'} \ \alpha_{L''} + \dots \} ,$$ (5)

$$\eta = 1/(1 + 1.387 \ a^{-1} + 3.290 \ a^{-2} + \dots),$$ (6)

$$a = \beta \ R_o ,$$ (7)

and

$$T(L,L',L'') = \int_{-1}^{1} dx \ P_L(x) \ P_{L'}(x) \ P_{L''}(x).$$ (8)

The α_o is not an independent function of α_L.
For the positive background, $N_p = (4\pi/3)\rho_{po}R_o^3$. The charge $Z = N_p - N_e$.
The function $T(L,L',L'')$ becomes zero if the sum of three angular momenta is odd.

3. Total Energy

We focus on the stability of a charged small particle. By using the local density functional approximation, the total energy is given as follows:

$$E_t = E_{ke} + E_{kec} + E_{xc} + E_{cor} + E_{ee} + E_{ep} + E_{pp} \quad . \tag{9}$$

From the left to the right, each term means a kinetic energy, a kinetic energy correction, an exchange correlation energy, a correlation energy except an exchange correlation energy, an electron-electron Coulomb interaction energy, an electron-positive background Coulomb interaction energy and a Coulomb interaction energy between positive backgrounds.

In the present treatment, we take them as follows:

$$E_{ke} = \frac{3}{10} (3\pi^2)^{2/3} \int dr \ [\rho_e(\mathbf{r})]^{5/3}, \tag{10}$$

$$E_{kec} = \frac{1}{72} \int dr \ |\nabla \rho_e(\mathbf{r})|^2 / \rho_e(\mathbf{r}) \ , \tag{11}$$

$$E_{xc} = -\frac{3}{4} (3/4\pi)^{1/3} \int dr \ [\rho_e(\mathbf{r})]^{4/3}, \tag{12}$$

$$E_{cor} = \int dr \ \rho_e(\mathbf{r}) \ [-0.0474 + 0.0155 \ \ln \rho_e(\mathbf{r})], \tag{13}$$

$$E_{ee} = \frac{1}{2} \iint drdr' \ \frac{\rho_e(\mathbf{r}) \ \rho_e(\mathbf{r}')}{| \mathbf{r} - \mathbf{r}'|} \ , \tag{14}$$

$$E_{ep} = -\iint drdR \ \frac{\rho_e(\mathbf{r}) \ \rho_p(\mathbf{R})}{| \mathbf{r} - \mathbf{R} |} \ , \tag{15}$$

$$E_{pp} = \frac{1}{2} \iint dRdR' \ \frac{\rho_p(\mathbf{R}) \ \rho_p(\mathbf{R}')}{| \mathbf{R} - \mathbf{R}'|} \ , \tag{16}$$

where the unit $m = \hbar = e = 1$ is used.

The E_{kec} is based on the result by Kirzhnits [5] and E_{cor} is taken from the form given by Noziere and Pines [6].

To derive the surface shape dependence of E_t, we expand $\rho_e(\mathbf{r})$ and $\rho_p(\mathbf{R})$ around $\beta\zeta(\theta) = 0$ as follows:

$$\rho_e(\mathbf{r}) = \rho_{eo}\{1/[1 + \exp \beta(r - R_0)] + (\beta\zeta + \beta^2\zeta^2 + \ldots \ )\ldots \} \ , \tag{17}$$

$$\rho_p(\mathbf{R}) = \rho_{po}\{\theta(R - R_0) + \beta\zeta \ \delta(R - R_0) - \frac{1}{2} \beta^2\zeta^2 \ \delta'(R - R_0) + \ldots \}. \tag{18}$$

From these expansions, we calculated each term of E_t. The integration is carried out within an error of $\exp(-a)$.

By these procedures, we obtain the following result for E_t.

$$E_t = E_0 + \sum C_L^{(1)} \ \alpha_L + \sum\sum C_{LL'}^{(2)} \ \alpha_L \ \alpha_{L'} + \sum\sum\sum C_{LL'L''}^{(3)} \ \alpha_L \ \alpha_{L'} \ \alpha_{L''} + \ldots \tag{19}$$

where C means an expansion coefficient. The parameter a defined by eq.(7) is determined so as to make E_0 be the minimum.

The coefficient $C_L^{(1)} = 0$ for all L and $C_{LL'}^{(2)} = 0$ for $L \neq L'$. The $C_{LL'L''}^{(3)}$ is not zero since the contributions from seven terms of E_t are different (calculation of the third order term is in progress and the result will be reported elsewhere).

4. Stability Condition

For a fixed Z (the number of charges), $C_{LL}^{(2)}$ becomes negative in the case of small N_p, which shows the instability of a charged small particle. Such a small particle explodes rapidly after ionization. Existence of the critical number of atoms in which the $C_{LL}^{(2)}$ is zero corresponds to the experimental result by Sattler et al [1]. For a fixed N_p, $C_{LL}^{(2)}$ becomes small as Z increases. These situations are shown in figs.2 and 3.

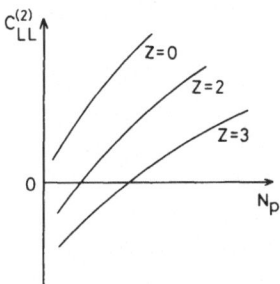

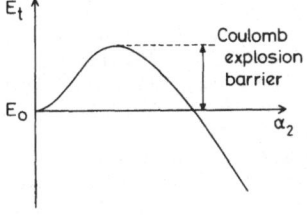

Fig.2 Schematic diagram of $C_{LL}^{(2)}$ as a function of N_p.

Fig.3 Schematic diagram of the total energy E_t as a function of α_2.

If $C_{LL'L''}^{(3)}$ is negative, the Coulomb explosion barrier locates in the region $\alpha_2 > 0$. In the third order term, there are contributions like $\alpha_2 \alpha_3^2$, $\alpha_2^2 \alpha_4$, ... These terms make the total energy surface asymmetric in the multi-dimensional space of the surface shape deformations. Therefore, there exists a fission barrier or an explosion barrier in the case that $C_{LL}^{(2)} > 0$. The Coulomb explosion barrier has an asymmetric shape because of the combination of plural surface shape deformations, e.g., α_2, α_3, α_4 and so on. The height determines the critical energy for the induced Coulomb explosion. When the system gains energy from external probes, the total energy increases and the system migrates in the multi-dimensional space of the surface shape deformations. The small particle shows the surface shape fluctuations and the shape changes transiently. This corresponds to the experimental result by Iijima and Ichihashi [4] on the dynamical behavior of an Au small particle on a large Si particle. In their experiment, the number of charges is thought to be fluctuating since the small particle suffers electron bombardment. Secondary electrons are likely to be emitted. As is mentioned above, the $C_{LL}^{(2)}$ depends on Z. When the excitation energy exceeds the explosion barrier, the Coulomb explosion occurs more frequently.

In the case of a neutral small particle, $C_{LL}^{(2)}$ is positive for all N_p as shown in fig.2. The small particle is stable against the surface shape deformations. It does not explode as is expected.

In the present treatment, material dependence of the stability is characterized by an interatomic spacing parameter r_s. As the r_s increases, the critical number of atoms for explosion becomes large. This is due to the trend that the Coulomb interaction energy becomes predominant for the low density system.

Comparing with previous studies by Sattler et al [1] and by Bennemann [2], our treatment is based on the spreading excessive charge, not on the localized holes. Moreover, the present theory does not include a fitting parameter.

5. Conclusion

By using LDF, the total energy of a charged small particle shows a strong dependence on the surface shape deformations. Recent experimental results on the stability of charged small particles are described as well as that against the surface shape deformations.

References

1. K. Sattler, J. Mühlbach, O. Echt, P. Pfau and E. Recknagel:
   Phys. Rev. Lett. 47, 160 (1981)
2. K. H. Bennemann: Z. Phys. B60, 161 (1985)
3. I. T. Iakubov, A. G. Khrapak, L. I. Podlubny and V. V. Pogosov:
   Solid State Commun. 53, 427 (1985)
4. S. Iijima and T. Ichihashi: Phys. Rev. Lett. 56, 616 (1986)
5. D. A. Kirzhnits: Soviet Phys. (JETP) 5, 64 (1957)
6. P. Noziere and D. Pines: Phys. Rev. 111, 442 (1958)

Part VII

Semiconductor Clusters

Production and Photofragmentation
of Semiconductor Clusters and Cluster Ions

L.A. Bloomfield[1], *M.E. Geusic*[2], *T.J. McIlrath*[†], *M.F. Jarrold*[2],
R.R. Freeman[2], *and W.L. Brown*[2]

[1]Department of Physics, University of Virginia, Charlottesville, VA 22901, USA
[2]AT&T Bell Laboratories, Murray Hill, NJ 07974, USA

1. Introduction

The study of atomic clusters dates from the 1950s, but it has been only in the past decade that advances in experimental and theoretical techniques have brought renewed interest to this exciting field, both as an area of fundamental research and for its potential industrial applications. Among the developments responsible for this increased activity have been the advent of laser vaporization, thermal stagnation, and supersonic expansion cluster sources, as well as powerful laser spectroscopic techniques, pulsed molecular beam nozzles, matrix isolation, and improvements in vacuum technology. These methods have finally made it possible to study clusters of materials which are highly refractory and in particular, the semiconductors.

Clusters of silicon and carbon were first observed by HONIG[1] in 1953 using thermal evaporation, however lasers have made it possible to study larger clusters at much colder internal temperatures than can be obtained conventionally[2]. Recent laser based studies of silicon[3-8], germanium[5,8], and carbon[6,9-17] clusters have greatly expanded the understanding of these tightly bound particles and prompted considerable theoretical interest.

Unfortunately, it is still not possible to obtain large quantities of homogeneous semiconductor clusters, as one might choose for direct, specific measurements of the cluster properties as a function of size, charge state, and temperature. Therefore, most of the current measurements of semiconductor cluster properties are indirect and subject to considerable uncertainty in interpretation. Mass selection, an essential step in simplifying experimental results, is virtually impossible for neutral clusters. Furthermore, it is quite difficult to detect neutral particles of most materials. To avoid these two problems, this work focuses on the study of charged cluster ions, permitting easy separation of the clusters throughout the experiments and facilitating their detection using charged particle techniques.

Because the samples involved are so minute, conventional absorption-based spectroscopic techniques are of little use. In addition, even very small clusters possess a bewildering number of excited states and often very short excited state lifetimes, limiting the usefulness of such techniques as two-photon ionization and laser-induced fluorescence. However photodissociation, which has been used for several years to investigate the spectroscopy and dissociation dynamics of

---

† on leave from Institute of Physical Sciences and Technology, University of Maryland, College Park, MD 20742, USA

ions[18], can be used to probe the properties of the larger clusters. Despite the indirect nature of this approach, much useful information about binding energies and cluster stabilities can be obtained.

In this paper we report on the production and photodissociation of mass-selected silicon and carbon cluster ions. Observations include a number of "magic" fragment particles which frequently remain following the photoinduced fission of a cluster ion. Also measured were photofragmentation cross sections and branching ratios at several wavelengths, and, in carbon, bracketing of the photofragmentation energy thresholds. These measurements are generally in agreement with recent theoretical calculations.

2. Experimental Details

The semiconductor cluster ions used in these experiments were produced via one of two related techniques. All of the clusters originated in a laser vaporization cluster source, with helium or argon as a carrier gas. However, in the earlier experiments, neutral clusters were photoionized to ions for further study while, more recently, ionized clusters produced directly in the vaporization plasma have been investigated.

The direct cluster ions have three significant advantages over photoionized neutral clusters: direct ions are likely to be much colder internally than photoions, they include negative as well as positive ions, and they form a more intense ion beam than can be obtained with photoions. For these reasons, all current work is done on the direct cluster ions emitted from the source.

A diagram of the experimental apparatus is shown in Fig. 1. To produce the clusters, a 0.3-0.5 mm spot on a sample rod (12.5 mm diameter) is exposed to the focused, second harmonic beam (532nm) of a Q-switched Nd:YAG laser (Quanta Ray DCR-1A) at an energy density of 5-50 J/cm². The higher energies yield the greatest number of cluster ions while the neutrals are best generated at lower energies. Cluster yield is fairly sensitive to the vaporization energy density, so that optimization is important. During the laser pulse, a carrier gas (helium for small to moderate sized clusters and argon for very large clusters) passes over the sample, through a 1 mm tube and then into a

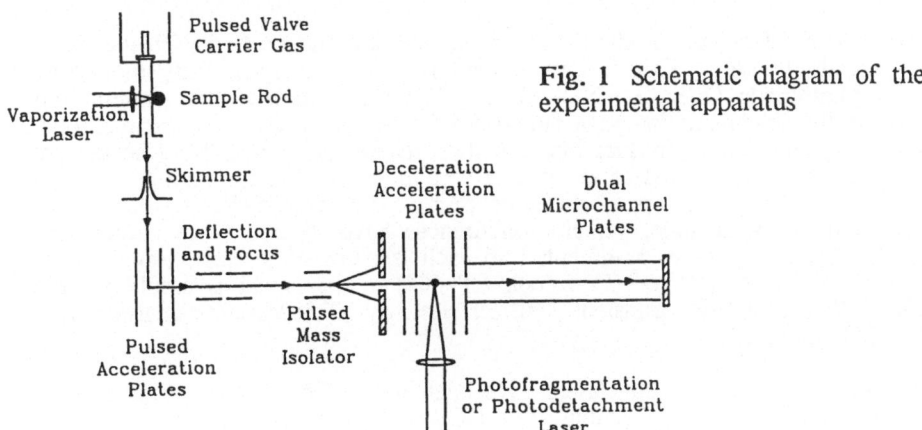

Fig. 1 Schematic diagram of the experimental apparatus

differentially pumped vacuum system. The carrier gas is pulsed (Lasertechnics LPV Valve) for 100 to 400 microseconds during the laser pulse, in order to reduce the load on the 3000 l/s untrapped source diffusion pump (Varian VHS-6).

Clusters are formed in the carrier gas during the several microseconds between the vaporization pulse and the passage of the sample/carrier gas mixture through the nozzle and into the vacuum. We are not alone in observing that semiconductor clusters form very easily and that, in contrast with metal clusters, they do not entirely require the carrier gas[7,9,13,17]. It is apparent, however, that very large clusters are more easily formed in the presence of the carrier gas and that colder clusters can be formed through the supersonic expansion of the gases into the vacuum. A further benefit of the supersonic expansion is a concentrating of the heavy clusters on the axis of the expanding carrier gas beam.

The clusters and cluster ions emitted from the source pass through a 1 mm diameter skimmer (Beam Dynamics Model 1) and into a second vacuum chamber. This chamber is pumped by a 2500 l/s, LN_2 cryotrapped diffusion pump (Varian VHS-6). Here the clusters are accelerated by a three-plate spectrometer system. In the case of clusters ions, as illustrated in Fig. 1, the three-plate system is held at ground potential while the particles enter it. A high voltage pulse, 3 μs in duration, is then applied to the left and middle plates at 2.25 kV and 1.8 kV respectively (Cober 606) to accelerate the ions. For studies of neutral particles converted to ions by photoionization, the plates are held at high voltage so that only neutral particles drift undeflected into the acceleration region. These neutrals are then photoionized using an excimer laser (Lambda Physik EMG101), not shown in Fig. 1, and are accelerated to 2 kV.

In either case, the now accelerated, ionized clusters pass through a series of ion optics and into the first time-of-flight region, pumped by a 1000 l/s water-baffled diffusion pump (Varian M-4). The different cluster sizes travel at different velocities and sufficient spatial separation occurs after 1 meter of travel to permit purification using a mass isolator. The isolator consists of two parallel plates, normally biased at 0 and 100V. However, while the cluster of interest passes through the plates, both plates are briefly biased at ground. This gating technique permits only the selected cluster ions (typically a packet of 10^3 to 10^4 ions) to continue through the apparatus.

The packet of clusters is then decelerated to 1 kV and crossed with the beam from a pulsed laser. This fragmentation laser fluence is carefully monitored with a photodiode (EG&G FND-100 or UV100BQ) so that the amount and nature of the fragmentation as a function of optical exposure can be recorded. The unfragmented and product ions are then reaccelerated to 5 kV (the neutral fragments are not detected) and enter a second time-of-flight region, pumped by a 2500 l/s LN_2 cryotrapped diffusion pump (Varian VHS-6). At the end of this second, 1 meter tube, velocity differences have separated the parent ions from the lighter daughter fragments and each is detected in turn, using a dual microchannel plate detector (Comstock CP-604). The signal is amplified and stored with a 200 MHz transient digitizer (LeCroy TR8828B) for analysis by a PDP 11/34 computer.

Several photofragmentation lasers have been used. The second harmonic (532 nm) of a Nd:YAG laser (Quanta Ray DCR-1A) was used for the studies of

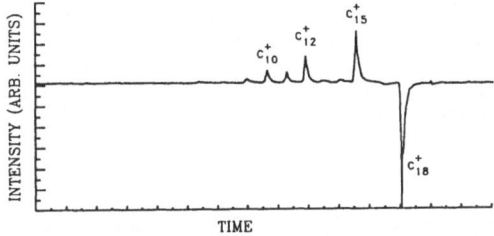

Fig. 2 Time-of-flight mass spectrum of the photofragments from C_{18}^+ photodissociation measured with 248nm light and a photodissociating laser fluence of approximately 12 mJ/cm²

silicon while an excimer laser (248 and 351nm) was used for carbon (Lambda Physik EMG101). A dye laser and other harmonics of the Nd:YAG laser have been tried but have not yielded results of sufficient merit for inclusion in this paper.

An example of the photofragmentation of C_{18}^+ cluster ions is shown in Fig. 2. The spectrum was measured with 248nm light and a photodissociating laser fluence of approximately 12 mJ/cm². Among the fragment ions which are observed are C_{10}^+, C_{11}^+, C_{12}^+, and C_{15}^+. The original C_{18}^+ peak is inverted due to the use of a background subtraction technique and represents the depletion of those cluster ions as their population is transferred to the fragments.

The relative populations of the various photofragment ions have been found to vary dramatically as a function of laser fluence. To illustrate this effect, we present, in Fig. 3, the relative abundances of C_{15}^+ and C_{12}^+ as a function of laser fluence. With low laser fluence, <4 mJ/cm², the C_{15}^+ population increases linearly with laser fluence. For higher fluences, the C_{15}^+ population stops increasing and for fluences greater than 10 mJ/cm², begins to decrease. In contrast, the population of C_{12}^+ fragments appears to depend quadratically on laser fluence, indicating that it involves a multiphoton process.

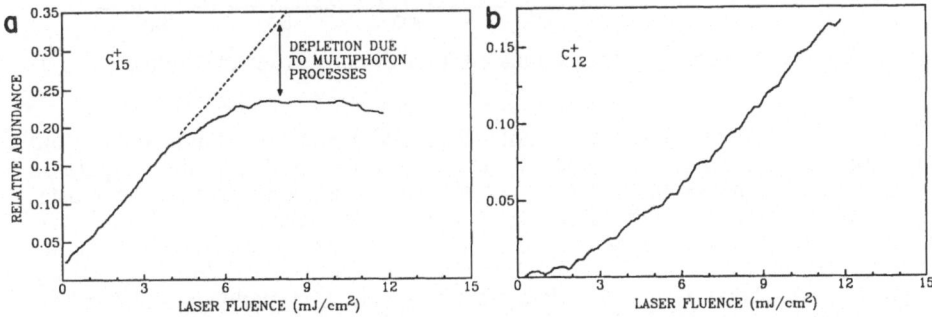

Fig. 3 Laser fluence dependence of the intensity of the (a) C_{15}^+ and (b) C_{12}^+ photofragments from C_{18}^+ photodissocation with 248nm light

These results suggest that C_{15}^+ is formed in a single step from C_{18}^+, whereas C_{12}^+ involves two photons in a sequential process during which C_{15}^+ appears as an intermediary. This two step process would require that the first photodissociation occur in a time scale short compared to the laser pulse (<10 ns). Such a dissociation time is not unreasonable, however it is also possible that C_{12}^+ could be formed directly from C_{18}^+ in a two-photon induced fragmentation with no intermediate cluster. Measurements of the depletion of possible intermediate states and comparisons with photofragmentation cross sections tend to support the stepwise theory. Nonetheless, there are cases where the intermediate clusters seem to deplete far too easily for their given cross sections. Such exceptions could be explained if the intermediate clusters are formed in excited states with large cross sections for further photofragmentation.

All of the results which follow involve the linear component of photofragmentation as a function of laser fluence. They were obtained from measurements of fragmentation at suitably low laser fluences and are essentially free of multiphoton effects.

3. Results

3.1 Carbon

The carbon positive cluster ions used in these studies were derived directly from the laser vaporization source. A typical spectrum of these positive ions is shown in Fig. 4. Intense peaks in this spectrum ("magic" numbers) appear with a periodicity of 4 at n=7, 11, 15, 19, and 23. The distribution of cluster ions produced in the source is quite different from that generated by photoionizing the neutral clusters. In the latter case, no odd clusters are observed for n>30, while they are quite evident among the ions which leave the source directly.

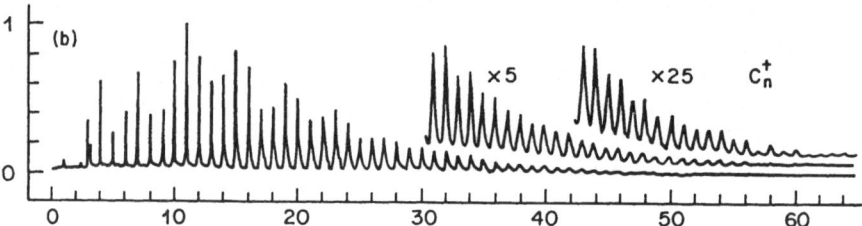

Fig. 4 Spectrum of carbon cluster ions emitted directly from the source

These carbon cluster ions were photodissociated using the 248nm and 351nm lines from an excimer laser, focused to a spot approximately 0.4 cm². Figure 5a shows a three-dimensional histogram of the photofragments for carbon cluster ion photofragmentation with 248nm light at low laser fluence (~2 mJ/cm²). At this low fluence, processes with intensities greater than 10% of the total are generally found to be linear with laser fluence. It is clear from the strong diagonal crest (shaded) that the dominant fragmentation pathway involves the emission of a neutral C_3 fragment. For some of the larger cluster ions, loss of C_5 is also an important fragmentation pathway. The branching ratios at both 248 and 351nm are tabulated in Fig. 5b. Results at the two wavelengths are very similar and differences of less than 5% of the total should not be considered significant. Similar results were also obtained with the second harmonic of a Nd:YAG laser, but are not reported here because the poor beam quality of the Nd:YAG laser made fluence dependence studies difficult.

By determining which cluster ions deplete linearly in laser fluence (single photon) at a given wavelength, it is possible to obtain bounds for the photodissociation energies of the cluster ions. The major problem with such

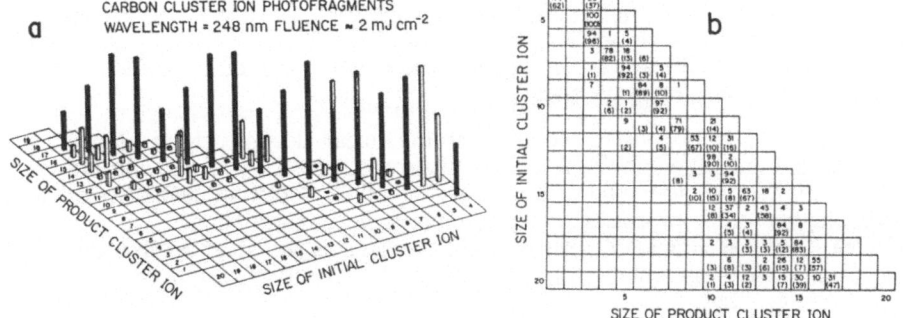

Fig. 5 Three-dimensional histogram (a) of the photofragments from carbon cluster ion photodissociation with 248nm light at a fluence of approximately 2 mJ/cm². The peaks corresponding to the loss of C_3 are shaded. Tabulated results are given in (b), with 351nm data in brackets

measurements is that two-photon processes can involve very different cross sections for the two excitation steps. The larger cross section saturates easily and the smaller one dominates. As a result, the process appears linear. To avoid such difficulties, we worked at very low laser intensities, small depletions, and consequently, low signal to noise ratios.

The results of such depletion measurements appear in Fig. 6. In three cases, C_4^+ at 351nm (3.53 eV) and C_5^+ at both 351 and 248nm (4.98 eV), the depletion was too weak to measure fluence dependence and is probably multiphoton. For all cluster ions with n=6-20, photodissociation appears one photon at 351nm. There is little experimental data with which to compare our measured binding energies although derived values, based on heats of formation[19] and ionization energies[20], are in general agreement with our results. Calculated values by RAGHAVACHARI[21] are also in reasonable agreement, with the exception of C_5^+, which is predicted to be bound by more than 5 eV. One explanation for such a discrepancy would be the existence of a high energy C_5^+ isomer in our

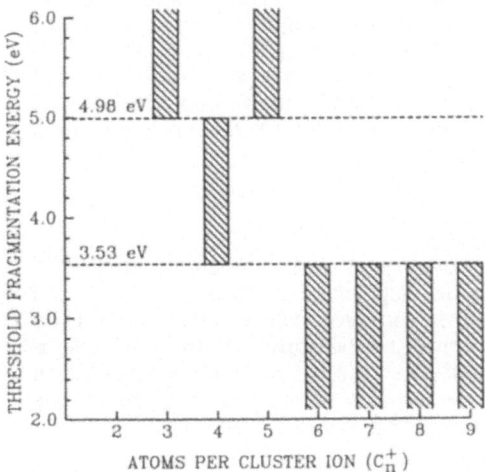

Fig. 6 Bracketed photofragmentation energy thresholds for C_{3-20}^+. C_5^+ value is derived from refs. [19,20]. C_{10-20}^+ are all less than 3.53 eV

beam, having a reduced photofragmentation threshold. Evidence for such an isomer has been found in FT-ICR (Fourier Transform Ion Cyclotron Resonance) measurements by MCILVANY, *et al.*[13], where only part of a sample of C_7^+ would react with some reactant gases.

Total photodissociation cross sections for the linear decomposition of carbon cluster ions are presented in Fig. 7. They were obtained by measuring cluster ion depletion at low laser fluences and low (<10%) depletion levels so as to minimize higher order processes. The absolute scale is approximately in unit of 10^{-17} cm^2. The accuracy of relative cross sections is about ±30% for nearby cluster sizes, while systematic errors reduce the accuracy for more widely separated cluster sizes.

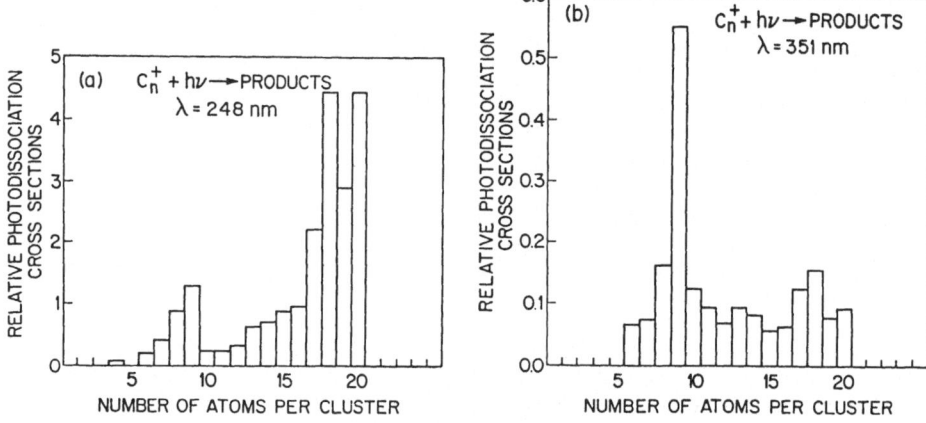

Fig. 7 Photodissociation cross sections for the photofragmentation of carbon cluster ions at (a) 248nm and (b) 351nm. Approximate absolute values for the cross sections can be derived from the relative values by multiplying the scale by 10^{-17} cm^2

It is interesting to speculate on the origin of the dramatic variation of the photodissociation cross sections with cluster size. While we reported previously[14] that the drop in cross section at n=10 might reflect a predicted change in structure from small linear chains to larger rings[22], more recent high level calculations[23,24] indicate that C_4 may already have a rhombic ground state. Further calculations and more detailed measurements of photodissociation spectra will be necessary for a definitive understanding of the cross sectional variations.

3.2 Silicon

Measurements in silicon predate the discovery, in this laboratory, that cold ionized clusters can be formed directly in the vaporization source[6]. Thus, the following results were obtained using silicon positive cluster ions derived by photoionizing neutral clusters. The internal temperature of the clusters is unknown, but is expected to be above 300° K. As a result, photodissociation threshold measurements could not be performed unambiguously, since the inhomogeneous thermal nature of the particles obscured any clear transition as a

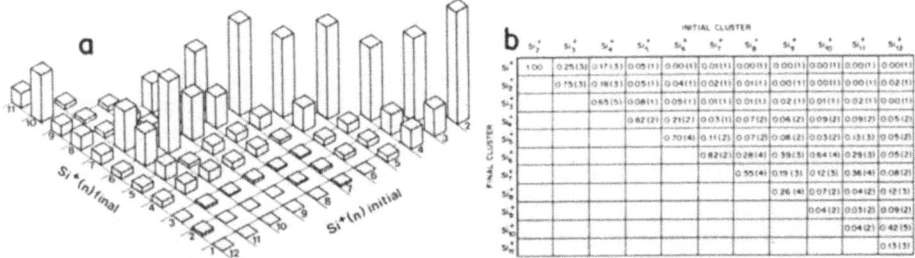

Fig. 8 Three-dimensional histogram (a) of the photofragments from silicon cluster ion photodissociation with 532nm light. Tabulated results are given in (b)

function of wavelength. Photofragmentation branching ratios, however, do not appear to vary significantly with changes in wavelength (see carbon results) and thus should be relatively insensitive to the ~100 meV thermal excitations of the cluster ions. Photofragmentation cross sections are also expected to be relatively unaffected at this low excitation level, though a small effect is likely.

The silicon cluster ions were photodissociated with the second harmonic beam (532nm) from a Nd:YAG laser. A small (2.5 mm²), relatively uniform portion of the unfocused beam was selected for use in these studies. As in the carbon measurements, the linear portion of photofragmentation branching ratios was extracted by analyzing the low intensity fluence dependence of fragmentation. A histogram of the branching ratios appears in Fig. 8a with the corresponding tabulation in Fig. 8b. It is evident that, unlike carbon, the "magic" fragments in silicon are not neutral clusters, but rather Si_6^+ and Si_{10}^+. These positive cluster ions appear as pronounced final state rows in the figure. These are the same clusters which appear as "magic" numbers in the spectrum of positive cluster ions emitted directly from the source (Fig. 9).

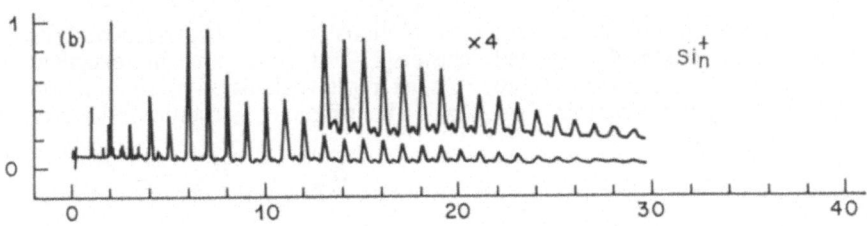

Fig. 9 Spectrum of silicon cluster ions emitted directly from the source

Further evidence for the special stability of Si_6^+ and Si_{10}^+ is given by the total photofragmentation cross sections at 532nm. The measured values for ions Si_2^+ to Si_{11}^+ appear in Fig. 10. Si_4^+, Si_6^+ and Si_{10}^+ are all local minima, suggesting that they have particularly compact or stable geometries.

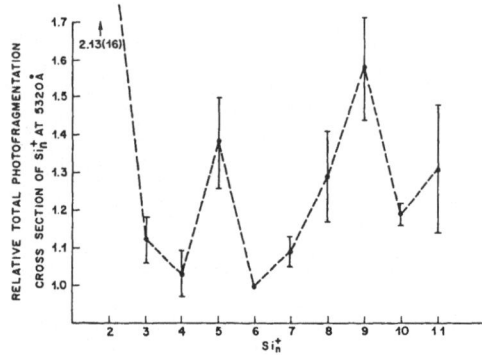

Fig. 10 Photodissociation cross sections for the photofragmentation of silicon cluster ions at 532nm. Approximate absolute values for the cross sections can be derived from the relative values by multiplying by 0.7 x 10^{-18} cm²

4. Discussion

There are two possible mechanisms for the appearance of "magic" sizes among the photofragments of carbon and silicon clusters. These "magic" fragments could be fundamental building blocks for the larger clusters and therefore the most easily detached pieces. We think this explanation unlikely. Instead, we believe that these special particles are unusually stable structures and first appear during the unimolecular decomposition process which occurs following the absorption of optical energy. A recent molecular dynamics study[25] for the silicon neutral clusters supports this view, predicting Si_4, Si_6, and Si_{10} among the common fragments of larger clusters. That study also finds that Si_4, Si_6, and Si_{10} are stable at greater internal temperatures than are their neighboring clusters.

Further support for the statistical dissociation process is found in the wavelength independence of the branching ratios. For a direct dissociation mechanism, the branching ratios would be expected to depend strongly on excitation energy. However, the similarity of branching ratios cannot be regarded as proof of statistical dissociation, since unimolecular reactions are not generally independent of energy[26].

Several high level, ground state energy calculations have been carried out for silicon neutral and positively ionized clusters[27-34], and are in excellent agreement with all of the silicon results. As example, calculated fragmentation energies[29] for Si^+_{2-7} are given in Fig. 11a. The predicted structure[29-31] of

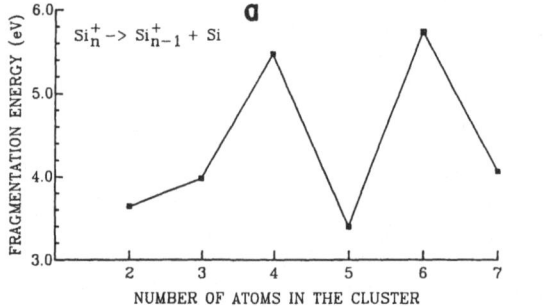

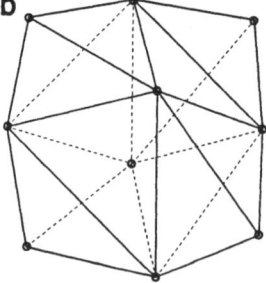

Fig. 11 a) Theoretical fragmentation energy for the reaction $Si^+_n \rightarrow Si^+_{n-1} + Si$ vs the number of atoms n for Si^+_2-Si^+_7. Obtain from ref. [29]. b) Calculated structure for Si_{10}, a tetra-capped octahedron [29-31]

Si_{10}, a tetra-capped octahedron, is shown in Fig. 11b. The high symmetry and compact nature of this particle are principally responsible for its stability. Similar calculations for carbon have also been undertaken[21,23,24] and are in general agreement with results of this study (apart from the C_7^+ exception, noted previously).

5. Conclusion

We have reported the results of detailed photofragmentation studies on carbon and silicon positive cluster ions. The most remarkable features of these measurements are the "magic" fragments Si_6^+, Si_{10}^+, and C_3. Measurements of the total photofragmentation cross sections have shown great variability with cluster size, and significant dependence on laser wavelength. Work with tunable dye lasers and cold cluster ions should further enhance our measurements and yield considerable information on the nature of these cluster ions.

1. R.E. Honig: J. Chem. Phys. 22, 126 (1954); R.E. Honig: J. Chem. Phys. 22, 1610 (1954)
2. D.E. Powers, S.G. Hansen, M.E. Geusic, A.C. Puiu, J.B. Hopkins, T.G. Dietz, M.A. Duncan, P.R.R. Langridge-Smith, R.E. Smalley: J. Phys. Chem. 86, 2556 (1982)
3. T.T. Tsong: Appl. Phys. Lett. 45, 1149 (1984)
4. L.A. Bloomfield, R.R. Freeman, W.L. Brown: Phys. Rev. Lett. 54, 2246 (1985)
5. J.R. Heath, Y. Liu, S.C. O'Brien, Q.-L. Zhang, R.F. Curl, F.K. Tittel, R.E. Smalley: J. Chem. Phys. 83, 5520 (1985)
6. L.A. Bloomfield, M.E. Geusic, R.R. Freeman, W.L. Brown: Chem. Phys. Lett. 121, 33 (1985)
7. M.L. Mandich, W.D. Reents, Jr., V.E. Bondybey: J. Phys. Chem. (in press)
8. Y. Liu, Q.-L. Zhang, F.K. Tittel, R.F. Curl, R.E. Smalley: To Be Published
9. N. Furstenau, F. Hillenkamp: Intern. J. Mass Spectrom. Ion Phys. 37, 135 (1981)
10. E.A. Rohlfing, D.M. Cox, A. Kaldor: J. Chem. Phys. 81, 3322 (1984)
11. H.W. Kroto, J.R. Heath, S.C. O'Brien, R.F. Curl, R.E. Smalley: Nature 318, 162 (1985)
12. Y. Liu, S.C. O'Brien, Q. Zhang, J.R. Heath, F.K. Tittel, R.F. Curl, H.W. Kroto, R.E. Smalley: Chem. Phys. Lett. 126, 215 (1986)
13. S.W. McElvany, W.R. Creasy, A. O'Keefe: J. Chem. Phys. 85, 632 (1986)
14. M.E. Geusic, T.J. McIlrath, M.F. Jarrold, L.A. Bloomfield, R.R. Freeman, W.L. Brown: J. Chem. Phys. 84, 2421 (1986)
15. M.E. Geusic, M.F. Jarrold, T.J. McIlrath, L.A. Bloomfield, R.R. Freeman, W.L. Brown: Z. Phys. D (in press)
16. S.C. O'Brien, J.R. Heath, H.W. Kroto, R.F. Curl, R.E. Smalley: (to be published)
17. R.D. Knight, R.A. Walch, S.C. Foster, T.A. Miller, S.L. Mullen, A.G. Marshall: Chem. Phys. Lett. (to be published)
18. R.C. Dunbar: In *Molecular Ions: Spectroscopy Structure and Chemistry*, ed. by T.A. Miller, V.E. Bondybey (North Holland, Amsterdam 1983);J. Moseley, J. Durup: J. Chim. Physique 77, 673 (1980)
19. J. Drowart, R.P. Burns, G. DeMaria, M.G. Ingram: J. Chem. Phys. 31, 1131 (1959)

20. H.M. Rosenstock, K. Draxl, B.W. Steiner, J.T. Herron: J. Phys. Chem. Ref. Data 6, Supplement No. 1 (1977)
21. K. Raghavachari: Private Communication
22. K.S. Pitzer, E. Clementi: J. Amer. Chem. Soc. 81, 4477 (1959); S.J. Strickler, K.S. Pitzer: In *Molecular Orbitals in Chemistry Physics and Biology*, ed. by B. Pullman, P.O. Lowden (Academic Press, New York 1964); R. Hoffman: Tetrahedron 22, 521 (1966)
23. R.A. Whiteside, R. Krishnan, D.J. DeFrees, J.A. Pople, P. von R. Schleyer: Chem. Phys. Lett. 78, 538 (1981)
24. B.K. Rao, S.N. Khanna, P. Jena: Sol. St. Commun. 58, 53 (1986)
25. B.P. Feuston, R.K. Kalia, P. Vashishta: (to be published)
26. M.F. Jarrold, L.M. Bass, P.R. Kemper, P.A.M. van Koppen, M.T. Bowers: J. Chem. Phys. 78, 3756 (1983)
27. K. Raghavachari: J. Chem. Phys. 83, 3520 (1985)
28. J.C. Phillips: J. Chem. Phys. 83, 3330 (1985)
29. K. Raghavachari, V. Logovinsky: Phys. Rev. Lett. 55, 2853 (1985)
30. S. Saito, S. Ohnishi, C. Satoko, and S. Sugano: J. Phys. Soc. Japan 55, 1791 (1986)
31. K. Raghavachari: J. Chem. Phys. 84, 5672 (1986)
32. K. Balasubramanian: Chem. Phys. Lett. 125, 400 (1986)
33. D. Tomanek, M.A. Schluter: Phys. Rev. Lett. 56, 1055 (1986)
34. G. Pacchioni, J. Koutecky: J. Chem. Phys. 84, 3301 (1986)

Dynamical Characteristics of Laser Vaporization Process in Group IV Elements

A. Kasuya and Y. Nishina

The Research Institute for Iron, Steel and Other Metals,
Tohoku University, Sendai 980, Japan

Emission characteristics of electrons and ions desorbed from the solid target of group IV elements have been investigated by nitrogen laser excitation in the intensity range from 5 to 200 MW/cm^2. The results on Si, for example, show the presence of a threshold in the laser intensity above which emission of both electron and positive ion increases sharply by more than three orders of magnitudes in an approximately equal rate, and then tend to saturate with further increase in the intensity. The observed features can not be described in terms of ordinary vaporization process in which each atom/ion is thermally ejected individually from isolated atomic sites on the surface. The presence of threshold and its characteristics indicate that optically excited atoms are desorbed not individually into vacuum but in various forms of large clusters as a result of electronic excitation above a critical density. The measured results by means of a 127° energy analyzer indicate that the subsequent fragmentation of clusters takes place in a time period between 1 to 10 μs. The major decomposition process toward the final few steps is represented by $Si_k^{a+} \rightarrow Si_{k-1}^{(a-1)+} + Si^+ + E_r$, where the released kinetic energy $E_r < 3$ eV.

1. INTRODUCTION

Our previous time-of-flight (TOF) mass analysis shows that ions emitted from surfaces of Si, Ge and Sn by nitrogen laser excitation are initially coagulated in large-size clusters, and not in the form of individual atomic ions [1,2]. These clusters, however, are chemically unstable and dissociate into their ultimate form of mono-atoms and -ions in the time scale of 10 μs. This paper reports the result of our analysis on the threshold characteristics of ion emission and subsequent decomposition process in a period ranging from 1 to 10 μs. The decomposition process is analyzed with the time variation of detector current of ions passing through an electrostatic energy analyzer (EA). The details of our experimental arrangements are given in our previous publications [1,2]. The charge-to-mass ratios of emitted ions have been analyzed with a TOF spectrometer consisting of a straight flight tube of 62 cm, allowing us to analyze varieties of decomposition processes. EA of the 127° cylindrical sector type has its mean radius of 35 mm. The energy resolution is 5 meV at the kinetic energy of 1 eV.

2. EMISSION YIELD FOR IONS AND ELECTRONS

Figure 1 shows the emission yields for ions and electrons from Si surface in the peak intensity range of nitrogen laser beam from 5 to 200 MW/cm^2. Both of the emission yields sharply increase at an approximately definite level of laser intensity, $I_{th} = \sim 50$ MW/cm^2 at their comparable rates. Below I_{th} only electronic emission is observed. Bensousson et al. find in

249

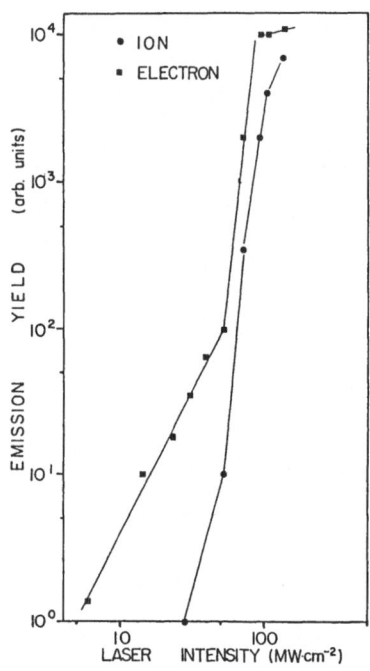

Figure 1. Emission yields of ions (circles) and electrons (squares) desorbed from silicon surface by nitrogen laser excitation in the peak intensity range from 5 to 200 MW/cm^2. The pulse width is 10 ns.

their measurements by nitrogen laser excitation in KW/cm^2 range that the electronic yield is proportional to the square of laser intensity [3]. Since the work function of Si is larger than the photon energy but is smaller than twice the energy, they interpret that the two-photon excitation process is the dominant mechanism for the electron emission. Our results of Fig. 1 also show a square-law dependence of the electron yield below I_{th}. More direct evidence of the two-photon process is found in our measurement of electron kinetic energy that the high-energy edge of its distribution plus the work function of silicon is equal to twice the photon energy of 7.6 eV. The above I_{th}, the electronic yield as well as the ionic increases sharply in comparison with the two-photon contribution of the electron yield, and tends to saturate with further increase in the laser intensity. In order to investigate the characteristic feature of desorption process, simultaneous detection of the emission yields vs. laser intensity is carried out for each laser shot. The following features are found: 1) Above I_{th} but below the saturation level for emission, both electronic and ionic yields fluctuate violently from shot to shot of laser pulses, even for a nearly equal peak intensity. 2) In this intensity range, the ionic emission is always accompanied by electronic emission. No ionic emission is observed if no electronic emission takes place. Here the minor background contribution from two-photon excitation process is neglected. Hence, the fluctuation comes from two causes. One is the random alternation of presence with absence of ionic and electronic emission for a given number of laser shot. The other is the fluctuation in emitted numbers of ions (electrons). Both of these fluctuations tend to diminish as the yields approach their saturation values for higher intensities.

The above emission characteristics lead to the following interpretation of the process of ion desorption: a) The presence of a definite threshold shows that the ion desorption takes place as a consequence of extreme electronic excitation of valence electrons into the conduction band in such a high degree as to reach a critical density above which surface atoms can no longer remain in the condensed phase of the bulk. b) Desorbed ions are not ejected individually from isolated lattice sites on the surface, but a shot of laser pulse above I_{th} induces desorption of a large number of neighboring atoms (ions) in the irradiated area. If the emission process were vaporization due to thermal energy transfer, neither the threshold nor the violent fluctuation of yields would take place near the threshold. The saturation above the threshold would not happen either. Furthermore, the ordinary vaporization process can not explain the fact that the electronic yield increases in the rate and magnitude approximately equal to those of ions. Such experimental results imply that the work function for positive ions, w_p is comparable in magnitude with that of electron, w_e. In general, $w_p > w_e$ [4]. The skin depth consideration suggests that the apparent saturation of the yields comes from the energy transfer process as follows: Namely, once the surface atoms are emitted, the residual laser energy above I_{th} is dissipated in the excitation of emitted clusters.

The above experimental observations can be explained consistently in the picture deduced from our TOF analysis [2] that the critical level of surface excitation corresponds to the change in the state of surface atoms from the condensed phase into ionized vapor of various sizes in cluster forms. These clusters are in highly excited states so that they begin to decompose through numerous fragmentation channels immediately after their formation.

3.   ANALYSIS OF DECOMPOSITION PROCESS BY MEANS OF ENERGY ANALYZER

Our method of TOF analysis gives information on decomposition process in the time period comparable to the acceleration time in the TOF spectrometer. This time period is about 1 μs for the acceleration voltage of 300 V in our previous measurement [2]. In principle, the decomposition process in a time scale longer than 1 μs can be analyzed by decreasing the acceleration voltage, thereby increasing the acceleration time, provided that the kinetic energy gained by acceleration is much higher than the initial kinetic energy of ions. Since the latter energy is in the order of 10 eV, lowering the former voltage below 300 V results in the degradation of spectral resolution. In order to overcome this difficulty, one introduces the emitted ions into EA and analyzes the decomposition process by measuring the variation in the flight time of ions through EA. The mean radius of EA is 35 mm. The total flight path from the sample to the exit slit is 95 mm so that it takes 15.8 μs for ions with their kinetic energy of 5 eV. If there is no decomposition during the flight, a sharp peak in the detector current appears in the flight time spectrum. If decomposition of ions takes place, it induces as a change in the line-profile, i.e. the time variation, of the detector current.

Figure 2 shows spectra obtained with Si for various settings of the path energy, E_p, of the analyzer. If E_p is set above 18 eV a relatively sharp peak is observed at the spectral position of M/q = 1 with the transit time of 8.3 μs through EA, where M is the mass in units of a Si atom and q its charge multiplicity. The peak is followed by a weak shoulder extending toward the longer flight time up to about M/q = 2. A sharp peak found at M/q = 1 indicates that most of ions are decomposed into Si^+ in a period much less than 8.3 μs after laser excitation. The presence of weak shoulder, on the other hand, implies that there remain some ions which decompose into Si^+ in a period comparable to their flight time of 8.3 μs.

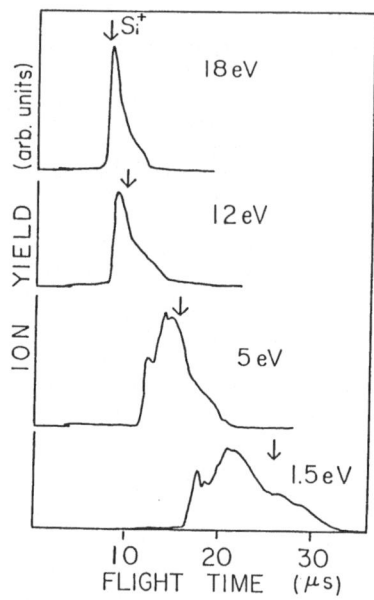

Figure 2. Time variations in the detector current of ions passing through a 127° electrostatic energy analyzer for various settings of the path energy of the analyzer. The arrows indicate the respective positions of flight time for Si^+ without decomposition during the flight.

The ions which contribute to the shoulder are those which initially fly into the analyzer with their M/q > 1 and then decompose into Si^+ in the middle of the flight path toward the exit slit of EA. Flight times of such ions are delayed with respect to those of ions which fly through the entire path in the analyzer with M/q = 1 (Si^+).

As E_p decreases below 18 eV, the intensity of shoulder grows up to the magnitude comparable with the main peak and contributes to the broadening of the peak for Si^+. At the same time several new peaks become prominent. In addition, the peak energy shifts toward the shorter flight time. This energy shift is one of the evidences that the decomposition is indeed taking place. In our measurements of TOF and of quadrupole mass analyzer (QM), ionic species with M/q < 1 (e.g. Si^{2+}) have not been found in any appreciable quantity. Hence those ions reaching the detector in a flight time shorter than that for M/q= 1 must have M/q > 1 initially, but flying into the analyzer with their kinetic energies higher than E_p. Then they decompose into Si^+'s with their kinetic energies comparable to E_p, and directions of their orbits are nearly parallel to the analyzer electrodes so that the decomposed ions pass through the exit slit without impinging upon the electrodes. Our calculation described in the following section (IV) shows that a sharp peak is observed only if the initial ions fly into the analyzer with their kinetic energy comparable to E_p and continue to fly, after decomposition, with their energies and directions nearly equal to those before decomposition. In other words, a sharp peak is observed only if the over-all flight path of ions is kept nearly circular. These conditions are satisfied if the kinetic energy release, E_k, at the time of decomposition into Si^+ is small in comparison to E_p. Accordingly, the sharp peak found for E_p greater than 18 eV indicates that $E_k \ll 18$ eV, while the substantial broadening of the peak below E_p < 5 eV comes from the fact that the flight path of ions deviates appreciably from the circular orbit because E_p is comparable in magnitude with E_k.

The above results of flight time distribution with EA and those of TOF measurements indicate that the desorbed ionic clusters begin to dissociate in less than 1 μs and continue to decompose through various fragmentation channels into their ultimate form of monoatomic ions in the time scale longer than 1 μs. In this time scale the last one or two steps of decomposition are identified with the process $Si_k^{a+} \rightarrow Si_{k-1}^{(a-1)+} + Si^+$. Incidentally, similar decomposition process for the neutral clusters of Si is predicted by Raghavachari and Logovinsky in their theoretical analysis [5]. This result with EA correlates the result of TOF with that of QM in a consistent manner, since QM measurement provides mass distribution in a period longer than 10 μs after laser excitation and shows that only the dominant speices of ions is Si^+ [2].

4. CALCULATION OF LINE-PROFILE

Consider a two-dimensional motion of a molecular ion in a 127° electrostatic analyzer as shown in Fig. 3. The ion enters through the slit A of analyzer with its velocity of v_o in the direction perpendicular to the electric field applied between the inner electrode, I, and the outer one, O. The ion, then, decomposes into Si^+ and the residual ion(s) at some point P through one of the following succesive fragmentation processes, $Si_k^{a+} \rightarrow Si_{k-1}^{(a-1)+} + Si^+$, a,k = 1, 2, 3,..., with a kinetic energy $E_r = Mv_r^2/2$ released to Si^+ with M=1. The fragment ion Si^+ continues to travel along another orbit $\overset{\frown}{PB}$ with a new velocity $\vec{v}_f$, where $\vec{v}_f = \vec{v}_p + \vec{v}_r$ at P, and reaches the exit slit B. Here, $\vec{v}_p$ is the velocity of the parent ion at P. If the Si^+ ion moves along one of the different orbits such as $\overset{\frown}{PC}$ or $\overset{\frown}{PD}$, it does not pass through B and is not detected by the charge-multiplying detector M. One can calculate the flight time of an ion leaving the sample surface S at t= 0 , entering the analyzer through the slit A. We treat the case where this ions decomposes into Si^+ at various positions in the analyzer, and escapes through the slit B. Both parent and fragment ions move in the radial electrostatic field in two dimension during the flight time through the analyzer. The flight time distribution can be obtained by plotting the number of ions passing through the slit B vs. their flight times. This plot represents the line-profile, i.e. the time variation, of detector current.

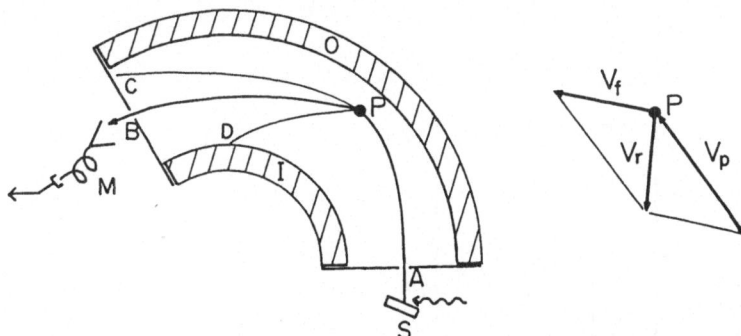

Figure 3. Schematic drawing for the flight path of an ion which experiences a single event of decomposition   once in the energy analyzer. At the point P, a Si^+ ion is ejected with its velocity $\vec{v}_r$ from the parent ion of velocity $\vec{v}_p$. The resultant velocity of Si^+ in the laboratory frame is given by $\vec{v}_f = \vec{v}_p + \vec{v}_r$.

In calculating the line-profile, the following assumptions are introduced: i) The initial velocity distribution of parent ions is constant. No restriction is imposed on the possible range of initial velocity. After all those ions having their initial energies appreciably larger or smaller than E_p impinge upon the electrodes and do not reach the slit B. The radii of the inner and outer electrodes are 26.5 mm and 43.5 mm, respectively. The slit width is 1 mm. ii) The parent ions decompose at any point P in the analyzer with an equal probability. At the point P the Si^+ fragment can be scattered toward any direction in the two-dimension with equal probability. iii) The range in k/a (mass-to-charge-ratio of the parent ions) is set in the range 1 < k/a < 2. Our TOF and QM measurements show that the ratio of the dominant chemical species before and after decomposition falls in this range. The uniform mass distribution is assumed within this range. Under these three conditions, the ratio of E_r/E_p = R becomes an independent variable which determines the line-profile.

Figure 4 shows the result of calculation under various settings of R. The flight time t_o corresponds to that of ions with M/q = 1 (Si^+) for each

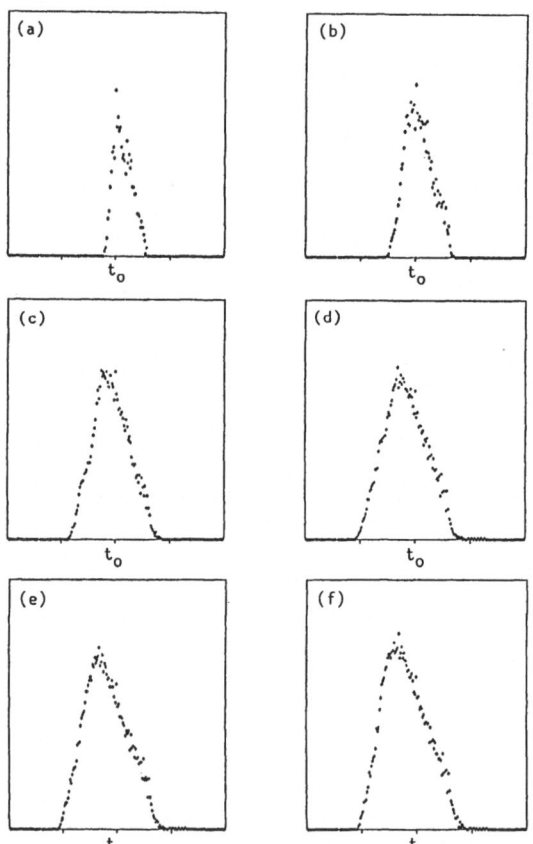

Figure 4. Results of calculation for the flight time distribution of the parent ions decomposing into Si^+ for various ratios, R's, of the released kinetic energy E_r to pass energy E_p, (a) 0 < R < 0.1, (b) 0 < R < 0.2, (c) 0 < R < 0.3, (d) 0 < R < 0.4, (e) 0< R < 0.5, (f) 0 < R < 0.6.

setting of E_p. Varieties of line-profiles are deduced for different choices of R. Figure 4 shows distributions of the number of detectable ions for six values of in R. In addition, the peak of the distribution shifts with increase R. In addition, the peak position of the distribution shifts to shorter flight time with respect to t_o. The result of calculation is consistent with the essential features of our experimental results shown in Fig. 2. For E_p = 18 eV a relatively sharp peak is observed at t_o. As E_p decreases, the linewidth broadens. The peak position shifts toward flight time shorter than t_o. Furthermore, in spite of such broadening a relatively sharp cut-off remains at both ends of the distribution. The tail, however, to the shorter flight time is steeper than that to the longer one. This asymmetry of the line-profile is predicted in the calculation, as shown in Fig. 4. Those parent ions which move in the orbit with their radial distances longer (shorter) than the mean radius of our analyzer, the orbit APB for example, reach the slit B in longer (shorter) flight time in comparison to t_o. Under the conditions imposed on k/a, more ions which decompose at a shorter radial distance have more chance of hitting the inner electrode rather than reaching the exit slit, B. From the above line-profile analysis, one finds the following: A) The dominant decomposition process in the time period of 1 μs is represented by $Si_k^{a+} \rightarrow Si_{k-1}^{(a-1)+}$ + Si^+, for 1 < k/a < 2. B) The kinetic energy, E_r, released to Si^+ at the time of decomposition is less than 3 eV. C) The maximum initial kinetic energy of parent ions is of the order of 10 eV.

5. DISCUSSION

The above experimental evidence on Group IV elements indicate that the laser desorption process in the intensity range of our measurement is not the ordinary vaporization of surface atoms in thermal equilibrium. There are two models of laser desorption which have been referred to frequently [4]. One is the thermal model in which energy of the laser beam is merely converted into heat in the lattice system so that surface atoms are thermally ejected. These atoms are partially ionized with the ratio given by the Langmuir-Saha equation which is valid only under thermal equilibrium. The other model considers the vaporization of ions in two steps. The earlier part of a laser pulse causes the vaporization of surface atoms according to the process referred to in the first model. The partially ionized vapor subsequently absorbs the latter part of the laser energy to increase the degree of ionization and to expand itself into vacuum in the form of high-energy atoms and ions. The former model has been applied to the case of low laser excitation with its peak intensity below 1 MW/cm^2 in the nanosecond range of pulse width. The extensive investigations on the laser annealing process in semiconductors indicate that the former model explains the gross feature of the desorption of electrons and neutral atoms up to the melting of the sample surface [6]. This model, however, does not predict any of the characteristics 1) and 2) in Sec. II found in our measurements. The latter model has been applied to cases of laser excitation of solid surfaces in the GW/cm^2 range where laser beam is so intense (>> I_{th}) that the resulting high density of ionized vapor continues to absorb the latter part of pulsive laser fluence. Consequently the increase in ionization and expansion leaves no more energy to excite the surface lying under the opaque plasma. This model corresponds to the case where the major portion of laser energy is given to the excitation of vaporized materials in gas phase. The mass, charge and kinetic energy distributions of ions reflect essentially the state of excitation in the gas phase. The latter model can explain the observation of high-energy ions, but not the threshold characteristics found in our measurements.

There has been no systematic study as to the continuity between the two models discussed above in terms of the level of laser fluence. The thermal model considers only the excitation process in the solid phase, while the other model deals with the ionized vapor in gas phase. Neither model takes explicit account of interaction of laser beam in the transition stage of the desorption process. Our experimental results clearly show the necessity of developing a new model which deals with the dynamics in the transition of solid surface into ionized vapor under the extreme excitation of the electronic system. There are a few models which consider the electron–phonon interaction in the critical state of a high density of electron–hole plasma [7–11]. These models predict the presence of a threshold intensity and desorption of surface atoms into highly ionized vapor. They do not refer, however, to the dynamics of energy transfer associated with coagulation and/or fragmentation of emitted particles.

6. CONCLUSIONS

The dynamical aspect of the emission characteristics for laser desorption process has been investigated by time-resolving analysis. Our four independent measurements, i.e. ion yield, time variations of TOF, QM and EM spectra,explain in a consistent manner the emission behavior of highly ionized vapor in coagulated forms of constituent particles.. The existing microscopic models can not provide satisfactory description of the essential features of dynamical desorption. The analysis of our experimental results strongly suggests the non–local behavior of desorption process which shows a high degree of correlation among emitted particles near the critical excitation.

Authors would like to thank Professor C. Horie, Professor S. Maekawa and Mr. M. Ashizawa for valuable discussions. This work was supported in part by the Mitsubishi Foundation.

1. A. Kasuya and Y. Nishina, Phys. Rev. B28, 6571 (1983).
2. A. Kasuya and Y. Nishina, Phys. Rev. Lett. 57, 755 (1986).
3. M. Bensoussan, L.M. Moison, B Stoesz, and C. Sebenne, Phys. Rev. B23, 992 (1981).
4. J.F. Ready, Effects of High Power Laser Radiation (Academic Press, New York, 1971) Chap. 4.
5. Raghavachari and Logovinsky, Phys. Rev. Lett. 55, 2853 (1985). This paper considers fragmentation process of neutral Si cluster and does not necessarily apply to the case of ionic clusters.
6. Laser Annealing of Semiconductors, edited by J.M. Poate and W. Mayer (Academic Press, New York, 1982).
7. J.A. Van Vechten, Mat. Res. Soc. Symp. Proc. 23, 81 (1984).
8. M. Combescot and J. Bok, Phys. Rev. Lett. 48, 1413 (1982).
9. N. Ito and T. Nakayama, Phys. Lett. 92A, 471 (1982).
10. R. Biswas and V. Ambegaokar, Phys. Rev. B26, 1980 (1982).
11. M. Wautelet, Surface Sci. 133, L437 (1983).

Localized Orbital Approach to Carbon Clusters

K. Terakura[1], T. Hoshino[2], and T. Asada[2]

[1]Institute for Solid State Physics, University of Tokyo,
Roppongi, Minatoku, Tokyo 106, Japan
[2]Department of General Education, College of Engineering,
Shizuoka University, Hamamatsu 432, Japan

1. Introduction

Recent progress in the experimental techniques of producing semiconductor
microclusters and of analyzing the distribution of cluster size has
revealed several interesting aspects in the stability problem of the microclusters. In particular, the work by ROHLFING et al.[1] on carbon clusters
has attracted our attention strongly. They showed that for very small
clusters, C_3 seems to be fairly stable and that for C_n with n ranging from
about 10 to 30, the cluster-size distribution exhibits a periodic alternation every four atoms with maxima at n=11, 15, 19, 23, and 27.
Furthermore, for larger clusters with n ranging from 40 to 200, only even
clusters seem to be stable. Although they detected the cluster by photoionization time-of-flight mass spectrometry, they claimed that the
stability problem is concerned with the neutral clusters rather than
ionized ones. There are some speculations to interpret the experimental
observation[1], but definite answer has not been given yet. Therefore, the
systematic understanding of the phenomenon based on the detailed
electronic structure calculations within the local density functional
framework is the ultimate goal of our study.

However, this is a very difficult problem, because the structure optimization of a given cluster becomes formidably difficult as the cluster size
increases. Furthermore, even if such a work may become possible with the
help of gigantic computers, huge computations may cause one to fail in
obtaining clear-cut ideas for the basic mechanisms. Therefore, we decided
to proceed in the following way. As the first step, we will perform the
electronic structure calculations as accurate as possible for relatively
small clusters, in order to check the validity of our fundamental tools,
i.e., the norm-conserving pseudopotential (NCPP)[2] and the local spin
density approximation (LSDA) in the spin density functional method. We
will show in the following that as far as the structural problems of carbon
clusters are concerned, careful calculations based on the above two tools
can give remarkably good results. In the second step, we try to find out a
simpler but sufficiently accurate way of doing the electronic structure
calculations. Our localized orbital theory [3,4] is associated with this
context. Although this approach will not reduce the amount of computations, the use of a good minimal basis set provided by our method will make
it easy to analyze the calculated results and also to develop model
Hamiltonians for dealing with larger clusters. The validity of the
approximations introduced in the second step can be checked with the
results obtained in the first step. Unfortunately, however, we have not
completed the whole project yet and can present only a limited number of
results. This article is organized as follows. Section 2 is devoted to a
brief description of the computational procedure and presentation of
calculated results for C_n(n=2, 3, 4 and 5) with linear chain structures and

also C_4 with a rhombus structure. Basic aspects of our localized orbital theory and an application to C_2 will be described in Section 3. Section 4 is devoted to concluding remarks.

2. Detailed Electronic Structure Calculations for Small Carbon Clusters

As mentioned in Introduction, the fundamental tools in our electronic structure calculations are the LSDA for the spin density functional theory and the NCPP. CEPERLEY and ALDER's results[5] parametrized by PERDEW and ZUNGER[6] is adopted for the LSDA. We use Gaussian-type orbitals (GTO's) as variational basis functions. The use of the NCPP enables us to save the number of basis functions and four GTO's for each of s- and p-type orbitals are sufficient. The actual values of exponents are (0.16000, 0.50000, 1.96655, 5.14773) for s-type and (0.16000, 0.50000, 1.14293, 3.98640) for p-type orbitals. The smallest two exponents in each case are modified from the original values by HUZINAGA[7]. The fast Fourier transform(FFT) is adopted to solve Poisson's equation.

In order to check the accuracy of the present calculations, the binding energies, interatomic distances and vibrational frequencies were calculated for the ground state $^1\Sigma_g^+$ and the excited state $^3\Pi_u$ of C_2 molecule and the results are shown in Table 1. The overall agreement with experimental data[8] is quite satisfactory. One should note that C_2 can be a very severe test case, as exemplified by the fact that the Hartree-Fock calculation even with a large basis set yields only 0.5 eV as the binding energy of the ground state[9]. The results of CI calculation by FOUGERE and NESBET[10] for C_2 agree with experimental data reasonably well but not so accurately as the present ones.

Table 1. Comparison between the present calculation and experiment for the binding energy E_b, interatomic distance R and the vibration frequency ω for C_2.

		E_b (eV)	R (a.u.)	ω (cm^{-1})
$^1\Sigma_g^+$	cal.	6.78	2.35	1808
	exp.	6.35	2.35	1856
$^3\Pi_u$	cal.	6.76	2.48	1605
	exp.	6.27	2.48	1641

The optimized atomic arrangements for linear chain form of C_n (n=3, 4, 5) and a rhombus structure of C_4 are shown in Fig.1. As for the bond length of C_3, the present result and some of the other theoretical results coincide with each other within 1% and agree with the experimental value of 2.42(in a.u.)[11] fairly well. Similarly, most of the calculations are mutually consistent with regard to the structure of C_5, but no direct experimental information is available. On the other hand, some confusion exists for C_4. According to the infra-red(IR) absorption measurements of the stretching modes of linear chain carbon clusters[12], the stretching force constant f_R of the end bond takes a fairly large value of 16.6 mdyn/A for C_4 compared with other cases (11 and 12 mdyn/A for C_5 and C_6). The experimental data has been regarded as an indication that the end bond of C_4 may have a strong character of triple bond, while all bonds in C_5 may be regarded as double bond. HOFFMANN's extended Hükel calculation[13] predicted a strong bond alternation for C_4 with triple end bonds, but such a

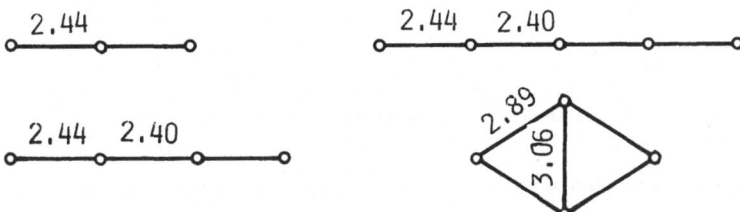

Fig.1. The optimized structures of linear chains of C_3, C_4, and C_5 and a rhombus of C_4. The figures along bonds denote bond lengths in a.u..

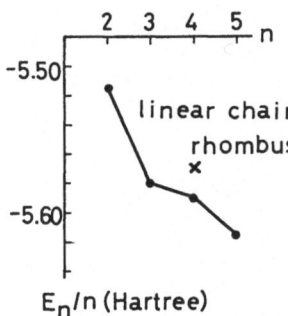

Fig.2. The total energies per atom of C_n clusters. The circles are for linear chain structures and the cross is for a rhombus structure.

bond alternation persists even up to C_{17}. Therefore, although his calculation seems to be qualitatively consistent with the IR data for C_4, it can not explain a rapid change in f_R when moving from C_4 to C_5. On the other hand, our calculation as well as the Hartree-Fock calculation[9,15] of C_4 yielded a very small bond alternation with an opposite sense, being quite similar to the situation in C_5 as can be seen in Fig.1. A consistent interpretation of the IR data is left as a future task. We then recon-firmed the interesting predictions by PITZER and CLEMENTI[14] about the even-odd alternation: for n odd, the ground state is singlet and the cluster is relatively stable, while for n even, the ground state is triplet (except for C_2) and the cluster is less stable. The total energy per atom in the NCPP framework is plotted against n in Fig.2, from which significant stability of C_3 is readily seen. The cross in Fig.2 corre-sponds to C_4 with a rhombus structure. In the present calculation, a linear chain structure is definitely more stable than a rhombus one for C_4. However, the CI calculations predict a rhombus as the most stable structure for C_4, but the energy difference between the two structures is generally only marginal[15,16].

3. Local Orbital Theory

Structural optimization by means of detailed ab initio electronic structure calculations may be possible only up to about n=10. Therefore, it is almost inevitable to adopt a simpler model Hamiltonian in order to deal with a larger cluster, and it is highly desirable to have a proper prescription to derive such a model Hamiltonian. The first important step toward this direction is to prepare a good minimal basis set with well localized orbitals. If such a minimal basis set is available for a given atomic configuration, we can calculate the corresponding Hamiltonian and overlap matrix elements. Repeating the procedure for some different atomic configurations and interpolating (or extrapolating) the results to an arbitrary configuration, one may obtain information necessary in deriving a model Hamiltonian. The aim of this section is to give a brief description of the local orbital (LO) theory which we have developed in order to prepare a minimal basis set[3,4].

The starting equation for determining the LO is a general ADAMS-GILBERT equation[3,17] expressed as

$$H|I,i> = E_{Ii}|I,i> + \sum_{(Jj)\neq(Ii)} |Jj>\lambda_{Jj,Ii} \tag{1}$$

where H is the Hamiltonian of the system and $|I,i>$ is the i-th LO associated with the I-th atom. The parameters E_{Ii} and $\lambda_{Jj,Ii}$ are in principle arbitrary. Equation(1) means simply that the LO's which we are solving for span a closed subspace with regard to a given H. If we set

$$\lambda_{Jj,Ii} = <J,j|V_J|I,i> \tag{2}$$

with V_J an atomiclike potential assigned appropriately to the J-th site, Eq.(1) is reduced to ANDERSON's chemical pseudopotential equation[18,19]. As the LO's are coupled nonlinearly in the chemical pseudopotential equation, no reliable solutions have ever been obtained. We therefore introduce a single-site approximation[3] in which the LO, $|I,i>$ is expressed as

$$|I,i> = \sum_n |I,n\} c^I_{n,i} \tag{3}$$

where $|I,n\}$ denotes a local basis orbital, for example, a GTO centered at the I-th site. The fundamental assumption here is that the LO at the I-th site is expressed only by a finite set of local basis orbitals on the same site. We call the LO obtained with this approximation the SSA-LO. Some details for solving Eq.(1) were described in Ref.3. We would like to mention one important aspect. The arbitrariness of E_{Ii} and $\lambda_{Jj,Ii}$ is utilized to optimize the single-site approximation. The E's, λ's and LO's are all self-consistently determined. The SSA-LO may be sufficient for most purposes, but further accuracy may be required sometimes. Then we simply add some contributions from the local basis orbitals centered at neighboring sites of I-th site[4]. The improved LO is called cluster LO. C_2 molecule is one of the cases where such an improvement is crucial[4].

The 2s- and $2p_z$-like SSA-LO's and cluster LO's for C_2 molecule along the molecular axis are shown in Fig.3. The corresponding atomic orbitals (AO) are also shown. Note that we are using the pseudopotential so that the orbitals shown here behave very differently from the ordinary ones in the core region. A noticeable aspect of the SSA-LO is that the LO of a given atom shrinks in the direction of another atom[3]. This may not be always true but seems to be expected in most cases from the analysis of the chemical pseudopotential equation. The cluster LO's behave quite differently. They have larger amplitudes in the interatomic region so that the bonding is enforced. The minimal-basis-set calculation with SSA-LO's for C_2 improves the corresponding one with AO's greatly but not satisfactorily. In the accurate calculation described in the previous section, the lowest π level is fully occupied and an empty σ level, the antibonding state of the dangling orbitals, is situated just above the π level. The minimal-basis-set calculation with AO's result in the reverse level order-

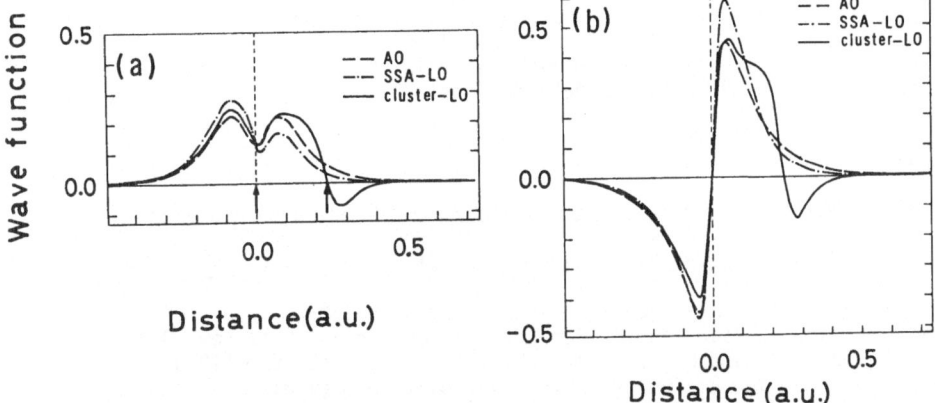

Fig.3. The (a) 2s- and (b) $2p_z$-like SSA- and cluster LO's for C_2 molecule along the molecular axis. The 2s- and $2p_z$-AO's are also shown. Arrows in (a) indicate the nuclear positions.

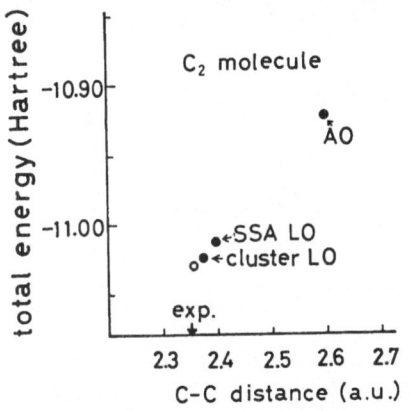

Fig.4. The total energy and interatomic distance of C_2 molecule obtained by minimal-basis-set calculations with different levels of approximations. The open circle corresponds to the result obtained by treating all GTO's as variational basis functions.

ing between these π and σ levels and this qualitatively wrong result is remedied by the cluster-LO but not by the SSA-LO. However, as mentioned earlier, C_2 molecule is one of the most severe test cases and therefore the shortcomings of the SSA-LO may be too much emphasized. Figure 4 shows how the calculated results for the total energy and the interatomic distance are improved as we do more elaborate calculations. Note that the electronic configuration is fixed as $\sigma_g^2\sigma_u^2\pi_u^4$ even in the minimal AO and minimal SSA-LO calculations in the results of Fig.4. The error in the total energy by the minimal AO calculation amounts to about half of the binding energy, while the corresponding one by the SSA-LO is reduced to only about 1/10. The cluster LO improves the result even further.

4. Conclusion

As concluding remarks, we just list up possible virtues of the use of minimal-LO set which our local-orbital theory will provide:
 1) reduction in the secular-matrix dimension,
 2) feasibility in analyzing the calculated results,

3) determination of LO's in a limited local space,
4) proper inclusion of local-environment effect included in LO's, and
5) usefulness in constructing a model Hamiltonian.
Demonstrations of these virtues,particularly in the physics of micro-clusters,will be made in future publications.

References

1. E. A. Rohlfing, D. M. Cox and A. Kaldor: J. Chem. Phys. 81, 3322 (1984)
2. G. B. Bachelet, D. R. Hamann and M. Schlüter: Phys. Rev. B26, 4199 (1982)
3. T. Hoshino, T. Asada and K. Terakura: Phys. Rev. B31, 2005 (1985)
4. T. Hoshino, T. Asada and K. Terakura: to appear in Phys. Rev. B34
5. D. M. Ceperley and B. J. Alder: Phys. Rev. Lett. 45, 566 (1980)
6. J. P. Perdew and A. Zunger: Phys. Rev. B23, 5048 (1981)
7. S. Huzinaga: J. Chem. Phys. 32, 1595 (1960)
8. E. A. Ballik and D. A. Ramsay: Astrophy. J. 137, 84 (1963)
9. D. W. Ewing and G. V. Pfeiffer: Chem. Phys. Lett. 86, 365 (1982)
10. P. F. Fougere and R. K. Nesbet: J. Chem. Phys. 44, 285 (1966)
11. N. H. Kiess and H. P. Bordia: Can. J. Phys. 34, 1471 (1956)
12. K. R. Thompson, R. L. DeKock and W. Weltner, Jr.: J. Am. Chem. Soc. 93, 4688 (1971)
13. R. Hoffmann, Tetrahedron: 22, 521 (1966)
14. K. S. Pitzer and E. Clementi: J. Am. Chem. Soc. 81, 4477 (1959)
15. R. A. Whiteside, R. Krishnan, D. J. Defrees, J. A. Pople and P. Von R. Schleyer: Chem. Phys. Lett. 78, 538 (1981)
16. P. Jena, the article in this volume
17. T. L. Gilbert, In Molecular Orbitals in Chemistry, Physics and Biology, ed. by P. O. Löwdin and B. Pullman (Academic, New York, 1964), p.405
18. P. W. Anderson: Phys. Rev. Lett. 21, 13 (1968)
19. J. D. Weeks, P. W. Anderson and A. G. H. Davidson: J. Chem. Phys. 58, 1388 (1973)

Electronic and Structural Theory of Group-IV Microclusters

S. Saito and S. Ohnishi

Fundamental Research Laboratories, NEC Corporation, Miyazaki 4-1-1, Miyamae-ku, Kawasaki 213, Japan

The non-spherical model-potential study on group-IV, sp^3-hybridized microclusters has revealed that there are two different series for sp^3 clusters, crystalline series and amorphous series. Starting from the model potential structures, we have investigated electronic states of several typical silicon clusters based on the Bond-Orbital Model and the LCAO-Xα method. The Bond-Orbital-Model study shows that surface-dangling-bond states concentrate on the Fermi level in every cluster. This may cause the reconstructions of the silicon clusters from the sp^3 structures. LCAO-Xα-force calculations on Si_6 and Si_{10} of crystalline series have shown the presence of the common reconstruction mechanism named "triangle contruction." On the other hand, the structural deviations of Si_8 and Si_{20} of amorphous series are found to be small as long as their symmetries are conserved.

1. Introduction

In recent years, several interesting experimental [1-5] and theoretical studies [6-11] on group-IV, silicon clusters have been done. Time-of-flight and photofragmentation experiments have shown the relative abundances of small silicon clusters [1-4], and Si_6 and Si_{10} are found to be relatively stable. Since silicon atoms are bound with covalent bonds which have mutually related orientations, the observed relative abundances are expected to reflect cluster geometries. There are some theoretical studies on silicon cluster geometries calculating their electronic structures [6-9]. Another theoretical approach, using the model potential for silicon atoms, has been also adopted [10,11].

In our previous work [12], a non-spherical model potential for sp^3-hybridized atoms was introduced and the cluster geometries and magic numbers were discussed. The obtained structures tend to have ring structures and have been classified into two groups and named "crystalline series" and "amorphous series" after their pair-distance distribution. The crystalline-series and amorphous-series clusters mainly consist of six- and five-membered rings, respectively.

In this paper, electronic structures of several typical clusters of both series are studied for the case of silicon clusters. The selected clusters from crystalline series are Si_6 and Si_{10}, and from amorphous series Si_8 and Si_{20} (Fig. 1). Si_6 is, of course, the smallest cluster among those consisting of six-membered rings. Si_{10} has a three-dimensional cage structure made up of four six-membered rings. Both are magic-number clusters. Similarly, Si_8 and Si_{20} are typical two- and three-dimensional magic-number clusters of only five-membered rings. Electronic states of these clusters are calculated using a simple tight-binding-approximation method, i.e. Bond-Orbital Model [13], and the Linear-Combination-of-Atomic-Orbitals

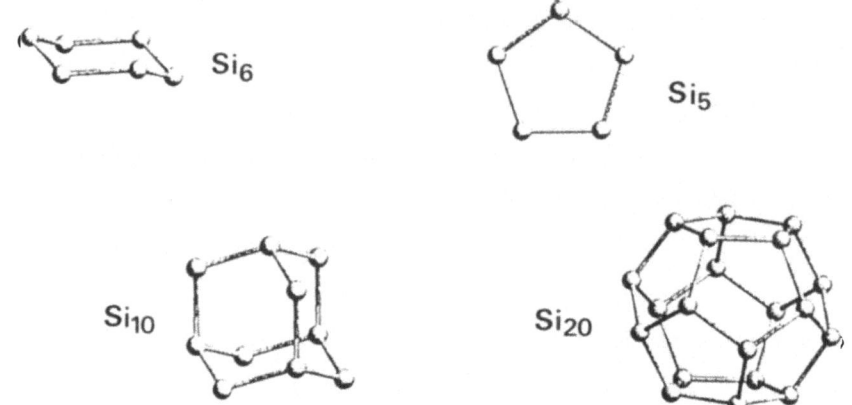

<u>Fig.1</u> Geometries of the clusters.

(LCAO)-Xα method [14,15]. In the LCAO-Xα calculation, the reconstruction mechanisms of these clusters are investigated under the condition of conserving their symmetries.

2. Bond-Orbital-Model Study

The Bond-Orbital Model (BOM), which has been introduced for the bulk materials with tetrahedral bonds, is simple but very useful to understand their physical properties. Hence, in order to know the general properties of sp^3-hybridized silicon clusters, BOM is applied for crystalline-series Si_6 and Si_{10}, and for amorphous-series Si_5 and Si_{20}. For transfer parameters between various orbital pairs, V_1-V_5 [13], the values for silicon bulk are used. (V_5 is known to be very small and ignored in the present calculation.) In the case of amorphous-series clusters, however, the transfer parameters only between orbitals which correspond to the bulk geometries are used. Calculated electronic energy levels for Si_6, Si_{10}, Si_5, and Si_{20} are shown in Fig. 2 with Gaussian-broadened density of states (DOS) for the schematic comprehension. It has been found from components of eigen states that dangling-bond states are concentrating near the Fermi level, E_F. Therefore, orbital energies for these clusters are expected to be very sensitive to the deformations, and the deformation can easily occur. Especially, the deformation which induces changes in transfers between dangling-bond orbitals is most preferable, since the levels near E_F are dangling-bond states. In other words, dangling-bond interactions cause deformations of the clusters.

In order to know the sensitivities of levels near E_F in regard to the transfer between dangling bonds, electronic states for clusters with additional transfer parameters between dangling bonds have been calculated tentatively (Fig. 3). For every cluster, new transfers (-0.6 eV) are introduced into the nearest pair of dangling bonds which have no transfers in BOM. In the case of Si_5 and Si_{10}, the additional transfer is introduced not only into the above dangling-bond pairs but also into the corresponding pair of bonding orbitals, because the distance between the pair orbitals is relatively short. It can be seen from the differences between Fig. 2 and

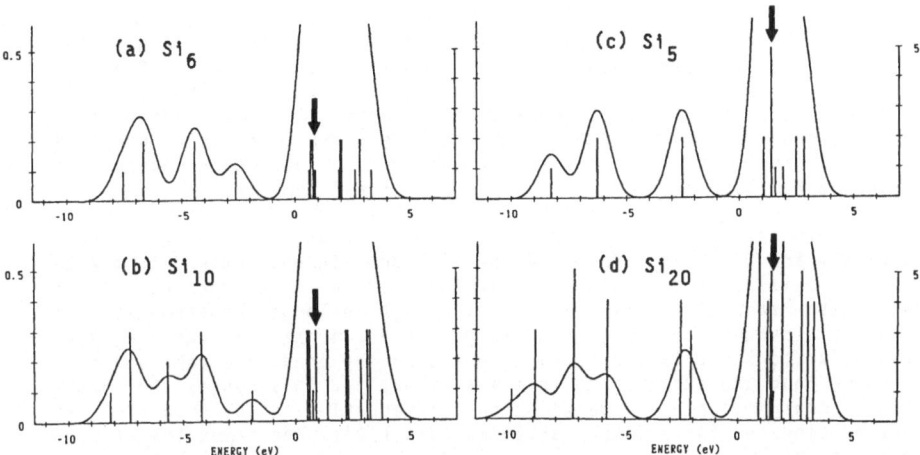

Fig.2 Energy levels and their Gaussian-broadened DOS by BOM. E_F for neutral clusters are indicated by arrows. The right scale represents degeneracies of levels, while the left scale is for DOS [1/eV·atom].

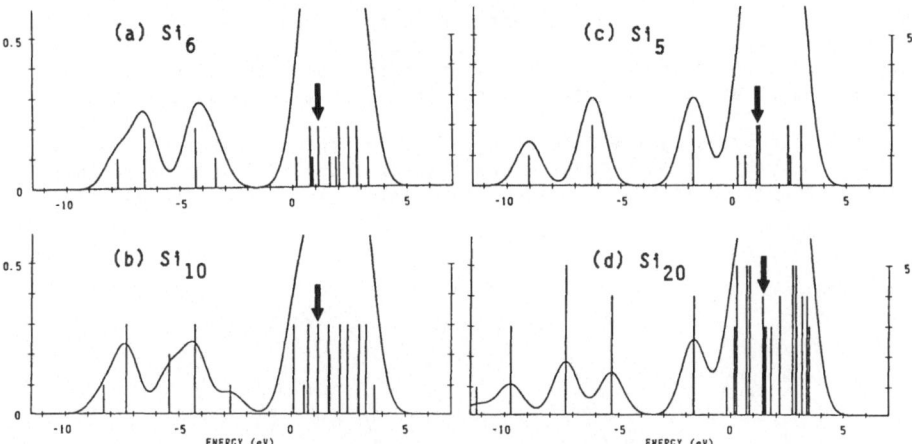

Fig.3 Energy levels with additional transfer parameters.

Fig. 3 that the dangling-bond states are very sensitive to the additional transfers. BOM is, however, too simple to discuss deformations of the clusters quantitatively. Hence, LCAO-Xα-force method has been adopted for further investigation.

3. LCAO-Xα-Force Study

Reconstructions of Si_6, Si_{10}, Si_5, and Si_{20} clusters are studied based on the local-density-functional scheme with the LCAO-Xα-force method. In the present calculation, numerical basis functions of 1s, 2s, 2p, 3s, and 3p orbitals of a silicon atom, obtained by the Xα calculation, are used. The parameter α is set equal to 0.7. The bond length in every cluster is chosen to be 4.444 a.u., the Si-Si covalent bond length. The structural deviation

for each cluster is explored so that all the forces acting on atoms may
vanish. Their symmetries, D_{3d}, T_d, D_{5h}, and I_h for Si_6, Si_{10}, Si_5, and
Si_{20}, respectively, are fixed throughout the calculation. The reason is
that, in each cluster, symmetry-preserving distortion modes are expected to
have strong couplings with electronic states. Moreover, so as to know the
general characters of crystalline-series and amorphous-series clusters, it
is favorable for each cluster to retain its original symmetry.

It has been found that there are some common features to the reconstructions of Si_6 and Si_{10} [16]. Both Si_6 and Si_{10} contain equilateral triangles
consisting of (111)-surface atoms of diamond lattice. After the reconstructions, these triangles have contracted by 20% in length compared to the
initial sp^3 structures. Each atom in the triangle has a dangling-bond
perpendicular to the triangle plane. These dangling bonds are parallel with
each other and are expected to interact strongly. This causes the above
"triangle contraction." The dangling-bond interaction on the triangle has
been ascertained by the electron distributions [16]. The community of the
reconstructions of Si_6 and Si_{10} can also be found from their electronic
states. In Fig. 4, Gaussian-broadened DOSs for Si_6 and Si_{10} obtained by
LCAO-Xα calculations are shown. It is found that DOSs for Si_6 and Si_{10}
resemble each other both before and after the reconstructions.

The structures of Si_5 and Si_{20} are a regular pentagon and a regular
dodecahedron, respectively. Hence, the breathing modes are only the
distortion modes which preserve the cluster symmetries. By calculating the
forces acting on atoms, the stable structure for Si_5 is found to be the
pentagon larger than the initial one by 6% in length. In the case of Si_{20},
the stable structure is the dodecahedron smaller than the initial one by
only 1% in length. In both cases, the highest occupied levels are only
partially occupied. In Fig. 5, electronic energy levels for Si_5 and Si_{20}

(a)

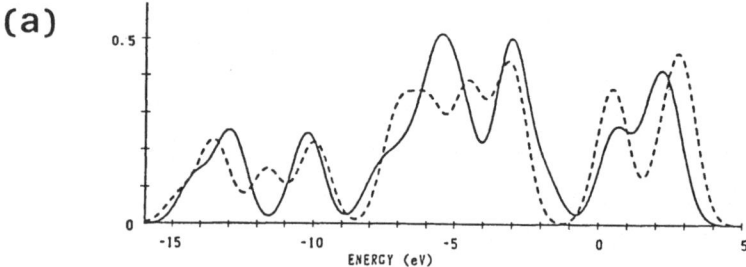

(b)

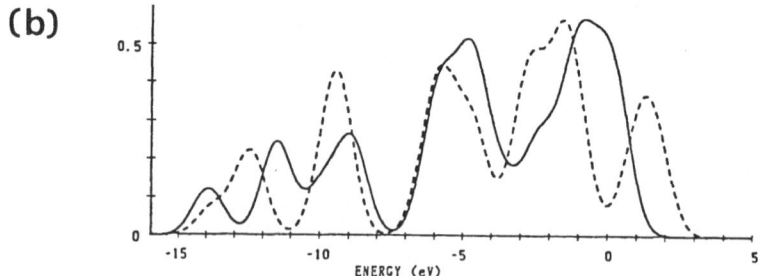

Fig.4 Gaussian-broadened DOSs for Si_6 (solid line) and Si_{10} (broken line).
(a) Before reconstructions. (b) After reconstructions.

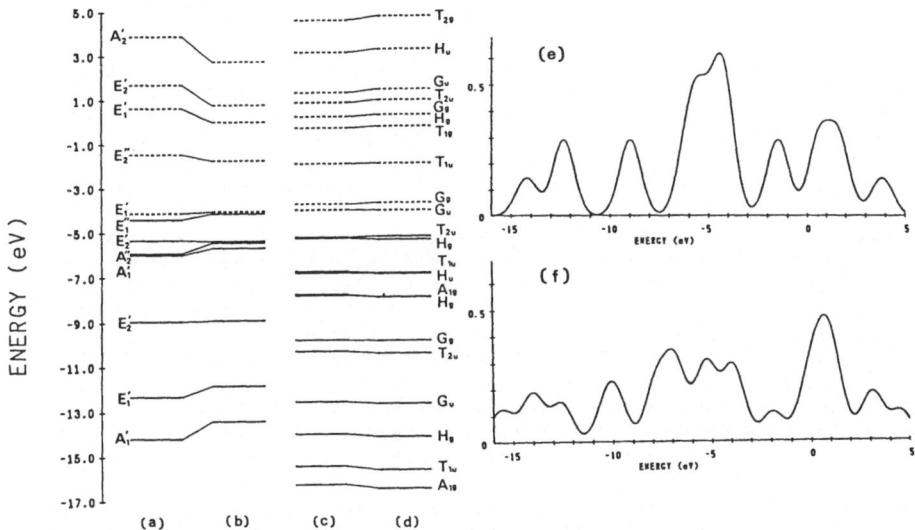

Fig.5 Electronic energy levels for (a) initial and (b) stable Si_5, and (c) initial and (d) stable Si_{20}. (e) Gaussian-broadened DOS for initial Si_5. (f) DOS for initial Si_{20}.

are shown. In contrast to the crystalline-series clusters, distortions of both Si_5 and Si_{20} are relatively small. Dangling bonds of these clusters are directed outwards from the clusters and will have a weak interaction between themselves. Hence, the distortions of the amorphous-series clusters will be small as long as their symmetries are conserved.

4. Discussion

It is interesting to compare DOSs obtained by LCAO-Xα calculations with those obtained by BOM. In the case of Si_6, a left shoulder of the lowest peak of BOM DOS without additional transfer (Fig. 2(a)) is not so clear as LCAO-Xα DOS before the reconstruction (Fig. 4(a)), but the lower two peaks correspond to each other. After the reconstruction, the lowest peak of LCAO-Xα DOS (Fig. 4(b)) has split into two peaks and the original second lowest peak has grown large. By introducing the additional transfer between dangling bonds, the lowest peak of BOM DOS (Fig. 3(a)) has shown the tendency to split and the second peak has also grown large. On the other hand, BOM DOS for Si_{10} without additional transfer (Fig. 2(b)) correspond well to LCAO-Xα DOS for the unreconstructed Si_{10} (Fig. 4(a)) in their lower parts. After the reconstruction, the original third lowest peak of LCAO-Xα DOS (Fig. 4(b)) has grown large. BOM DOS with additional transfer (Fig. 3(b)) has also shown this tendency. As a whole, in the case of crystalline-series clusters, introducing the transfer parameter between (111)-surface dangling bonds in BOM reproduce the change in lower parts of LCAO-Xα DOSs due to the reconstructions. This fact supports our interpretation of their reconstruction mechanisms, i.e. triangle construction attributed to the (111)-surface dangling-bond interaction. It is also found that Gaussian-broadened LCAO-Xα DOSs for Si_5 and Si_{20} before reconstruction also show correspondence to BOM DOSs without additional transfers.

In order to know whether the electronic structures of silicon clusters possess the characteristics of those of bulk silicon or not, LCAO-Xα DOSs for unreconstructed clusters are compared to those for bulk silicon. In the case of Si_6 (Fig. 4(a)), the lower two peaks correspond well to crystalline-silicon DOS [17]. Si_{10} DOS has also peaks at nearly the same positions as the lower two peaks of Si_6 DOS, although an additional small peak appears between two peaks. In both cases, there are dangling-bond states at the position corresponding to the energy gap for crystalline silicon. On the other hand, LCAO-Xα DOS for Si_5 (Fig. 5(e)) has bonding-state peaks at the positions which those for Si_6 do not possess. Therefore, the merger between five- and six-membered-ring units will make the lower side of DOS rather structureless. Amorphous silicon DOS has only one shallow high peak and no other distinct peak in the valence band [18]. This also indicates that there are many five-membered-ring units in amorphous silicon. LCAO-Xα DOS for Si_{20} (Fig. 5(f)) is also distinct from crystalline-silicon DOS. The structure of the lower half of the bonding states is gentle. Hence, the existence of Si_{20} unit can also broaden amorphous-silicon DOS. This fact supports the assignments "crystalline series" and "amorphous series," which have been named afer their pair distribution functions.

In the model-potential calculation [12], amorphous-series clusters generally have lower energies than crystalline-series clusters. The present work, however, shows that the structural deviations for crystalline-series clusters are large. Therefore, the crystalline series is expected to become stable in comparison with the amorphous series, on account of the energy gain due to the reconstruction. In other words, the crystalline series of the model potential reflects the geometries of the real silicon clusters better than the amorphous series.

Acknowledgements

The authors would like to thank Prof. S. Sugano (Institute for Solid State Physics) for his continual guidance. They are also grateful to Dr. C. Satoko (Institute for Molecular Science) for his helpful comments.

1. L. A. Bloomfield, R. R. Freeman, W. L. Brown: Phys. Rev. Lett. 54, 2246 (1985)
2. T. P. Martin, H. Schaber: J. Chem. Phys. 83, 855 (1985)
3. L. A. Bloomfield, M. E. Geusic, R. R. Freeman, W. L. Brown: Chem. Phys. Lett. 121, 33 (1985)
4. J. R. Heath, Y. Liu, S. C. O'Brien, Q.-L. Zhang, R. F. Curl, F. K. Tittel, R. E. Smalley: J. Chem. Phys. 83, 5520 (1985)
5. A. Kasuya, Y. Nishina: to be published in Phys. Rev. Lett.
6. G. Pacchioni, J. Koutecky: J. Chem. Phys. 84, 3301 (1986)
7. K. Raghavachari, V. Logovinsky: Phys. Rev. Lett. 55, 2853 (1985)
8. K. Raghavachari: J. Chem. Phys. 84, 5672 (1986)
9. D. Tomanek, M. A. Schluter: Phys. Rev. Lett. 56, 1055 (1986)
10. E. Blaisten-Barojas, D. Levesque: Phys. Rev. B34, 3910 (1986)
11. B. Feuston, R. Kalia, P. Vashishta: (private communication)
12. S. Saito, S. Ohnishi, S. Sugano: Phys. Rev. B33, 7036 (1986)
13. S. T. Pantelides, W. A. Harrison: Phys. Rev. B11, 3006 (1975)
14. C. Satoko: Chem. Phys. Lett. 83, 111 (1981)
15. C. Satoko: Phys. Rev. B30, 1754 (1984)
16. S. Saito, S. Ohnishi, C. Satoko, S. Sugano: J. Phys. Soc. Japan 55, 1791 (1986)
17. J. R. Chelikowsky, M. L. Cohen: Phys. Rev. B14, 566 (1976)
18. L. Ley: The Physics of Hydrogenated Amorphous Silicon II, Electronic and Vibrational Properties, ed. by J. D. Joannopoulos and G. Lucovsky, 61 (Springer-Verlag, Berlin, Heiderberg 1984)

Effects of Electrostatic Field
on the Electronic Structure of Microclusters

S. Ohnishi and Y. Chiba

Fundamental Research Laboratories, NEC Corporation, Miyazaki 4-1-1, Miyamae-ku, Kawasaki 213, Japan

Electrostatic field effects on the electronic structure of microclusters are studied by means of Weyl-Titchmarsh theory with a complex-energy parameter which is convenient especially for investigating resonance states. Resonance states for a spherically symmetric quantum well under the influence of strong electrostatic field are analyzed by numerical calculations of the energy spectra of the model system, an embedded GaAs semiconductor cluster in AlGaAs, within the effective-mass approximation. Stability of an electron and a hole is discussed from a point of view of the resonance widths of the energy spectra. The excitonic state inside the GaAs sphere, particularly the influence of its Coulomb interaction, is also discussed.

1. Introduction

Microclusters are of interest as "elementary particles" for new material phases and new material designs. It is getting possible to design newer and more artificial materials by using microclusters as a kind of "atomic" element consisting of an aggregate of real atoms. The "giant atom" concept for clusters of alkali and noble metals,based on the assumption of the electronic shell structure,has been successfully established and is one of the most striking features of cluster physics [1,2,3]: the quantum size effect is found to be the most important property for a free electron-like system. We apply that to the system of the heterostructured quantum well of GaAs semiconductor cluster embedded in AlGaAs, because a recent progress of microfabrication techniques has made it possible to enchase a microcluster in bulk crystal [4] and the similar situation to the above-mentioned metallic clusters is expected in the system. For GaAs/AlGaAs superlattice systems reduced to one- or two-dimensional problems, intensive discussions have been held, e.g. about the quantum-confined Stark effect on an excitonic state, the exciton field ionization being inhibited by an effect of the quantum well [5,6]. Compared with the one-dimensional cases, three-dimensional systems of microclusters have a lot of varieties to be designed, therefore if various arrangements of clusters can be realized, many interesting physical phenomena can be expected to occur, such as the interaction between excitons, field-induced tunnelings of the electron or the hole from one cluster to the other, non-linear optical phenomena related with these excitons, etc..

It is of great importance to investigate the electrostatic field, in particular strong field, effects on the electron-hole states and the confined excitonic states inside the GaAs microcluster in order to resolve the field ionization mechanism. Figure 1 shows 1s eigen values in no-field case of the electron, heavy-hole, and light-hole states for the system of GaAs sphere in AlGaAs as functions of the radii of the quantum wells with split ratio 85:15 between conduction- and valence-band within the

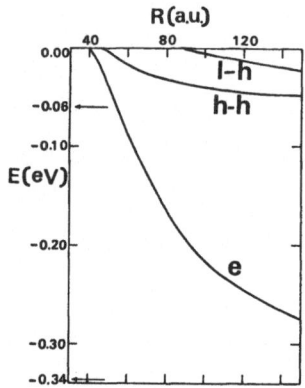

Fig. 1 1s eigen values of electron (e), heavy-hole (h-h), and light-hole(1-h) in spherical quantum well of GaAs cluster embedded in AlGaAs.

effective-mass approximation. There exists a critical radius of about 40 (a.u.) for both electron and heavy-hole under which no eigen state exists. In this paper, we focus on the electron-heavy-hole system in the spherical wells with radii of 60(a.u.) and 120(a.u.). We study two kinds of effects, one of the electrostatic field and the other of Coulomb interaction between the electron and the hole inside the GaAs cluster, on the stability of the electron-hole resonance states by Weyl-Titchmarsh theory [7].

2. Calculational Method

We consider the following Hamiltonian for the electron (hole) with the effective mass m^*_e (m^*_h) under the constant electrostatic field F of absolute strength in the z direction

$$H = - \frac{1}{2m^*_e} \Delta + (V_{well}(r) + Fz) \qquad (1)$$

$$V_{well}(r) = V_{0e} \, \theta (R-r) \quad (V_{0e}<0) \qquad (2)$$

where $\theta (t)$ is the step function. (For the hole, change m^*_e into m^*_h, V_{0e} into V_{0h}, and F into -F). The eigenvalue problem is $H\Psi=E\Psi$ with the boundary conditions, Ψ being regular at origin and vanishing at infinity.

As under the influence of the electrostatic field the potential energy is no longer spherical symmetric, but z-axial symmetric, we use conventional parabolic coordinates which are well known to be convenient in case of the presence of homogeneous electrostatic field: $x=(\xi \eta)^{1/2}\cos\phi$, $y=(\xi \eta)^{1/2}\sin\phi$, and $z=(\xi-\eta)/2$ ($\xi, \eta \geq 0$, $0\leq\phi<2\pi$). Here we introduce a shape approximation for the spherical well potential in order to make a variable separation be possible as

$$(\xi+\eta)V_{well}(r) \sim (\xi+\eta)V_{ap}(r) = U_1(\xi) + U_2(\eta) \qquad (3)$$

where $U_1(\xi)=V_0\xi \, \theta (R_{ap}-\xi)$, $U_2(\eta)=V_0 \eta \, \theta (R_{ap}-\eta)$. R_{ap} is determined by the suitable condition that the volume of the deepest part (i.e. such domain as $V_{ap}=V_0$) of the approximate well is the same as that of spherical well. Althouth we adopt the approximate well potential, the field term can be separated with exactness. Putting $\Psi(\vec{r})=u(\xi)v(\eta)\exp(im\phi)$ (m being magnetic quantum number) and $u(\xi)=\xi^{-1/2}u_1(\xi)$ and $v(\eta)=\eta^{-1/2}v_1(\eta)$, we obtain the one-dimensional second-order differential equations coupled by a constant α

$$[- \frac{1}{2m^*_\alpha} \frac{d^2}{d\xi^2} + P(\xi)] u_1(\xi) = \frac{E}{4} u_1(\xi) \qquad (4.a)$$

$$[- \frac{1}{2m^*_\bullet} \frac{d^2}{d\eta^2} + Q(\eta)] v_1(\eta) = \frac{E}{4} v_1(\eta) \qquad (4.b)$$

where $P(\xi)$ and $Q(\eta)$ are defined as

$$P(\xi) = - \frac{1-m^2}{8m^*_\bullet \xi^2} + (+\frac{F}{8} \xi - \frac{\alpha}{\xi} + \frac{U_1(\xi)}{4\xi})$$

$$Q(\eta) = - \frac{1-m^2}{8m^*_\bullet \eta^2} + (-\frac{F}{8} \eta + \frac{\alpha}{\eta} + \frac{U_2(\eta)}{4\eta})$$

As the intervals of equations are semi-infinite, we use Weyl-Titchmarsh theory which is a useful tool in investigating, particularly of the continuous spectrum in scattering problem. Let us summarize briefly Weyl's limit-point limit-circle theory [7]. The differential operator L is defined by $L[y(t)]=-(p(t)'y(t))'+q(t)y(t)$ (✳) where $p(t)$ and $q(t)$ are assumed to be real and continuous and $p(t)>0$ on interval $[0, \infty)$. Consider the equation with a complex-energy parameter $\lambda=E+i\varepsilon$ (let $\varepsilon>0$) instead of real E and let $\Phi(t)=\Phi(t;\lambda)$, $\Psi(t)=\Psi(t;\lambda)$ be the two linearly independent solutions of (✳) with the real initial conditions such that $\Phi(0;\lambda)=\sin\beta$, $p(0)\Phi'(0;\lambda)=-\cos\beta$, $\Psi(0;\lambda)=\cos\beta$, $p(0)\Psi'(0;\lambda)=\sin\beta (-\pi/2 \leqq \beta < \pi/2)$. Then there exists at least one monomorphic function m_w of λ, which has the following propaties; (1) $\chi(\lambda)=\Phi(\lambda)+m_w\Psi(\lambda)\in L^2(0,\infty)$ (2) either (a) there is a limit circle at infinity: $\Phi(\lambda)\in L^2(0,\infty)$ and $\Psi(\lambda)\in L^2(0,\infty)$, in which case m_w is not unique, or (b) there is a limit point at infinity: $\Phi(\lambda)\notin L^2(0,\infty)$ and $\Psi(\lambda)\notin L^2(0,\infty)$, in which case m_w is unique. This classification depends only on the operator L and dosen't depend on a specific value of λ. It is proved that if there exists some positive constant k such as $q(t)\geqq-k$ and $\int \{p(t)\}^{-1/2}dt=\infty$, then L is in the limit point case. The quantum systems we treat in this paper all satisfy this condition. In the limit point case Weyl's function m_w is related to the spectral density of L:

$$\frac{d\rho(E)}{dE} = \lim_{\varepsilon \to 0} \frac{1}{\pi} Im[m_w(E+i\varepsilon)] \qquad (5)$$

The nature of the spectrum is studied by Titchmarsh in detail [7]. For the equation $[-d^2/dt^2+V(t)]\Psi(t)=E\Psi(t)$ $(0\leqq t<\infty)$, there are four cases on account of the behavior of $V(t)$ in the limit $t\to\infty$: (a) $V(t)\to+\infty$: point spectrum $(-\infty, +\infty)$. (b) $V(t)\to0$: continuous spectrum $(0, +\infty)$ and point spectrum $(-\infty, 0)$. (c) $V(t)\to-\infty$, $\int |V(t)|^{-1/2}dt$ is divergent: continuous spectrum $(-\infty, +\infty)$. (d) $V(t)\to-\infty$, $\int |V(t)|^{-1/2}dt$ is convergent: point spectrum $(-\infty, +\infty)$. Equation (4.a) is in case (c), having continuous spectrum in $-\infty<E<+\infty$ and (4.b) in case (a), having point spectrum. For the determination of resonance states as functions of the field F, it is necessary to determine such a set of the quantum number α and the peak of the continuous spectrum or the point spectrum that satisfies (4.a) and (4.b) simultaneously. Figure 2.a and 2.b illustrate typical examples of

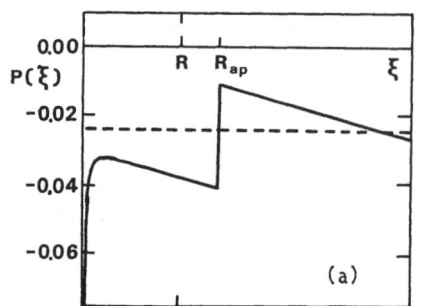 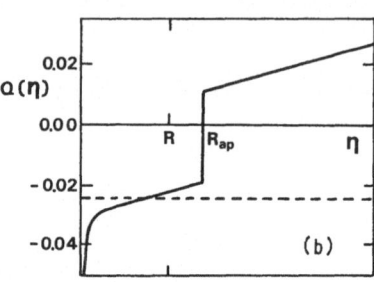

Fig. 2   Typical potential curves for $P(\xi)$ (a), and $Q(\eta)$ (b) in (4.a) and
(4.b) for the hole with parameters of m=0, R=120(a.u.), and F=50(kV/cm).
Dashed line is resonance energy (E/4).

$P(\xi)$ and $Q(\eta)$ in (4.a) and (4.b), which correspond to continuous and
point spectrum cases, resepectively.

3. Results

Figure 3.a and 3.b show the energy spectra for the electron and hole states
in the GaAs sphere with R=120 (a.u.). From Figure 3.b for the hole state it
is seen that the peaks are broadened by the strong fields indicating its
instability and the short life time evaluated by $\hbar/\Delta E$ (ΔE being the half
width): about 1(pc) for F=50(kV/cm), while that of electron is about five
times as long as it. Figure 4.a and 4.b show resonance energy shifts of the
electron and hole states resepectively with R=60(a.u.) and 120(a.u.).   The
strong non-linearity with respect to F should be noted. Eigen values of
them at F=0(kV/cm) determined by (4.a) and (4.b) in case of the approximate
well potential agree quite well with values of exact spherical wells shown
in Figure 1: for the electron about 19% and 2% and for the hole 10% and 2%
deviations with R=60 (a.u.) and R=120 (a.u.), respectively.

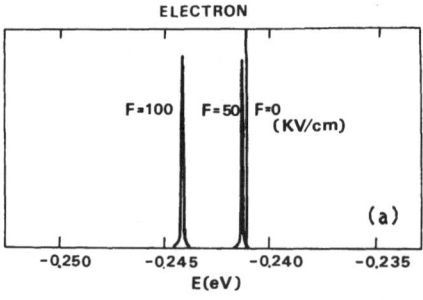 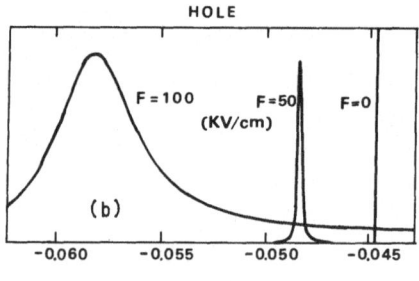

Fig. 3 Energy spectra for electron (a), and hole (b), states.

By introducing the electron-hole Coulomb interaction $V_c=-1/\varepsilon\ |\vec{r}_e-\vec{r}_h|$,
the total Hamiltonian of the electron-hole system is expressed

$$H = - \frac{1}{2\mu}\Delta_{\vec{r}} - \frac{1}{2M}\Delta_{\vec{R}} + V'_{well} + Fz + V_c(\vec{r}) \qquad (6)$$

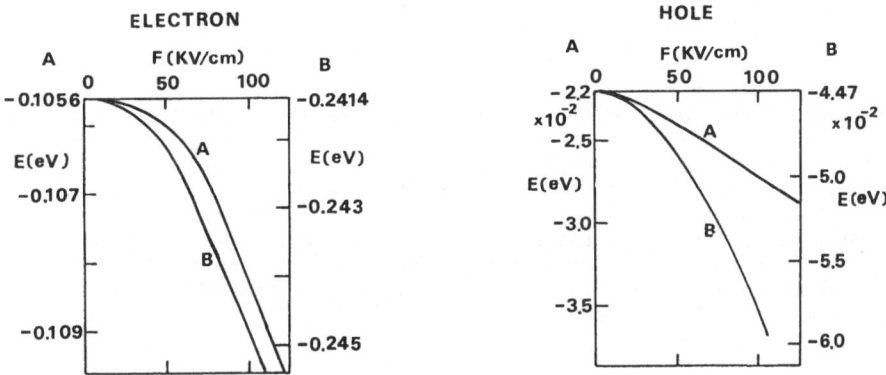

Fig. 4 Energy level shifts vs. field strength F for electron (a), and hole (b) with R=60(a.u.) (A), and R=120(a.u.) (B).

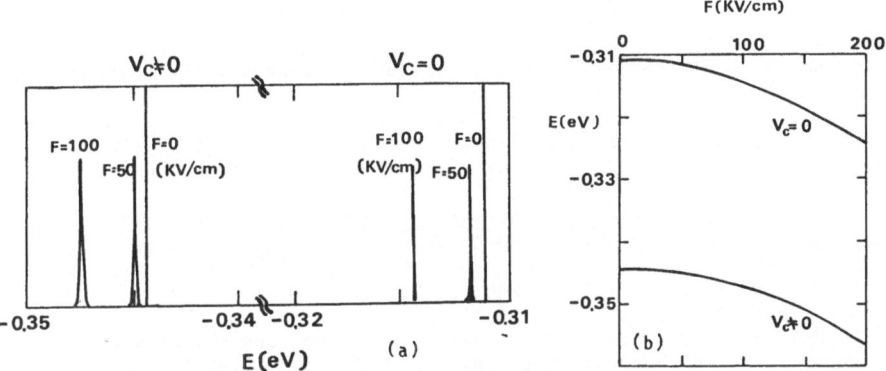

Fig. 5 Energy spectra of excitonic states (a) and Energy shifts vs. F (b).

where $V'_{well} = V_e(|\vec{R} + (m*_h/M)\vec{r}|) + V_h(|\vec{R} - (m*_e/M)\vec{r}|)$, $\vec{r} = \vec{r}_e - \vec{r}_h$, $\vec{R} = (m*_e\vec{r}_e + m*_h\vec{r}_h)/M$, $M = m*_e + m*_h$, and $\mu = m*_e m*_h/(m*_e + m*_h)$. We solve excitonic states only within an adiabatic approximation by neglecting terms including $\vec{R}$ on the assumption of $\vec{R} = \vec{0}$. Figure 5.a and 5.b show the energy spectra with F=0, 50, 100 (kV/cm) and the resonance energy shifts due to the electrostatic field in cases of including and not-including the Coulomb interaction V_c. Resonance peaks are very sharp, which indicates the stability of the excitonic states under the influence of the strong field. The Coulomb binding energy, which is given by the difference between the resonance energies obtained from Hamiltonian with and without V_c, is about 30 meV. This is a significant result which indicates that the exciton in the spherical well has almost ten times the Coulomb binding energy of the GaAs bulk crystal (about 4 meV) and several times that of the GaAs/AlGaAs superlattice.

4. Discussion

As the Coulomb binding energy may be influenced by the motion of the center of gravity of the electron-hole pair, we are afraid that our estimate for it has no validity any more. It needs further analyses. However, the Coulomb binding energy by the first order perturbation given by

$$E_c = \langle \Psi_e \Psi_h | V_c | \Psi_e \Psi_h \rangle / \langle \Psi_e \Psi_h | \Psi_e \Psi_h \rangle$$

is about 27(meV) at F=0(kV/cm), which coincides quite well with that obtained from the adiabatic approximate Hamiltonian (6). This confirms the appropriateness of the approximation. Also, since the electronic structure of the systems we treat in this paper depends rather strongly on the external field, standard perturbative treatments for them are not considered to be valid, especially for the electron and hole with small effective masses in semiconductor, and our method is found to be very effective to analyze the quasi-eigen states and the stability given by the broadened peaks of the spectra. We should emphasize that the most important fact found in our calculations is the extreme strong quantum-confined effects for the exciton inside GaAs cluster.

References

1. W. D. Knight, K. Clemenger, W. A. Saunders, M. Y. Chou, and M. L. Cohen:
   Phys. Rev. Lett. **52**, 2141 (1984)
   M. Y. Chou, A. Cleland, and M. L. Cohen: Solid State Commin. **52**, 645 (1984)
2. I. Katakuse, I. Ichihara, Y. Fujita, T. Matsuo, T. Sakurai, and
   H. Matsuda: Int. J. Mass Spect. Ion Proc. **67**, 229 (1985)
3. Y. Ishii, S. Ohnishi, and S. Sugano: Phys. Rev. B **33**, 5271 (1986).
   S. Saito and S. Ohnishi: to be published in The Physics and Chemistry of Small Clusters, ed. by P. Jena, K. Rao, and N. Khanna (Plenum, New York 1987)
4. H. Watanabe: In The Physics and Fabrication of Microstructures and Microdevices, ed. by M. J. Kelly and C. Weisbuch (Springer-Verlag, Heidelberg 1986)
5. E. E. Mendez, G. Bastard, L. L. Chang, L. Esaki, H. Morkoc, and
   R. Fisher: Phys. Rev. B **26**, 7101 (1982)
6. D. A. B. Miller, D. S. Chemla, T. C. Damen, A. C. Gossard, W. Wiegmann, T. H. Wood, and C. A. Burrus: Phys, Rev. Lett. **53**, 2173 (1984),
   Phys. Rev. B **32**, 1043 (1985)
7. H. Weyl: Ann. Math. **68**, 20 (1910)
   E. C. Titchmarsh: In Eigenfunction Expansions Associated with Second-Order Differential Equations (Oxford U.P., London, 1946)
   E. A. Coddington and N. Levinson: In Theory of Ordinary Differential Equations (McGraw-Hill, New York, 1955)
   M. H. Hehenberger, H. V. McIntosh, and E. Brandas: Phys. Rev. A **10**, 1494 (1974)

Quantum Box: Gallium Arsenide Microclusters Embedded in Aluminum Arsenide

A. Oshiyama

Fundamental Research Laboratories, NEC Corporation, Miyazaki 4-1-1, Miyamae-ku, Kawasaki 213, Japan

1. INTRODUCTION

Since the first proposal on semiconductor superlattices by Esaki and Tsu[1], there has been a considerable expansion in the amount and variety of both theoretical and experimental research on semiconductor heterostructures[2,3]. This has been caused by the characteristic phenomena of quasi-two-dimensional systems constructed in semiconductor quantum wells or superlattices on one hand[4]. On the other hand, this system has attracted much interest from the view point of application to, for example, optoelectronic devices[5]. Recent developments in technique of the molecular-beam-epitaxy or the metal-organic-compound-vapor-deposition have enabled us to construct the new artificial superlattices with atomic layer-by-layer controllability[6]. Even the atom-by-atom manipulation in growth of the artificial materials is now discussed in a realistic manner[7] with the combination of sophisticated technique of the electron-beam-deposition and the atomic-layer-epitaxy. A new realm of condensed matter physics would emerge[8] from this expansion of technological developments.

One possibility for the novel heterostructures which could be produced from such developments in technology is the semiconductor microcluster embedded in the matrix of another type of semiconductor. When the electron affinity of the semiconductor constituting the embedded cluster is larger and the energy gap is narrower than the matrix semiconductor, the quantum box or the quantum ball is expected to be constructed (Fig. 1): i.e. the cubic or the spherical potential well resulting from the band offsets for the conduction and the valence electrons. A peculiar sequence of electron energy-levels is formed in this type of potential well [9]. The electron states whose wavefunctions have no nodes are energetically favorable because of the lack of the deep attractive region in the vicinity of the center of the potential wells. Thus, we have the unique level sequence of the electron states such as 1s, 2p, 3d and so forth. This distinguished feature might cause new phenomena in this new type of semiconductor heterostructure .

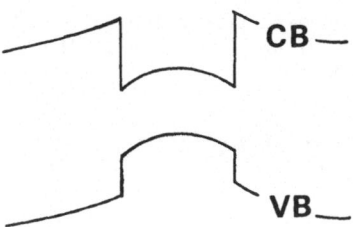

Fig.1 Schematic diagram of valence- and conduction-band edge of the quantum box.

When the size of the embedded semiconductor cluster is reduced to the order of tens of angstroms in its diameter, however, the above picture based on the effective-mass-approximation is not assured to be valid. The extreme case of the small size of the microcluster is one foreign atom (e.g. Ga atom) in the matrix semiconductors (e.g. AlAs). In a usual case of the semiconductor heterostructure, the chemical species of the foreign atom is similar to those of the host atoms. Thus, it is not likely that the foreign atom induces the deep energy level, which corresponds to the band offsets in case of the large cluster, in the energy gap of the matrix semiconductors. Therefore, the aim of the present paper is to perform the electronic structure calculation from the first principles for the small embedded microcluster, and clarify the distinguished features of this new type of the heterostructures. This calculation will reveal the validity and the limitation of the effective-mass-approximation, and further imply the possibility of microscopic assessment of the band offsets in the semiconductor heterostructures. Moreover, the calculated total energy of this system would provide us with rich information about the alloying of two different compound semiconductors, which are important materials in application point of view.

The most promising method to study the embedded microcluster from the first principles is the scattering-theoretic or the Green's function method which has been succesfully applied to the point defects in semiconductors[10-12]. Recent developments in the processor and the capacity in the supercomputer could enable us to apply the first principles Green's function method to the case of the embedded microcluster beyond the point-defect regime. In this paper, I have done the Green's function calculation for three GaAs microclusters embedded in AlAs: first, one Ga atom in AlAs. Second, the planer GaAs cluster containing 5 Ga atoms and thirdly, the GaAs cluster containing 13 Ga atoms.

2. GREEN'S FUNCTION METHOD

This section is concerned with the description of the Green's function method. We start with the effective single-particle equation within the density functional framework:

$$H \Psi_i = \varepsilon_i \Psi_i, \tag{1}$$

or, more explicitly (in atomic units),

$$(- \Delta/2 + V[n(r)]) \Psi_i = \varepsilon_i \Psi_i, \tag{2}$$

where the electron charge density is given by

$$n(r) = \sum_i | \Psi_i(r) |^2 \Theta(\varepsilon_F - \varepsilon_i). \tag{3}$$

with $\Theta(x)$ being 1 for $x>1$ and zero otherwise. The effective potential V consists of the pseudopotentials v_{ps} of the constituting ion cores, the Coulomb potential and the exchange-correlation potential:

$$V(r) = \sum_\tau v_{ps}(r-\tau) + \int dr' \frac{n(r')}{|r-r'|} + \mu_{xc}[n(r)]. \tag{4}$$

The exchange-correlation potential here is given within the local density approximation using the analytic form[13] which is fitted to the results for the electron gas by Ceperly and Alder[14].

In the present calculation, the non-local norm-conserving pseudopotentials[15] of the ion core are constructed from the results of the all-electron atomic calculation following the scheme by Hamann et al[16]. In this scheme, the pseudopotential is obtained numerically, and then fitted to the analytic long- and short-range components:

$$v_{ps}(r) = v_{ps}^{lr}(r) + v_{ps}^{sr}(r),$$

(5)

where

$$v_{ps}^{lr}(r) = - Z_v \, erf(\, r/r_0 \,)/r$$

(6)

and

$$v_{ps}^{sr}(r) = \sum_{l} \sum_{k=1}^{3} (\, a_k^l + b_k^l r^2 \,) \, exp(\, -d_k^l r^2 \,) \, \hat{P}_l.$$

(7)

Here Z_v is the atomic valence charge, and $\hat{P}_l$ is a projection operator on the angular momentum state l. r_0, a_k^l, b_k^l and d_k^l are the fitting parameters. The resulting pseudopotential has the norm-conserving properties which assure that the pseudo- and real-valence charge density are identical to each other outside the core region.

The wavefunction Ψ_i is expanded in a linear combination of gaussian orbitals Φ_α situated at atomic sites τ:

$$\Psi_i(r) = \sum_{\alpha,\tau} C_{i\alpha} \, \Phi_\alpha(r-\tau)$$

(8)

with

$$\Phi_\alpha(r) = r^l exp(\, - \zeta r^2 \,) \, Y_{lm}(r).$$

(9)

In terms of this expansion, the differential equation (2) is transformed to the secular equation,

$$\sum_{\beta} (\, H_{\alpha\beta} - \varepsilon_i S_{\alpha\beta} \,) \, C_{i\beta} = 0,$$

(10)

where $H_{\alpha\beta}$ and $S_{\alpha\beta}$ are the hamiltonian and the overlap matrices, respectively.

We are now in the position to introduce the Green's function $G_{\alpha\beta}(z)$ at the energy variable z. It is defined through

$$\sum_{\beta} (\, zS_{\alpha\beta} - H_{\alpha\beta} \,) \, G_{\alpha\beta}(z) = \delta_{\alpha\gamma}.$$

(11)

More explicit definition for $G_{\alpha\beta}(z)$,

$$G_{\alpha\beta}(z) = \sum_{i} C_{i\alpha}C_{i\beta}^{*} / (\, z-\varepsilon_i \,),$$

(12)

is readily verified by substituting (12) to (11) and using the properties of the expansion coefficients $C_{i\alpha}$. In general, the physical quantity A (the corresponding operator $\hat{A}$),

$$A = \sum_i \Theta(\varepsilon_F - \varepsilon_i) \int dr \ \Psi_i^*(r) \ \hat{A} \ \Psi_i(r),$$ (13)

is expressed as a trace over a density matrix ρ:

$$A = \sum_{\alpha\beta} \rho_{\alpha\beta} A_{\alpha\beta},$$ (14)

and the density matrix $\rho_{\alpha\beta}$ can be written as

$$\rho_{\alpha\beta} = \sum_i C_{i\alpha}^* C_{i\beta} \ \Theta(\varepsilon_F - \varepsilon_i).$$ (15)

By using the Green's function (12), this density matrix is expressed as

$$\rho_{\alpha\beta} = (1/2\pi i) \int_{c(\varepsilon_F)} dz \ G_{\alpha\beta}(z),$$ (16)

where the contour $c(\varepsilon_F)$ encloses the poles of $G_{\alpha\beta}(z)$ corresponding to the occupied states of the system.

Our objective is to construct the Green's function $G_{\alpha\beta}(z)$ for the system of the microcluster embedded in the perfect crystal. Denoting the hamiltonian matrix, the overlap matrix and the corresponding Green's function for the perfect crystal by $H_{\alpha\beta}^0$, $S_{\alpha\beta}^0$ and $G_{\alpha\beta}^0(z)$, respectively, we obtain the Dyson equation for the perturbed Green's function $G_{\alpha\beta}(z)$:

$$G_{\alpha\beta}(z) = G_{\alpha\beta}^0(z)$$
$$+ \sum_{\gamma,\delta} G_{\alpha\gamma}^0(z) \ (\ \Delta H_{\gamma\delta} - z \Delta S_{\gamma\delta}\ ) \ G_{\delta\beta}(z),$$ (17)

where $\Delta H_{\alpha\beta} = H_{\alpha\beta} - H_{\alpha\beta}^0$ and $\Delta S_{\alpha\beta} = S_{\alpha\beta} - S_{\alpha\beta}^0$ indicate matrix differences between the unperturbed and the perturbed systems. The perturbation $\Delta H - z \Delta S$ includes not only the perturbation potential induced by embedding the cluster in the perfect crystal, but also the difference in the basis set employed in the expansion of the wavefunctions for the perturbed and the unperturbed systems. When the perturbation is restricted around the region of the embedded microcluster, the gaussian orbitals situated at the atomic sites in the microcluster and in its neighboring shells is sufficient in the expansion (8). In that case (17) is just the matrix equation with finite dimension so that it can be easily solved as (we omit the suffix α for brevity),

$$G = [\ 1 - G^0(\ \Delta H - z \Delta S)\ ]^{-1} \ G^0.$$ (18)

Once the Green's function G and then the density matrix are obtained, any quantities can be calculated by the expression (14). Further, the change in the density of states $\Delta D(z)$ is given by

$$\Delta D(z) = (1/2\pi i) \ (\ \mathrm{Tr}\{G(z)S\} - \mathrm{Tr}\{G^0(z)S^0\ \}$$
$$= (1/2\pi i) \ d \ \xi(z) / dz,$$ (19)

where $\xi(z)$ is a phase shift defined by

$$\xi(z) = -\tan^{-1}\{\ \mathrm{Re}\ \det[\ 1 - G_0^0(\Delta H - z \Delta S)\ ]$$
$$/ \ \mathrm{Im}\ \det[\ 1 - G^0(\Delta H - z \Delta S)\ ]\ \}.$$ (20)

278

The total change in the number of states $\Delta N(\varepsilon)$ below the energy ε is then written as

$$\Delta N(\varepsilon) = \int_{c(\varepsilon)} dz \quad \Delta D(z). \tag{21}$$

3. RESULTS AND DISCUSSION

In the actual calculation, I use 22 gaussian orbitals per atom as the basis set. The exponents of the gaussian orbitals are { 0.15, 0.4, 0.8 } for the s and p orbitals and { 0.15, 0.4 } for the d orbitals of each atom. For the unperturbed system (e.g. perfect AlAs crystal), the calculated conduction-band energies measured from the top of the valence bands are 2.7 eV for the Γ point, 1.5 eV for the X point and 2.2 eV for the L point. The overall features of the energy bands are well reproduced by the present basis set.

In the case of Ga substitutional atom in AlAs, no electron state has been found in the energy gap of AlAs. I put the gaussian orbitals as the basis set at the Ga site and at the 4 nearest neighbor sites first, and then enhanced the basis set with the orbitals at the 12 second neighbor Al sites. The results have not changed by this enhancement of the basis set. The calculated difference ΔE in total energy is -0.28 Ry. Then the energy cost to substitute one Al atom by a Ga atom, $\Delta E + \varepsilon$ (Al atom) $- \varepsilon$ (Ga atom), equals 0.6 eV[17]. This value is comparable with the cohesive-energy difference of GaAs (6.7 eV) and AlAs (7.7 eV)[18].

When the 5 neighboring Al atoms in a [001] plane are replaced by Ga atoms, there is still no state in the energy gap of AlAs. Next, the calculation is performed for the GaAs cluster, consisting of the central Ga atom, the 4 neighboring As atoms and the next-shell 12 Ga atoms. It is found that a singlet A_1 state appears at about 0.4 eV above the top of the valence band. In this calculation the basis set consists of the gaussian orbitals at the atomic sites only in the GaAs microcluster. Thus, there remain some ambiguities on the completeness of the basis set. Nevertheless, this result suggests that there is some critical number of Ga atoms above which the electron states appear suddenly in the energy gap of AlAs. More detailed analysis is now in progress.

REFERENCES

1. L. Esaki and R. Tsu: IBM J. Res. Develop. 14, 61 (1970).
2. See, for example, L. Esaki: in Recent Topics in Semiconductor Physics, ed. by H. Kamimura and Y. Toyozawa (World Scientific, Singapore 1983).
3. T. Ando, A. Fowler and F. Stern: Rev. Mod. Phys. 54, 437 (1982).
4. See, for example, Proc. Int. Conf. Electronic Properties of Two-Dimensional Systems (Kyoto 1985) = Surf. Sci. 170, 1986.
5. See, for example, Proc. Int. Conf. Modulated Semiconductor Structures (Kyoto 1985) = Surf. Sci. 174, 1985 and Extended Abstracts of 18-th Int Conf. Solid State Devices and Materials (Tokyo 1986).
6. A. Ishibashi, Y. Mori, M. Itabashi and N. Watanabe: J. Appl. Phys. 58, 2691 (1985), T. Baba, T. Mizutani and M. Ogawa: Jpn. J. Appl. Phys. 22, L627 (1983) and articles in ref. 5.
7. H. Watanabe: in The physics and Fabrication of Microstructures and Microdevices ed. by M.J. Kelly and C. Weisbuch (Springer-verlag, Heidelberg 1986).

8. T. Inoshita, S. Ohnishi and A. Oshiyama: Phys. Rev. Lett. $\underline{57}$, 2560 (1986).
9. A similar sequence of the levels is found in the shell model for the alkali metal clusters. W.D. Knight, K.Clemenger, W.A. de Herr, W.A. Saunders, M.Y. Chou and M.L. Cohen: Phys. Rev. Lett. $\underline{52}$, 2141 (1984).
10. J. Bernhol, N.O. Lipari and S.T. Pantelides: Phys. Rev. Lett. $\underline{41}$, 895 (1978), G. A. Baraff and M. Schluter: Phys. Rev. Lett. $\underline{41}$, 892 (1978).
11. A.R. Williams, P.J. Feibelman and N.D. Lang: Phys. Rev. B$\underline{26}$, 5433 (1982).
12. The recent successful applications of the Green's-function method are found in R. Car, P.J. Kelly, A. Oshiyama and S.T. Pantelides: Phys. Rev. Lett. $\underline{52}$, 1814 (1984) and Phys. Rev. Lett. $\underline{54}$, 360 (1985), G. A. Baraff and M. Schluter: Phys. Rev. Lett. $\underline{55}$, 2340 (1985), H. Katayama-Yoshida and A. Zunger: Phys. Rev. Lett. $\underline{55}$, 1618 (1985).
13. J. Perdew and A. Zunger: Phys. Rev. B$\underline{23}$, 5048 (1981).
14. D.M. Ceperly and B.J. Alder: Phys. Rev. Lett. $\underline{45}$, 566 (1980).
15. A. Oshiyama and M. Saito: to be published.
16. D.R. Hamann, M. Schluter and C. Chiang: Phys. Rev. Lett. $\underline{43}$, 1494 (1979). see also G.B. Bachelet, D.R. Hamann and M. Schluter: Phys. Rev. B$\underline{26}$, 4199 (1982).
17. The total energy of the pseudoatom constructed in this paper is -3.54 Ry for Al and -3.86 Ry for Ga, respectively.
18. J. Ihm and J.D. Joannopoulos: Phys. Rev. B$\underline{24}$, 4191 (1981) and references therein.

Electronic Structure of the Superatom

T. Inoshita and H. Watanabe

Fundamental Research Laboratories, NEC Corporation, Miyazaki 4-1-1, Miyamae-ku, Kawasaki 213, Japan

1. Introduction

Semiconductor microfabrication technology has seen a series of revolutionary innovations in the past decade and has reached the point where it no longer seems fanciful to discuss 'single atom manipulation'. Such single atom manipulation techniques will make it possible to realize various new semiconductor structures which are not possible with conventional fabrication techniques. This situation motivated one of the authors (H.W.) to propose recently a new spherically symmetric modulation-doped heterostructure named superatom[1,2]. The purpose of this paper is to clarify the general features of the electronic structure of superatoms on the basis of a self-consistent local-density-functional calculation[3].

2. What is a superatom?

A superatom is a semiconductor heterostructure consisting of (i) a spherical domain, called 'core', doped with a controlled number (z) of donors and (ii) a surrounding impurity-free matrix with electron affinity larger than that of the core (Fig. 1a). If the conduction band offset at the interface is sufficiently large, all z donors in the core are ionized and give rise to nearly centro-symmetric Coulomb potential outside the core. The electrons released from the donors are kept outside the core but are bound by this potential. If the core diameter is larger than the electron de Broglie wavelength, the

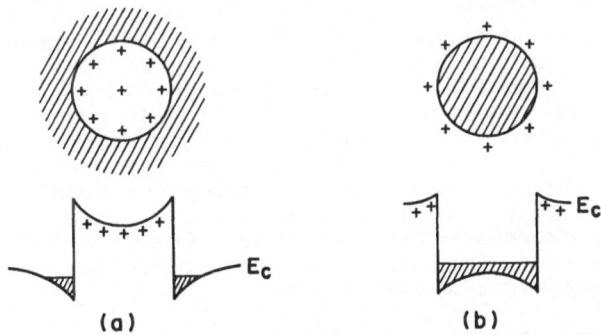

(a) (b)

Fig. 1 Sketch of a superatom (a) and an inverse superatom (b) in the real space. The electrons are shown by shades, and E_c denotes the conduction band edge

system has a quasi-two-dimensional electronic spectrum. But when the core diameter is on the order of or smaller than the electron wavelength, the electrons are quantized not only in the direction perpendicular to the interface but also parallel to it to form a quasi-atomic electronic structure. If the matrix is indeed impurity-free, the system is electrically neutral at T=0, binding all z electrons. Thus, z is the equivalent of the atomic number in ordinary atoms. This is an n-type superatom with bound electrons, not holes. It is obvious that, by doping acceptors, one can as well form a p-type superatom with bound holes. Also, if one performs doping in the matrix immediately outside the core, which is kept impurity-free, one has an 'inverse superatom' (Fig. 1b). In this paper, however, we shall confine ourselves to n-type superatoms.

One of the most promising material combinations for the superatom is AlGaAs(core)-GaAs(matrix), because of the close lattice matching. AlGaAs-GaAs heterostructures as well as superlattices have been extensively studied and are known for their excellent interface properties[4].

3. Method of Calculation

Our calculation is based on the effective mass approximation (EMA) and the standard self-consistent non-relativistic scheme for atomic structure calculations. By assuming the potential to be spherically symmetric, an eigenfunction for Schroedinger equation can be separated into radial and angular parts as $X(r)/r \cdot Y_{lm}(\theta\phi)$ and the equation for the radial wavefunction $X(r)$ reads (in atomic units)

$$\left[-\frac{1}{2m^*} \cdot \frac{d^2}{dr^2} + \frac{l(l+1)}{2m^*r^2} + V(r)\right] X(r) = E\ X(r), \qquad (1)$$

where the effective mass m^* takes on different values depending on whether r is inside or outside the core. The potential $V(r)$ is expressed as

$$V(r) = E_{offset}\ \Theta(R-r) + V_{core}(r) + V_H(r) + V_{xc}(r). \qquad (2)$$

The first and the second terms represent the band offset ($\Theta(r)$: step function) at the interface (R: core radius) and the Coulomb potential of the ionized donors, respectively. In obtaining the latter, the donor charge distribution is averaged within the core and modeled as a uniformly charged sphere of total charge +ze. The third term represents the Hartree potential, and the fourth term, the exchange-correlation potential, is treated in the local-density-functional scheme using an analytical expression[5] for the Ceperley-Alder potential[6]. Numerical calculations are performed on $Al_{0.35}Ga_{0.65}As$(core)-GaAs(matrix) superatoms.

EMA has been known to give excellent results for the electronic structure of AlGaAs-GaAs single heterostructures, and, therefore, it should also be valid for AlGaAs-GaAs superatoms when the core diameter is sufficiently large. Its validity becomes doubtful as the core size is reduced, and the lower limit for the core diameter is estimated to be ~ 30 A. This is based on the fact that the electronic structure of $(AlAs)_n (GaAs)_n$ superlattices is well described by EMA down to $n=10$ $(AlAs$ ^{n} layer thickness ~ 30 A) [7]. We treat the cases where the core is significantly larger than this critical size.

4. Results and Discussions

The general characteristics of the electronic structure of superatoms are seen in the typical result shown in Fig. 2, which presents the self-consistent potential, orbital energy and the radial wavefunctions X_{nl}/r for the ground state configuration $1s^2 2p^6 3d^{10} 2s^2$ of a superatom with z=20 and R=120 A. There are three features to be noted. First, the level sequence is quite distinct from that of the ordinary atoms. To be more specific, the lowest three levels have no radial node. This dominance of the no-radial-node states is due to the lack of $1/r$ singularity in the potential. In case of ordinary atoms, the $1/r$ singularity at the origin favors low angular momentum states which have a larger amplitude around the origin. This is the reason that the electronic structure of ordinary atoms is governed by s and p states. In contrast, this singularity is absent in superatoms, which favors states with fewer radial nodes. We actually found that the electronic structure of superatoms is dominated by no-radial-node states (1s, 2p, 3d, 4f etc.) over a wide range of parameter values. Therefore large angular momentum states are possible, which might give rise to interesting magnetic properties.

This characteristic level sequence is also evident in Fig. 3 where the z dependence of the orbital energy is plotted for R=120 A. Up to z=18, electrons fill consecutively the no-node sequence 1s, 2p and 3d. The next shell 2s has one radial node,

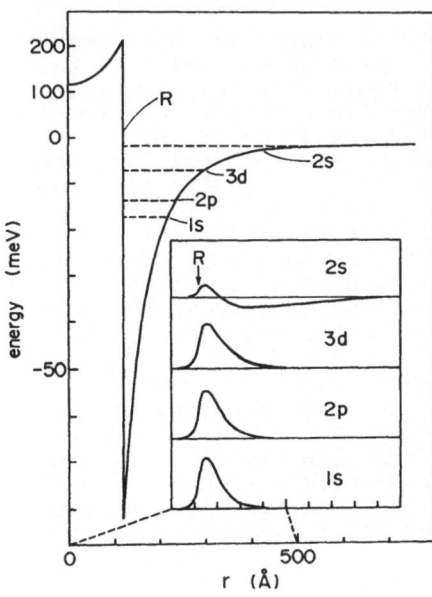

Fig. 2   The self-consistent potential (solid line), orbital energies (dashed line) and orbital wavefunctions (inset) for the ground state electron configuration of an $Al_{0.35}Ga_{0.65}$ As-GaAs superatom with z=20 and R=120 A.

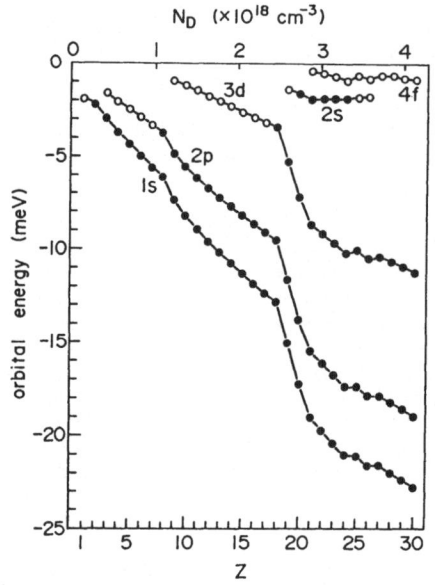

Fig. 3 Calculated orbital energy as a function of z. The donor concentration in the core is shown in the upper abscissa. The open and closed circles represent partially-occupied and fully-occupied shells, respectively.

but after that again comes a no-node shell 4f. Although the level sequence depends on the system parameters, this dominance of no-radial-node states is found to be a universal feature of superatoms.

The largest possible z value is limited by the requirement that the donor concentration cannot exceed the donor solubility limit, which usually is on the order of 10^{19} cm^{-3}. In all the cases considered here, the donor concentration is in the lower 10^{18} cm^{-3} range (see the upper abscissa in Fig. 3) and well satisfies this requirement[8].

The second salient feature seen in Fig. 2 is the large wavefunction extent of the highest occupied state, which is 2s in this case. In Fig. 4 the atomic radii, as defined by the maximum position of the radial wavefunction of the outermost orbital, is shown as a function of z (again R=120 A). The atomic radius is almost twice as large as R for small z and gradually decreases as z is increased. The case where the outermost orbital is 2s has a particularly large atomic radius because 2s has a radial node in contrast to 1s, 2p etc , which has no radial node. In any event the atomic radius is reasonably large over a range of z, and this indicates the validity of the atomic orbital picture for superatoms and also the feasibility of forming superatom molecules or crystals using superatoms as building blocks.

The third feature to be noted is the binding energy of the outermost orbital being on the order of meV. Thus the neutral ground state configuration can be realized by cooling the system down to temperatures on the order of 1 K. To be more precise, we should calculate the total energy for various electron configurations and compare them in order to see the stability of the ground state. The result of this calculation

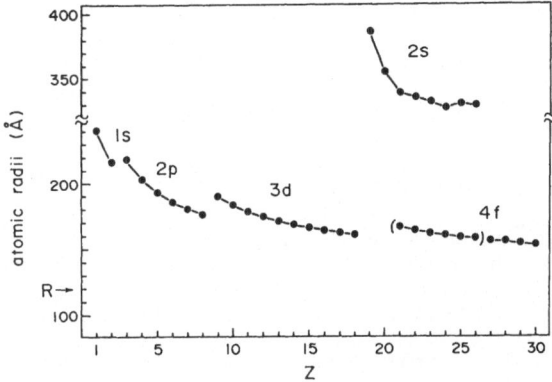

Fig. 4 Atomic radii versus z of $Al_{0.35}Ga_{0.65}As$–GaAs superatoms ($R=120$ A). The characters of the outermost orbitals are indicated. For $21 \leqq z \leqq 26$, the 2s orbital has the largest radii, although 4f has the smallest binding energy

also reveals that the excited state configurations are separated from the ground state by energy on the order of meV except near a configuration crossover, which takes place for certain parameter values.

The main effect of changing R is to induce such a configuration crossover. For example, for z=20, the ground state configuration is $1s^2 2p^6 3d^{10} 4f^2$ when R is around 180 A. With a reduction in R, $1s^2 2p^6 3d^{10} 4f^1 2s^1$ comes down in energy and becomes the ground state, transferring one of the 4f electrons to 2s, for R<162 A. With a further decrease in R the remaining 4f electron is again transferred to 2s at R=145 A, and below this the ground state becomes $1s^2 2p^6 3d^{10} 2s^2$ more strongly dominated by no-radial-node-states) for larger R and more ordinary atom-like for smaller R. (In particular, when z=1 and R=0, the system reduces to the ordinary hydrogenic donor in GaAs.) In practical situations, R is bounded from below again by the donor solubility limit. Therefore, although in the limit of vanishing R, one should recover the ordinary atomic level sequence, one can never approach this limit for z>1. Note, however, that, even for small R, the outermost orbital remains shallow, due to the small effective mass and the large dielectric constant of GaAs.

5. Summary

The electronic structure of the superatom is studied by a self-consistent density-functional-calculation using the effective mass approximation. Numerical results on AlGaAs–GaAs superatoms revealed that the atomic orbital picture can be successfully applied and that the ground state configuration is stable at T~1 K. The level sequence, however, is quite different from ordinary atoms due to the absence of 1/r singularity in the potential, and the electronic structure is dominated by no-radial-node states.

Acknowledgemet

The calculation was performed in collaboration with S. Ohnishi and A. Oshiyama.

References

1. H. Watanabe: In *The Physics and Fabrication of Micro-structures and Microdevices*, ed. by M. J. Kelly and C. Weisbuch (Springer, Berlin, Heidelberg 1986) p. 158
2. H. Watanabe, T. Inoshita: Optoelectronics: Devices and Technologies $\underline{1}$, 33 (1986)
3. T. Inoshita, S. Ohnishi, A. Oshiyama: Phys. Rev. Lett. $\underline{57}$, 2560 (1986)
4. H. Kroemer, Wu-Yi Chien, J. S. Harris, Jr., D. D. Edwall: Appl. Phys. Lett. $\underline{36}$, 295 (1980)
5. J. Perdew, A. Zunger: Phys. Rev. B$\underline{23}$, 5048 (1981)
6. D. M. Ceperley, B. J. Alder: Phys. Rev. Lett. $\underline{45}$, 566 (1980)
7. A. Ishibashi, Y. Mori, M. Itabashi, N. Watanabe: J. Appl. Phys. $\underline{58}$, 2691 (1985)
8. Strictly speaking, there is another factor that limits the range of z (and also R). See [3]

Index of Contributors

The 1st NEC Symposium on Fundamental Approach to New Material Phases
Microclusters Oct, 20~23 1986, Tokyo

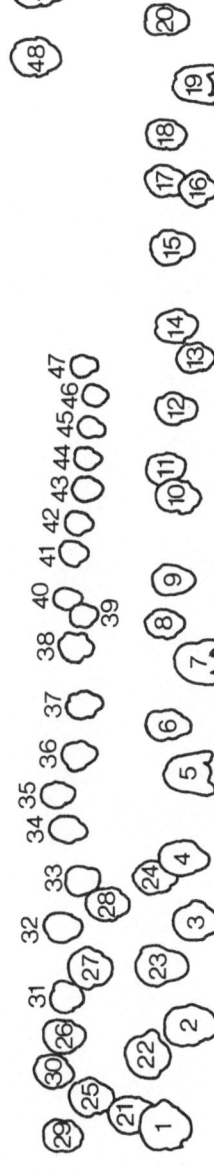

1   S. Sugano
2   M. Tsukada
3   T. Arai
4   P. Jena
5   Y. Chiba
6   T. Shinjo
7   Mrs. Cohen
8   T.P. Martin
9   M.L. Cohen
10   R. Kubo
11   D.E. Ellis
12   R. Car
13   H. Kamimura
14   C. Brechignac

15   K.H. Bennemann
16   Mrs. Kaldor
17   L. Wöste
18   L.A. Bloomfield
19   Miss K. Tanaka
20   K. Sattler
21   O. Sugino
22   H. Tatewaki
23   T. Kondow
24   A.J. Freemann
25   Y. Ishii
26   S. Ohnishi
27   S. Sawada
28   C. Satoko

29   S. Tanaka
30   S. Saito
31   A. Tamura
32   T. Fujiwara
33   Y. Nishina
34   D.G. Pettifor
35   N. Hamada
36   K. Terakura
37   A. Kasuya
38   A. Oshiyama
39   R. Uyeda
40   A.P. Kaldor
41   I. Yamada
42   K. Tanaka

43   R.S. Berry
44   Y. Saito
45   Y. Achiba
46   I. Katakuse
47   S. Iijima
48   D. Shinoda
49   M. Uenohara

D. Bäuerle

Chemical Processing with Lasers

1986. 88 figures. IX, 245 pages. (Springer Series in Materials Science, Volume 1). ISBN 3-540-17147-9

Contents: Introduction. – Fundamental Excitation Mechanisms. – Laser-Induced Chemical Reactions. – Experimental Techniques. – Material Deposition. – Surface Modifications. – Alloying, Compound Formation. – Etching, Cutting, Drilling. – Comparison of Processing Techniques, Applications of Laser Chemical Processing. – List of Abbreviations and Symbols. – References. – Subject Index.

Chemical Processing with Lasers is the first monograph to appear on this new, interdisciplinary and rapidly-expanding field which is increasingly attracting the attention of scientists, engineers and manufacturers alike. This book is intended as an introduction which caters for the very broad range of interests different groups of readers will have. Some of the chapters give a deeper insight into the fundamental mechanisms of laser-induced chemical reactions at or near solid surfaces, while other chapters have a more general approach. The characteristics of selective chemical processing within dimensions ranging from a few tenths of a micrometer up to several centimeters are discussed for single-model systems in chapters on material deposition, etching, surface modification, compound formation and synthesis. Actual and potential applications of two- and three-dimensional manufacturing techniques in micromechanics, metallurgy, integrated optics, semiconductor manufacture and chemical engineering are given brief separate treatment. Extensive tables will provide the reader with a quick and comprehensive overview of the "state-of-the-art" and immediately point him to the original literature.

Springer-Verlag
Berlin Heidelberg
New York London
Paris Tokyo

M. von Allmen

Laser-Beam Interactions with Materials

Physical Principles and Applications

1987. Approx. 200 pages. (Springer Series in
Materials Science, Volume 2).
ISBN 3-540-17568-7

Contents: Introduction. – Absorption of laser
light. – Heating by laser light. – Melting and solid-
ification. – Evaporation and plasma effects. –
Appendices. – References. – Subject Index.

Lasers are becoming popular tools and research
instruments in materials research, metallurgy,
semiconductor technology and engineering. This
text treats, from a physicist's point of view, the
processes that lasers can induce in materials. In
contrast to earlier books or reviews on related
issues, a broad view of the field and its perspec-
tives is given: physical topics covered range from
optics to shock waves, and applications range from
semiconductor annealing to fusion-plasma
production. Intuitive analytical models are used
whenever possible, in order to foster creative
thinking and facilitate access to newcomers and
nonspecialists. While basic ideas and their connections are emphasized more than technological
details, a sufficient amount of material data and
application information is included to make the
book useful to the practical experimentalist.

Springer-Verlag
Berlin Heidelberg
New York London
Paris Tokyo

Springer